KV-219-262

Mechanics of FIBRE COMPOSITES

V.K. TEWARY
Birla Institute of Technology & Science
Pilani

WILEY EASTERN LIMITED
New Delhi/Bangalore/Bombay

Copyright © 1978, Wiley Eastern Limited

This book or any part thereof may not be reproduced in any form without the written permission of the publisher

This book is not to be sold outside the country to which it is consigned by Wiley Eastern Limited

Price: Rs. 60.00

ISBN 0 85226 860 2

Published by Vinod Kumar for Wiley Eastern Limited, AB 8 Safdarjang Enclave, New Delhi 110016, and printed by Krishna Avtar Rastogi at Prabhat Press, 20/1 Nauchandi Grounds, Meerut 250002. Printed in India.

Preface

This book contains a detailed theoretical analysis of mechanics of fibre composites with emphasis on carbon fibre composites. The theory is discussed with an eye on the practical aspects of design and testing of these materials. S.I. units have been used throughout the book.

The development of carbon fibre composites is a major achievement in the field of materials science during the past ten years. In spite of considerable technological progress in the design and applications of these materials and a substantial number of published books in this area, only a few books are available which deal with the theory and analysis of mechanical properties of fibre composites, based upon their fundamental physical properties. This book is intended to fill this particular gap in the literature.

Many bulk properties of the fibre composites can be understood in terms of the simple laws of mixture and the continuum model of a solid. However, those properties of fibre composites which are sensitive to their fibrous nature such as the initiation and the propagation of cracks, propagation of elastic waves of wave length comparable to interfibre spacing etc. cannot be explained with the help of the simple laws of mixture and the continuum model. For the solution of such problems, it is essential to take the fibre-fibre interaction into account which requires a more fundamental approach. Fortunately, very advanced and highly sophisticated techniques have already been developed in solid state physics which can be used to great advantage in the field of fibre composites. One of the objectives of this book is to illustrate the use of such techniques as for example the Born-von Kármán model and Green function method which have been used here for studying the propagation of elastic waves and the calculation of the fracture energy due to debonded fibres in fibre composites (Chapters 5 and 6).

This book should be of interest to analysts and research and development engineers and also to solid state physicists who may want to apply their expertise and experience in solid state physics to the technologically oriented problems concerning the fibre composites. The book should be useful to senior undergraduate and graduate students of materials science and also to design engineers since it contains several formulae which can be applied to real cases.

About 650 references have been given in this book. This list is by no means complete but it does give a fair survey of the published work in the field. In addition to a review of the published work, this book also contains some of my own research work—published as well unpublished (mainly in Chapters 5 and 6).

I have to apologize for non-uniformity of notation in the book. Most symbols have been defined "locally". A particular symbol may refer to different quantities in different sections; for example G denotes the crack extension force in Chapter 2, a force constant in Chapter 5 and Green function in Chapter 6. In view of a large number of symbols involved, the system of local notations was found to be more convenient than the uniform notation and I hope that it will not cause any confusion.

Most of the work on this book was completed during my stay at the Atomic Energy Research Establishment, Harwell, U.K. where I had a very enjoyable and fruitful time for which I wish to thank all my colleagues and friends at Harwell. In particular, I am grateful to Dr. R. Bullough with whom I had the privilege to work and collaborate and to Dr. A.B. Lidiard for their constant encouragement and help and many useful discussions. During my present assignment at BITS, Pilani, this book has benefited not only in terms of actual research work but also in terms of a certain attitude to research which has been evolved at this Institute through a synthesis of science and technology and their social and environmental relevance. This, I trust, would be reflected in the book and for this I specially wish to express my gratitude to Dr. C.R. Mitra.

I am grateful to all the authors and their publishers whose work has been quoted in this book—in particular to Drs. R. B. Abarcar, T.S. Chow, G.A. Cooper, P.F. Cunniff, P.R. Goggin, Z. Hashin, M.D. Heaton, J.J. Hermans, R. Hill, W.N. Reynolds and B.W. Rosen and the Institute of Physics and the publishers of *Journal of Mechanics & Physics of Solids*, *Journal of Applied Mechanics*, *Journal of Composite Materials* and *Journal of Materials Science*. I should also thank Academic Press, New York for permission to use the material given in my paper, reference Tewary and Bullough (1972).

I should thank Miss Rosemary Rosier and Mr. G. R. Verma for the excellent job of typing the manuscript and secretarial assistance. I would also like to thank all my friends and family members who contributed to the effort behind this book in various ways. In particular I wish to thank Miss Anita Misra for doing the difficult job of organising the manuscript, references and indexes; my friends Mr. A. Kishore and Dr. Y. Singh and my sister Suman who are my great well wishers, my father Mr. H.N. Tewary for his efforts and sacrifices to bring me to this level, my two very sweet and equally naughty daughters Kanupriya and Anuranjita for not tearing off the manuscript and to my wife Sharad for her valuable companionship. Finally, for whatever it is worth, I dedicate this book to the single inspiration and the driving force in my life.

Pilani, India
December 1977

V.K. Tewary

Contents

5. PROPAGATION OF ELASTIC WAVES IN FIBRE COMPOSITES

6. EFFECT OF DEFECTIVE FIBRES ON WAVE PROPAGATION IN COMPOSITES

APPENDICES

Mechanics of Fibre Composites

CHAPTER **1**

Introduction

1.1 HISTORICAL SURVEY

That we live in a materialistic society is literally true. Our whole civilization—indeed our very existence—depends upon the availability of suitable materials. From prehistoric days right up to modern times there has always been a great demand for strong materials and therefore human society has always been in search of stronger and stronger materials.

In fact some of the materials used in ancient days can be regarded as technological miracles even by modern standards. The Ashoka Pillar in New Delhi (India) is one example of such a remarkable achievement of the past. This pillar was erected in commemoration of the victories of Emperor Ashoka, who wanted an indestructible symbol of his greatness. Indestructible it certainly seems to be, as it has withstood the onslaught of more than 2000 years without any corrosion or other apparent damage.

Today, with the increasing demand for 'suitable' materials with specific properties, it has been clear that naturally occurring materials cannot fulfil all the requirements. Developments in alloying techniques have gone a long way towards fulfilling these requirements but even that is not enough.

From the application point of view, an ideal material should be strong, tough and light. Metals and their alloys come close to satisfying these requirements. They are strong and tough but not very light. Some covalent materials are strong but not tough. The plastics invented in this century are light but lack stiffness, strength and toughness. An obvious approach to attaining an ideal material, therefore, would be to combine two materials with complementary properties. Such composite materials should have the combined advantages of their constituents.

The idea of composite materials itself is not new. During the days of Pharaoh in Egypt it was a common practice to use chopped straws in bricks, which prevented them from cracking. For the same purpose, plant fibres were used in ancient Inca and Maya potteries. The Egyptian mummy cases were made of papier-mâché, a kind of composite material containing sheets of papyrus which was used as writing material in Egypt. In fact, the moulding techniques used in making these composites have been used in Europe until quite recently.

The use of moss for strengthening ice by the Eskimos is another example of the fabrication of a composite material. Ice is quite hard but very brittle. The freezing of moss incorporates the fragments of cellulose into ice, which prevents the propagation of cracks. The use of reinforced icebergs as aircraft carriers was suggested by Geoffery Pyke during

World War II. This project, code-named *Habakkuk*, was very seriously considered, although never actually put into effect.

Two naturally occurring composite materials, namely, bone and bamboo, may also be mentioned here. Bone contains apatite crystals incorporated into a matrix of collagen fibres. It is a fine example of a natural composite which is very strong but unfortunately quite brittle. Bamboo is a cellulose reinforced by silica which makes it a remarkably hard material with a high impact strength.

In more recent times glass-fibre-reinforced plastics have been developed. It was found that glass fibres are much stronger than bulk specimens of glass. A big advantage in the use of glass fibres is the ease and cheapness with which long and strong fibres can be made. However, the fibres cannot be used on their own for structural applications—they have to be embedded in a matrix. A plastics matrix was found to be a suitable choice on account of its light weight and good adhesion properties with glass. Glass-reinforced plastics have been widely used in the manufacture of various things, such as kitchen sinks, bathtubs, motor-car bodies, swimming pools and even aeroplanes. Glass fibres, however, have many disadvantages, such as, for example, the fact that they lose their strength at high temperatures.

The invention of carbon fibre composites is a great step forward in the quest for strong materials. It is certainly a major achievement in the field of materials science during the past decade. In these composites carbon fibres of very high strength are used to reinforce plastics—usually an epoxy resin. These materials are strong as well as light. 'Stronger than steel yet lighter than aluminium' is the catch phrase which summarizes the very attractive properties of carbon fibre composites.

Perhaps the first industrial application of carbon fibres was made by Edison in his electric light bulbs. These fibres were quite fragile and were obviously not intended for structural applications. It has, however, been known for a long time that the covalent bond between two carbon atoms is very strong. Carbon, therefore, has always been a potential candidate for the fabrication of a strong structural material. Now, with the advent of carbon fibres, the strength of a carbon–carbon bond can be fully utilized.

In the UK a technique for the manufacture of strong carbon fibres was developed at the Royal Aircraft Establishment, Farnborough, by Watt, Phillips and Johnson (Watt et al., 1966; Johnson et al, 1968). In the USA the first strong carbon fibre was manufactured under the trade name Thornel 25 by Union Carbide. A fairly high-strength carbon fibre was produced by La Carbone-Lorraine in France. A strong fibre from polyacrylo nitride yarn was also made in Japan by Akio Shindo but it was not exploited for structural applications (quoted by Langley, 1970).

The fibres of other materials, such as boron, silicon carbide, ceramics, etc., have also been developed and are being investigated for possible structural applications. On considerations of strength and modulus boron fibres seem to be the most serious rivals of carbon fibres. At present, however, they are even more expensive than carbon fibres.

Fibre composites have been tipped to be the material of the future and a large effort has been devoted by several laboratories in the UK as well as abroad to the study of their properties. In this book we shall be interested only in the mechanical properties of fibre composites. However, a brief description of the preparation, general properties and some applications of carbon fibre composites has been given in Section 1.3. For details of carbon fibres and their composites reference may be made to the numerous review articles and books on this subject—for example, Langley (1970, 1973), Prosen (1970), Watt (1970) and Gill (1972). The strength of solids and composite materials in general has been discussed by, among others, Zackay (1965), Kelly and Davies (1965), Kelly (1966), Burke et al. (1966), Rauch et al. (1966), Holiday (1966), Broutman and Krock (1967) and Felbeck (1968). For a supplementary reading on the mechanical properties of composite materials see Rosen (1965, 1970a), Kelly (1970), Cooper (1971), Spencer (1972) and Hashin (1972).

In addition to those given above, the following books may also be consulted: Parrat (1972), Garg et al. (1973) and the recent series on composite materials, viz. Metcalfe (1974), Sendeckyj (1974), Noton (1974), Kreider (1974), Broutman (1974), Plueddemann (1975) and Chamis (1975).

A review of several properties of the composite materials has been given in Dexter and Singer (1974). This report, among other things, describes artificial and natural high temperature composites (article by W. Bunk) and theoretical and experimental stress analysis of composite pressure vessels (article by A. Segal and others). A comprehensive review of ceramic matrix composites has been given by Donald and McMillan (1976). Many useful papers on the mechanical properties of composite materials are available in the *Proceedings of the International Conference on Mechanical Behaviour of Materials* held at Kyoto (Aug. 15–20, 1971) and published by Society of Materials Science 1972.

1.2 STRENGTH OF SOLIDS

In this section we shall discuss the theoretical strength of solids and see how theoretical considerations of strength could have or should have naturally led to the development of carbon fibre composites.

First, it should be emphasized that strength and toughness are two distinct properties of a solid. The strength of a solid is measured by the maximum stress which it can sustain without breaking. The toughness of a solid can be described as its resistance to crack propagation. A material may be weak but tough, or strong but brittle. For example, ordinary glass is strong but very brittle. A structural material, as mentioned before, should be tough as well as strong. In this section, however, we shall consider only the strength of a solid.

Let us consider a solid bar on which a tensile stress σ is applied along its length. Let σ_c denote the value of σ when the bar breaks into two pieces. σ_c is called the critical stress and, according to the discussion given in the preceding paragraph, is a measure of the strength of the solid.

The bar will break only when the atomic bonds at the surface are broken. The strength of the atomic bond arises from the atomic interactions in the solid. The breaking of the atomic bonds requires some energy. A rough estimate of this energy can be obtained in terms of U_v, the formation energy of a vacancy. This energy expressed per unit volume is given by

$$U_v' = \frac{3U_v}{4\pi r_0^3} \tag{1.2.1}$$

where $4\pi r_0^3/3$ is the volume of the vacancy expressed in terms of r_0, the radius of the Wigner–Seitz cell. (See, for example, Kittel, 1976, for the definition of the Wigner–Seitz cell.)

The energy U_v' required to break the atomic bonds must be supplied by the stress σ_c. Assuming the solid bar to behave elastically right up to the breaking point, the strain energy per unit volume in the solid at the breaking point is given by

$$U_s = \tfrac{1}{2}\frac{\sigma_c^2}{E} \tag{1.2.2}$$

where E denotes the Young's modulus of the solid.

The solid will break when $U_v' = U_s$. Thus, we obtain from Eqs. (1.2.1) and (1.2.2)

$$\sigma_c = \left(\frac{3EU_v}{2\pi r_0^3}\right)^{1/2} \tag{1.2.3}$$

Equation (1.2.3) can be written in the more familiar form by defining γ, the vacancy formation energy per unit area as follows:

$$\gamma = \frac{U_v}{4\pi r_0^2} \tag{1.2.3a}$$

so that

$$\sigma_c = \left(\frac{6E\gamma}{r_0}\right)^{1/2} \tag{1.2.4}$$

The parameter γ in Eq. (1.2.4) can be identified with the surface energy of the solid.

In most solids U_v, the vacancy formation energy, is of the order of an electronvolt (1.6×10^{-19} J) and r_0 about a few angstroms (10^{-10} m). Thus we get from Eq. (1.2.3a) that γ is about 1 J/m². E, the Young's modulus for most solids, is about 100 GN/m². Thus we infer from Eq. (1.2.4) that the strength of most materials should be about 100 GN/m².

Equation (1.2.4) has been derived in a very crude manner on the basis of certain approximations which are not really justified. For example, the assumption that the solid behaves elastically right up to the breaking point is not valid. Moreover, the surface energy γ is not really given by the simple formula in Eq. (1.2.3a) in terms of the vacancy formation energy, although it does yield a correct order of magnitude estimate of the surface energy. However, these corrections, if incorporated into Eq. (1.2.4), will not change drastically the estimated value of σ_c.

A more rigorous approach (Kelly, 1966) leads to an equation of the same form as Eq. (1.2.4) but without the factor $(6)^{1/2}$. The proper calculation of σ_c will require a detailed knowledge of the interatomic potential and will involve the use of quantum mechanics (for a recent review of the theoretical work on the ideal strength of solids see MacMillan, 1972). For our purpose of a qualitative discussion of the theoretical strength of solids, the order of magnitude estimate of σ_c as given by Eq. (1.2.4) is sufficient.

TABLE 1.1

Approximate tensile strengths of materials

Material	Tensile strength, GN/m²
Steel	0.4–3.0
Wrought iron	0.15–0.3
Cast-iron	0.07–0.3
Copper	0.15
Aluminium	0.07
Magnesium alloys	0.2–0.3
Titanium alloys	0.7–1.4
Aluminium alloys	0.15–0.6
Brasses	0.15–0.4
Ordinary glass	0.03–0.15
Cement	0.004
Cotton	0.3
Silk	0.4
Spider's thread	0.2
Bone	0.15

The actual values (approximate) of the tensile strength of some typical materials are quoted in Table 1.1. We see from this table that none of the solids reach anywhere near their theoretically expected strength of 100 GN/m^2. Even the strength of steel is 2–3 orders of magnitude lower than the theoretical value.

The discrepancy between the theoretical and the observed strengths of materials was for some time a puzzle. Now it is known that real materials do not achieve their theoretical strengths owing to various defects which are nearly always present in them. The most important of these defects which substantially affect the strength of materials are dislocations, grain boundaries, cracks and notches, and also perhaps voids.

There are two main mechanisms which cause a solid to break at a stress lower than its theoretical breaking stress. These are plastic flow and cleavage. In some solids, and particularly in metals, the dislocations are quite mobile. The strength of these solids is mainly determined by the movements of the dislocations. On the other hand, in some solids such as diamond or silicon the dislocations are not very mobile. The strength of such solids is controlled mainly by the surface steps, cracks and notches.

From the preceding discussion we infer that in order to achieve the theoretical strength, we have to produce a solid which is free from the above mentioned defects. This, however, is not an easy task. The solids invariably contain dislocations and microcracks, in addition to several other defects.

It should be obvious intuitively that, in general, the defects in a solid will be randomly distributed. Moreover, the number of defects in a solid will increase with the size of the solid. Thus, our best hope for the production of a perfect solid lies with small, thin-shaped materials: it has, indeed, nearly been achieved with the production of whiskers and, to some extent, fibres.

Whiskers are single crystals which are completely free of dislocations or at most contain a single screw dislocation along their axes. Very-high-strength whiskers have been produced which are reasonably near the ideal strength predicted by Eq. (1.2.4). Whiskers of graphite, Al_2O_3, SiC, etc., are now quite common and are used for reinforcing plastics.

In spite of their high strengths, whiskers are not ideally suited for reinforcing purposes because of their short lengths. Long and continuous fibres are more desirable for the purpose of reinforcement, since they impart better creep and crack stopping properties to composites. The use of long glass fibres for the reinforcement of plastics has already been mentioned.

Thus, we have made a case for the production of long fibres of a strong material. Next, the question arises regarding the choice of the fibre material. We know that the carbon–carbon bond is quite strong. The trouble is that apart from diamond, which is expensive and brittle, most of the naturally occurring forms of carbon, such as graphite, lamp-black or charcoal, are either amorphous or very weak in certain directions. They are, therefore, not suitable for structural applications.

The solution of this problem lies in the use of polymers, which are organic molecules containing long chains of covalently bonded carbon atoms. First, let us estimate the ideal strength of a polymer with the help of a qualitative argument given by Frank (1970). Frank's argument yields the ideal stiffness (elastic modulus) rather than the strength of the polymers. The strength and the stiffness are quite different quantities but they are positively correlated with each other, as would be expected from Eq. (1.2.4).

To estimate the stiffness of polymers following Frank (1970), let us consider the Young's modulus of a diamond crystal in the $\langle 110 \rangle$ crystallographic direction. In this direction the carbon atoms are arranged in a zigzag fashion along the line $\langle 110 \rangle$, a structure which is quite similar to that of a typical polymer such as polyethylene. The bonds in the diamond lattice which are perpendicular to the $\langle 110 \rangle$ line will obviously not resist any stretching of the crystal in the $\langle 110 \rangle$ direction and therefore will have a zero contribution to the Young's modulus in this direction.

Consider one chain of carbon atoms along the $\langle 110 \rangle$ direction. As explained in the preceding paragraph, the stretching of each chain is independent of its coupling with other parallel chains. We can, therefore, say that the Young's modulus, E, of each chain is equal to 1160 GN/m², the value for the diamond crystal in the $\langle 110 \rangle$ direction.

If a denotes the average area of cross-section of each chain then the net force in a C—C bond when the chain is subjected to unit strain is equal to Ea. Assuming that the strength of a C—C bond in a polymer is the same as in diamond, we find that the stress produced in a polymer chain for unit strain will be Ea/A, where A is the area of cross-section of the polymer chain. Taking $a = 4.48 \times 10^{-19}$ m² and A (for polyethylene) $= 18.2 \times 10^{-19}$ m², we obtain that the stiffness of a polymer should be about 285 GN/m². This value has been obtained on the basis of very crude arguments but is reasonably close to the values derived by more sophisticated methods, such as 180 GN/m² (Treloar, 1960) and 235 GN/m² (Sakurada et al., 1964).

Thus, we see that the Young's modulus of a polymer should be quite large. The predicted value of the Young's modulus of polyethylene is comparable to that of high-tensile steel, which is about 210 GN/m². It immediately suggests the use of polymers for the production of strong fibres.

However, the observed values of the Young's moduli of polymers are much less than the theoretical values. For example, the observed value of an unoriented polyethylene molecule is about 1 GN/m^2, which is two orders of magnitudes smaller than the theoretical value. Even for a highly oriented polyethylene, the observed value is 24 GN/m^2, which is still an order of magnitude smaller than the theoretical value.

The discrepancy between the theoretical and the observed stiffness of the polymers can again be attributed to various types of defects. For example, a large amount of back-folding is known to occur even in the apparently well-aligned polymers which will reduce the resistance of the polymer to an applied tensile stress and, hence, its stiffness. Several other factors also contribute to the stiffness of the polymers (see Frank, 1970, for details).

Although the observed values of the elastic moduli of the polymers are much less than their theoretical values, it should be possible to attain them with the help of a suitable technique. This is indeed found to be the case. The Young's modulus of the carbon fibre which can be produced from the polymeric materials is of the order of 100 GN/m^2, which is in the region predicted theoretically.

In this section we have considered only the strength and the stiffness of the materials and we have seen how these considerations lead to the choice of carbon fibre composites. There are, of course, many other factors, such as toughness, creep and fatigue properties and also, of course, the cost, which determine the overall usefulness of a material. Although carbon fibres seem to have the most attractive overall properties from a practical point of view, several other fibres, such as boron, ceramic, silicon carbide and various metallic fibres and their composites, are also being developed. The relevant physical and mechanical characteristics of some fibres are given in Table 1.2, alongwith those of high-tensile steel for the sake of comparison.

We notice from Table 1.2 that the whiskers have superior mechanical properties. Their strength is quite close to the theoretical strength as predicted by Eq. (1.2.4). Unfortunately, as mentioned earlier, they are not very suitable for use in composite materials. Their short length (about 3 cm) creates fabrication and alignment problems. In a misaligned composite the strength of the reinforcing phase is not fully utilized. Moreover, a composite containing whiskers or discontinuous fibres will have poor creep resistance.

The strengths of the commercially produced fibres, as quoted in Table 1.2, are much lower than the theoretical values Eq. (1.2.4). However, they are still comparable to the strength of the high-tensile steel. In fact, if the low density of the fibres is taken into account, the fibres are found to be far superior to steel, as can be seen by comparing the specific strengths given in Table 1.2.

We shall now give a brief comparison of glass, boron and the carbon fibres. As we see from Table 1.2, the strength of all three is nearly the

TABLE 1.2

Physical and mechanical characteristics of some whiskers, fibres and high-tensile steel

Material	Density (ρ) 10^3 kg/m^3	Young's Modulus (E) 10^2 GN/m^2	Specific Young's Modulus (E/ρ) 10^8 J/kg	Tensil Strength (σ) GN/m^2	Specific Tensile Strength (σ/ρ) 10^6 J/kg
High-tensile steel	7.9	2.1	0.27	1.9	0.24
Whiskers:					
Graphite	2.2	6.9	3.1	19.6	8.9
Al_2O_3	4.0	5.3	1.3	15.4	3.9
SiC	3.2	7.0	2.2	21.4	6.7
Carbon Fibres:					
Grafil–HM	2.0	4.0	2.0	2.0	1.0
Grafil–HT	1.7	2.1	1.2	2.6	1.5
Rigilor–AG	2.1	2.8	1.3	0.9	0.4
Rigilor–AC	1.8	1.8	1.0	2.1	1.2
Thornel–25	1.5	1.8	1.2	1.4	0.9
Thornel–50	1.6	3.5	2.2	2.0	1.3
Boron Fibres	2.5	4.2	1.7	3.5	1.4
E-glass Fibre	2.6	0.7	0.3	1.8–3.5	0.7–1.3
S-glass Fibre	2.5	0.8	0.3	$\leqslant$4.6	$\leqslant$1.8

same but in stiffness (modulus) the glass fibres fall much behind the other two. Boron and carbon fibres can be up to six times more rigid than glass fibres.

Glass fibres have the great advantage of low cost as compared with the other two. Whereas glass fibres cost much less than a pound sterling per kilogram carbon fibres may cost up to ten pounds sterling per kilogram. Eventually, of course, the carbon fibres may become cheaper. Boron fibres are most expensive—up to about forty pounds sterling per kilogram. In fact, their largescale production may not be feasible at all.

Unlike glass fibres, both boron and carbon fibres are susceptible to oxidation. Moreover, in contact with certain active elements the structure of boron and carbon fibres may change. An example is the reaction of carbon fibres with nickel. This will have a disastrous effect on their strengths. Glass fibres are quite safe in this respect.

The biggest disadvantage of glass fibres is that they lose their strength with an increase in the temperature. Both boron and carbon fibres retain their strengths at relatively high temperatures. In fact, the strength of carbon fibres even increases with temperature.

Boron fibres seem to be especially useful for reinforcing the metals. For this purpose, borsic (boron coated with silicon carbide) fibres are preferable to uncoated boron fibres. For example, a borsic fibre/aluminium matrix composite has very good fatigue properties and a high resistance to corrosion.

To summarize, the relative advantages of glass, boron and carbon fibres are quite evenly balanced. The choice of a particular material will depend upon the particular purpose for which the fibres are to be used. From an overall point of view and from considerations of cost-effectiveness, carbon fibres appear to be the most promising material. However, if the cost of boron fibres can be brought down, the balance may be tipped in their favour.

1.3 CARBON FIBRES AND THEIR COMPOSITES: PREPARATION, PROPERTIES AND APPLICATIONS

First, we shall indicate the principle of the process used for the manufacture of carbon fibres.

As suggested in the previous section, carbon fibres can be produced from polymers by appropriately orientating and aligning the molecular chains and then carbonizing them. This is done by squirting a solution of the polymer as a fine jet into a coagulating bath. It results in the precipitation of the polymer in fibre form. In this fibre the molecules are randomly oriented and therefore the fibre has poor mechanical properties.

The fibre is then wound onto a frame under a carefully controlled tension. The frame is heated in an air oven at approximately 250°C, where the fibre is oxidized. The fibre, originally white, now turns black and shrinks. The shrinkage of the fibres and the thermal expansion of the frame exert a tension on the fibres which orientates and aligns the polymer molecules. The fibre shrinks by about 10% in length and about 40% in thickness. This large shrinkage in thickness indicates that the polymer chains are pulled closer together.

In the next stage the fibres are removed from the frame and carbonized in a furnace at a temperature of approximately 1000°C. In this furnace the fibres are reduced to pure carbon fibres and lose about half their weight in the process. Finally, the carbon fibres are put into another furnace for appropriate heat treatment in order to improve their mechanical properties. The temperature of the furnace in the last stage is chosen according to the type of fibre required. For a high-modulus carbon fibre this temperature is taken to be about 2500°C, whereas for a high-strength fibre a temperature of about 1500°C may be enough.

In Europe carbon fibres are usually made from acrylic polymers. The fibres made by the British companies are produced from polyacrylonitrile (PAN) polymer, whereas cellulose materials such as rayon are used in the USA for this purpose. The structure and the overall properties of these two fibres are quite similar. The main difference between the two processes is that it is probably easier to make a high-modulus fibre from cellulose and a high-strength fibre from PAN. Another difference is in their cross-sectional shape, which is more uniform in the case of PAN fibres.

Both PAN and rayon fibres are quite expensive. It is, therefore, of great interest to exploit the possibility of producing carbon fibres from cheaper raw materials, such as resins or hydrocarbon pitches. Kawamura and Jenkins (1970) have shown that the fibres produced from phenol-hexamine can be as strong as the PAN fibres but with a much lower rigidity. The low rigidity is a great disadvantage. However, in the future it may be possible to produce cheap carbon fibres from one of the above mentioned non-textile raw materials.

Next we shall describe the materials which can be used as matrices in carbon fibre composites. The following types of materials can be used for this purpose: (i) metals, (ii) glass, (iii) ceramics, (iv) carbon and (v) polymers or plastics.

Metals have good strength and toughness but they are heavy and do not wet the carbon, which results in a poor strength of carbon fibre/metal matrix composite. The wetting properties of metal matrices can be improved by adding certain impurities (for example, chromium in copper) but it may result in a degradation of the fibre properties.

Glass and ceramic matrices tend to be too brittle. However, the toughness of the carbon fibre/glass matrix composite can be substantially increased by special processing. An advantage with the glass (silica) matrix is that its thermal expansion coefficient is not very different from that of the fibres, and it has been possible to quench in water a silica glass reinforced by 50% fibres from 1200°C without cracking (Crivelli-Visconti and Cooper, 1969). The main disadvantage with a glass matrix is that it loses its strength at high temperatures.

On the other hand, ceramics retain their strength to very high temperatures and have very good thermal shock resistance. However, in the case of a carbon-fibre-reinforced ceramic, difficulty arises because of the mismatch between the thermal expansions of the two constituents.

A carbon matrix reinforced by carbon fibres appears to be a rather attractive idea because of several advantages, such as, for example, the fact that it can be produced in a variety of forms with diverse engineering properties. It has a low density, good lubricating properties and a reasonably high electric conductivity. In pyrolytic and vitreous forms it has a high strength and a good resistance to oxidation. The disadvantage with a carbon matrix is that it is too brittle, and the fabrication of a carbon fibre/carbon matrix composite does not appear to be an easy task. However, these problems may eventually be overcome and carbon/carbon composites may come into wide use.

At present the polymer matrix materials are most widely used in carbon fibre composites. The main reason for the popularity of the polymer matrices is the ease of fabrication combined with a large strength-to-weight ratio of the composites. These composites have good resistance to creep and fatigue, and can also be made reasonably tough. A detailed description of polymer matrices has been given by Molyneux (1973).

The following polymers can be used as matrices in carbon fibre composites: (i) polyesters, (ii) phenolics, (iiii) polymides, (iv) thermoplastics and (v) epoxies.

There are strong restrictions on the range of temperatures over which these materials can be used. These restrictions are imposed by the degradation of mechanical properties of the polymers at higher temperatures. Polyesters and thermoplastics can be used only up to about 100°C, although the former can be used for a short period at temperatures up to 150°C. Epoxies are most suitable at temperatures below 150°C, although the range of temperature can be extended up to 250°C at the expense of toughness. Phenolics can be used up to 250°C but they have poor mechanical properties. Polymides and some other similar polymers retain their toughness and strength at high temperatures and therefore are very useful for high-temperature (250–400°C) applications. The problem with these materials is that their processing and the fabrication of their composites are rather difficult.

In addition to the organic polymer matrices, an inorganic substance—silicone—can also be used in carbon fibre composites. The mechanical properties of silicone are quite similar to those of phenolics, and is also unsuitable for high-temperature applications.

Of all the polymers described above, epoxies are most commonly used in carbon fibre composites. This is because their mechanical properties are superior to those of other polymers and they have a good adhesion to carbon fibres. They also have a good chemical resistance and their dimensional stability is good in stressed conditions.

Finally, in this section we shall briefly describe some of the present and the possible future applications of carbon fibre composites. A very detailed account of mechanical engineering applications has been given by Bedwell (1973) and of structural engineering applications by Tetlow (1973). Carbon fibre brushes in electrical machines have been discussed by Bates (1973).

What makes carbon fibre composites very useful for structural applications is their high specific tensile strength and specific modulus. This advantage is hampered by their low toughness—particularly in shear—and their large anisotropy, which creates additional design problems. At present the very high cost of carbon fibres is an extra disadvantage.

It is obvious that the high specific strength of a material is particularly useful in aerospace applications, such as spacecraft, aircraft and tactical weapons. In these cases any reduction in weight will immediately pay off in terms of fuel savings and improved efficiency. The main disadvantage of carbon fibre composites in the aerospace application, therefore, is not their high cost but low toughness.

Carbon fibres have been used in the fan blades of the compressor stage of the Rolls Royce RB–211 engine. They suffered, however, from the 'bird impact' problem, which reflected their poor resistance to interlaminar shear fracture. The possibility of using carbon fibres in strike aircraft wings, glider wing spars and helicopter rotor blades is also being investigated.

Another important area in which the high specific strength and modulus of carbon fibres can outweigh the disadvantage of their high cost is the construction of gas centrifuges for uranium enrichment. In this case carbon fibre composites have an extra advantage over metals and glass fibre composites. Since both metals and glass fibre composites react with the fluorine in uranium hexafluoride which is involved in the uranium enrichment process, they have to be given a protective coating. This increases their effective weight without any increase in their strength. No such protective coating is required for carbon fibre composites (Bedwell, 1973).

A device similar to the gas centrifuge is the analytical centrifuge, which is used for medical purposes such as the analysis of blood samples by mass chromatography. Because of the asymmetric stresses produced in the centrifuge, a material of large specific stiffness is required in order to reduce the inertia of the rotor. Carbon fibre composites should, therefore, serve a useful role in this field.

Carbon fibre composites have also been used in the construction of support frames for severely handicapped people, particularly thalidomide children, the victims of another branch of science. The reduction in the weight of these support frames made possible by the use of carbon fibre composites without any loss in strength should be particularly welcome by these children. The high specific strength of carbon fibres should also be useful in the construction of artificial limbs, but their brittleness will be a serious problem in this case.

According to Tewary and Bullough (1971) (see also Chapter 5) carbon fibre composites can be used as collimator and filter elastic waves. It may, therefore, be possible to construct an ultrasonic focusing device from carbon fibres (Bedwell, 1973), which should be very useful for medical and surgical applications. The possible use of the above mentioned elastic wave filtering property of carbon fibres in noise reduction devices has also been discussed by Bedwell (1973).

Some other areas in which carbon fibre composites have or can be used are the following: design of pressure vessels, radar discs, antennae; generator end-bells; sports equipment, such as sailing masts, dinghy centreplates and bicycle frames; and some automobile parts, such as transmission shafts and flywheels.

There seems to be no immediate possibility of using carbon fibre composites as the material for motor-car bodies. Two of the main requirements for a car body material are toughness and low cost. These are precisely the qualities lacked by carbon fibre composites. However, carbon fibres were used to reinforce the glass fibre body of a Ford GT racing car, which resulted in a 42% reduction of the weight of the car (from 69.4 kg to 40.2 kg). This carbon fibre car was twice the winner of the coveted Le Mans motor racing championship.

CHAPTER 2

Mechanics of Solids a brief review

2.1 INTRODUCTION

A solid is an assembly of microscopic objects such as atoms or molecules which interact with one another through various kinds of forces. In a crystalline solid the atoms (or molecules) are arranged in a regular geometric pattern which defines the lattice structure of the solid. No such geometrical order exists in an amorphous or disordered solid.

Most of the characteristic properties of a solid are determined by its lattice structure and its interatomic forces. For example, any attempt to distort a solid or displace an atom from its equilibrium site will be resisted by the forces between the atoms. Thus, the strength of a solid and its response to an external force will be governed by the interatomic forces in the solid and the distribution of these forces in space, which is determined by the lattice structure. Similarly, the atomic vibrations and therefore the passage of an elastic wave in a solid will also depend upon the interatomic forces in the solid and its lattice structure.

Many quantities of physical interest depend only upon the bulk or macroscopic behaviour of the solid. For a calculation of such quantities it is, in general, possible to ignore the discreteness of the lattice structure and treat the solid as a continuum of matter. This model of a solid is called the continuum model and is quite adequate for a calculation of elastic properties and overall strength of the solid. However, the continuum model may be totally unreliable for a representation of those quantities which are sensitive to the microscopic structure of the solid. For example, the propagation of a wave with wavelength comparable to the interatomic spacing in a solid is poorly described by the continuum model. For such calculations the discrete structure of the lattice cannot be ignored and a suitable lattice model has to be constructed.

The basic features of the lattice and the continuum theories have been briefly reviewed in Sections 2.2 and 2.3, respectively. As an illustrative example, the application of these theories to a simple cubic lattice model has been given in Section 2.4. For further details and various applications of these theories, the reader is referred to the excellent monographs by Love (1944) and Landau and Lifshitz (1970) on the continuum theory and by Born and Huang (1954) and Maradudin et al. (1971) on the lattice theory. A brief account of the fracture mechanics and the finite element method for a numerical study of the mechanics of solids have been given in Sections 2.5 and 2.6, respectively.

2.2 LATTICE THEORY

The lattice structure of a solid is defined by three basis vectors $\mathbf{a}_1$, $\mathbf{a}_2$ and $\mathbf{a}_3$ such that any linear combination of these vectors is a lattice vector. As an example, the three basis vectors $\mathbf{a}_i$ ($i = 1, 2, 3$) for the case of a simple cubic lattice structure have been shown in Figure 2.1. In this particular case, the vectors $\mathbf{a}_1$, $\mathbf{a}_2$ and $\mathbf{a}_3$ lie along the x, y and z axes, respectively. Other examples can be found in standard texts on crystallography or solid state physics such as Kittel (1976).

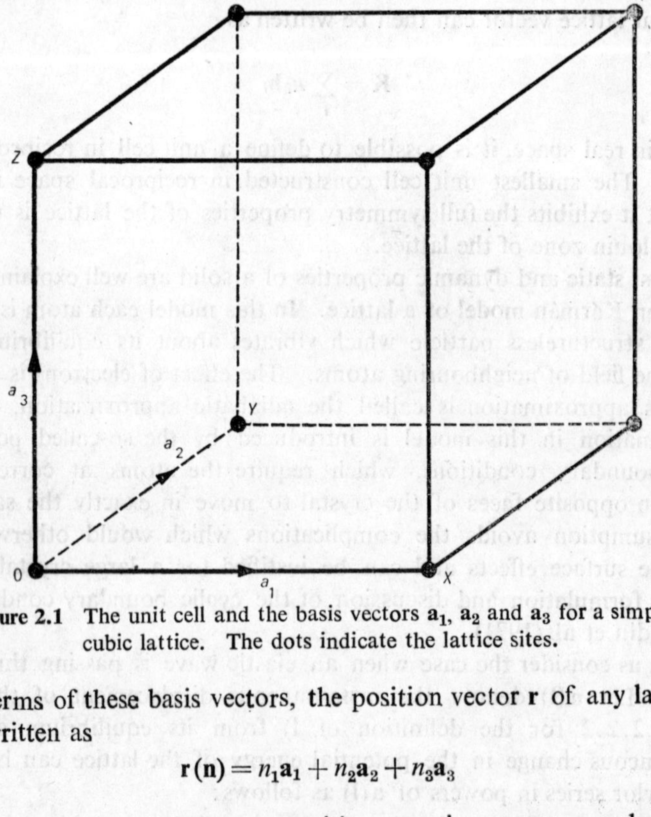

Figure 2.1 The unit cell and the basis vectors $\mathbf{a}_1$, $\mathbf{a}_2$ and $\mathbf{a}_3$ for a simple cubic lattice. The dots indicate the lattice sites.

In terms of these basis vectors, the position vector $\mathbf{r}$ of any lattice site can be written as

$$\mathbf{r}(\mathbf{n}) = n_1\mathbf{a}_1 + n_2\mathbf{a}_2 + n_3\mathbf{a}_3 \tag{2.2.1}$$

where n_1, n_2, n_3 are any integers—positive, negative or zero—and $\mathbf{n}$ can be regarded as a vector index with three components n_1, n_2 and n_3. It is, however, more convenient to express the vectors $\mathbf{a}_i$ in terms of their Cartesian components and label the lattice sites by vector indices $\mathbf{l}$ with components l_1, l_2 and l_3, i.e,

$$\mathbf{r}(\mathbf{l}) = l_1\mathbf{f} + l_2\mathbf{g} + l_3\mathbf{h} \tag{2.2.2}$$

where $\mathbf{f}$, $\mathbf{g}$ and $\mathbf{h}$ denote unit vectors along the x, y and z axes, respectively.

The periodicity and the translational invariance of the lattice allows us to divide the whole lattice into unit cells which are all equivalent to one another. In general, each unit cell may contain several atoms. In this book, however, we shall be interested only in monatomic lattice structures in which each unit cell contains only one atom.

It is convenient to define a reciprocal lattice or a lattice in reciprocal space with the help of the basis vectors $\mathbf{b}_1$, $\mathbf{b}_2$ and $\mathbf{b}_3$, which have the property ($i, j = 1, 2$ or 3)

$$\mathbf{a}_i \cdot \mathbf{b}_j = \delta_{ij} \tag{2.2.3}$$

where δ_{ij}, the Kronecker delta, is zero unless $i = j$, when it is unity. A reciprocal lattice vector can then be written as

$$\mathbf{K} = \sum_i n_i \mathbf{b}_i \tag{2.2.4}$$

As in real space, it is possible to define a unit cell in reciprocal space as well. The smallest unit cell constructed in reciprocal space in such a way that it exhibits the full symmetry properties of the lattice is called the first Brillouin zone of the lattice.

Most static and dynamic properties of a solid are well explained by the Born–von Kármán model of a lattice. In this model each atom is assumed to be a structureless particle which vibrates about its equilibrium lattice site in the field of neighbouring atoms. The effect of electrons is separated out: this approximation is called the adiabatic approximation. Another approximation in this model is introduced by the so-called periodic or cyclic boundary conditions, which require the atoms at corresponding points on opposite faces of the crystal to move in exactly the same way. This assumption avoids the complications which would otherwise arise from the surface effects and can be justified for a large crystal. For a rigorous formulation and discussion of the cyclic boundary conditions see Maradudin et al. (1971).

Let us consider the case when an elastic wave is passing through the lattice. Let $\mathbf{u}(\mathbf{l})$ denote the instantaneous displacement of the atom $\mathbf{l}$ (see Eq. 2.2.2 for the definition of $\mathbf{l}$) from its equilibrium site. The instantaneous change in the potential energy of the lattice can be written as a Taylor series in powers of $\mathbf{u}(\mathbf{l})$ as follows:

$$\phi = \sum_{\alpha,\mathbf{l}} \phi_\alpha(\mathbf{l})\, u_\alpha(\mathbf{l}) + \tfrac{1}{2} \sum_{\substack{\alpha,\mathbf{l} \\ \beta,\mathbf{l}'}} \phi_{\alpha\beta}(\mathbf{l}, \mathbf{l}')\, u_\alpha(\mathbf{l})\, u_\beta(\mathbf{l}') + \ldots \text{higher powers of } u \tag{2.2.5}$$

where α, β denote the Cartesian components and $\phi_\alpha(\mathbf{l})$ and $\phi_{\alpha\beta}(\mathbf{l}, \mathbf{l}')$ are, respectively, the first and second Taylor coefficients. There are several restrictions on the form of these Taylor coefficients, which are imposed by the conditions of symmetry and stability of the lattice (see Maradudin et al. 1971). Some of these restrictions are given below:

1. For a lattice free of initial stresses, which we shall assume to be the case, $\phi_\alpha(\mathbf{l}) = 0$ for all $\mathbf{l}$ and α.

2. On account of the translational symmetry, $\phi_{\alpha\beta}(\mathbf{l}, \mathbf{l}')$ depends upon $\mathbf{l}$ and $\mathbf{l}'$ only through their difference and therefore can be labelled by a single index $\mathbf{l} - \mathbf{l}'$. This is equivalent to the statement that since all lattice sites are equivalent, any of them can be chosen as origin and one of the indices $\mathbf{l}$ or $\mathbf{l}'$ in $\phi_{\alpha\beta}(\mathbf{l}, \mathbf{l}')$ can be taken as zero.

3. Invariance of the crystal potential energy against rigid body motion requires

$$\sum_{\mathbf{l}'} \phi_{\alpha\beta}(\mathbf{l}, \mathbf{l}') = 0 \quad \text{for all } \alpha, \beta \text{ and } \mathbf{l} \tag{2.2.6}$$

4. If S represents a symmetry operator in the point group of the lattice, i.e. an operation against which the lattice is invariant, then

$$\phi_{\delta\nu}(\mathbf{L}, \mathbf{L}') = \sum_{\alpha,\beta} S_{\delta\alpha} S_{\nu\beta} \phi_{\alpha\beta}(\mathbf{l}, \mathbf{l}') \tag{2.2.7}$$

where

$$r_\delta(\mathbf{L}) = \sum_{\alpha} S_{\delta\alpha} r_\alpha(\mathbf{l}) \tag{2.2.8}$$

and

$$r_\nu(\mathbf{L}') = \sum_{\beta} S_{\nu\beta} r_\beta(\mathbf{l}') \tag{2.2.8a}$$

i.e. $\mathbf{L}$ and $\mathbf{L}'$ are the lattice sites reached from $\mathbf{l}$ and $\mathbf{l}'$, respectively, with the help of the operator represented by the 3×3 matrix **S**, **S**, for example, can denote a proper or improper rotation about the origin.

5. $\phi_{\alpha\beta}(\mathbf{l}, \mathbf{l}') = \phi_{\beta\alpha}(\mathbf{l}', \mathbf{l}) = \phi_{\beta\alpha}(\mathbf{l}, \mathbf{l}')$

where the last equation is valid only for lattices with inversion symmetry which will be satisfied in the cases considered here.

We shall assume that the harmonic approximation is valid, so that the terms containing cubic and higher powers of u in Eq. (2.2.5) can be neglected. This approximation in the lattice theory corresponds to the assumption of linear elasticity in the continuum theory.

Since the first term in Eq. (2.2.5) is zero according to the first restriction given above, we obtain the following expression for the force on lattice site $\mathbf{l}$ due to the displacement $u(\mathbf{l}')$ of the atom $\mathbf{l}'$

$$F_\alpha(\mathbf{l}) = -\frac{\partial\Phi}{\partial u_\alpha(\mathbf{l})} = -\sum_{\beta} \phi_{\alpha\beta}(\mathbf{l}, \mathbf{l}')\, u_\beta(\mathbf{l}') \tag{2.2.9}$$

We notice from Eq. (2.2.9) that $\phi_{\alpha\beta}(\mathbf{l},\mathbf{l}')$ gives the negative of the α

component of the force on the atom **l** when the atom **l′** is displaced by unity in the β direction. The numbers $\phi_{\alpha\beta}(\mathbf{l}, \mathbf{l}')$ are, therefore, appropriately called the force constants. For fixed **l** and **l′** or $\mathbf{l}-\mathbf{l}'$, $\phi_{\alpha\beta}(\mathbf{l}-\mathbf{l}')$ can be regarded as a 3×3 matrix which is referred to as the force constant matrix between the atoms **l** and **l′** or between the origin and the atom $\mathbf{l}-\mathbf{l}'$.

The Fourier transform of the force constant matrix is defined as follows:

$$\phi(\mathbf{k}) = \sum_{\mathbf{l}} \phi(\mathbf{l}) \exp 2\pi i\, \mathbf{k}\cdot\mathbf{r}(\mathbf{l}) \tag{2.2.10}$$

where **k** is a vector in reciprocal space. It can be verified from Eqs. (2.2.3) and (2.2.4) that the addition of a reciprocal lattice vector **K** to **k** will not change the value of the exponential in Eq. (2.2.10). It is, therefore, sufficient to restrict the values of **k** within the first Brillouin zone of the lattice. The matrix $\phi(\mathbf{k})$ is a 3×3 matrix like $\phi(l)$. The vector **k** in Eq. (2.2.10) is also called the wave vector. Its magnitude is equal to the inverse wavelength of the elastic wave. In lattice theory only certain discrete values of **k** are allowed in the Brillouin zone (Maradudin et al., 1971).

The dynamical matrix $\mathbf{D}(\mathbf{k})$ is defined as

$$\mathbf{D}(\mathbf{k}) = \phi(\mathbf{k})/m \tag{2.2.11}$$

where m is the mass of an atom. The dynamical matrix contains almost all the information concerning the dynamics of the lattice. For example, the eigenvalues of $\mathbf{D}(\mathbf{k})$ are $\omega_j^2(\mathbf{k})$, where $\omega_j(\mathbf{k})$ ($j = 1$, 2 or 3) are the three possible frequencies when a wave of wave vector **k** is passing through the lattice. Similarly, the eigenvector $\mathbf{e}(\mathbf{k}j)$ of $\mathbf{D}(\mathbf{k})$ corresponding to the eigenvalue $\omega_j^2(\mathbf{k})$ defines the polarization vector of that particular mode. The necessary and sufficient condition for the stability of the lattice is that all the eigenvalues of $\mathbf{D}(\mathbf{k})$ are positive for all **k**.

The phase and group velocities of the wave can also be obtained from the eigenvalues of $\mathbf{D}(\mathbf{k})$ as follows:

$$C_{pj}(\mathbf{k}) = \omega_j(\mathbf{k})/k \tag{2.2.12}$$

and

$$C_{gj}(\mathbf{k}) = d\omega_j(\mathbf{k})/d\mathbf{k} \tag{2.2.12a}$$

Another important matrix concerning the dynamics of the lattices is the Green function matrix. Its Fourier transform is defined by

$$\mathbf{G}(\mathbf{k}; \omega^2) = \frac{1}{m}[-\mathbf{I}\,\omega^2 + D(\mathbf{k})]^{-1} \tag{2.2.13}$$

where **I** is a unit matrix and ω is the frequency. The Green function matrix for atoms **l** and **l′** is obtained by the inverse Fourier transform **G(k)** as follows:

$$\mathbf{G}(\mathbf{l}, \mathbf{l}'; \omega^2) = \frac{1}{N} \sum_{k} \mathbf{G}(\mathbf{k}; \omega^2) \exp 2\pi i \mathbf{k} \cdot [\mathbf{r}(\mathbf{l}) - \mathbf{r}(\mathbf{l}')] \qquad (2.2.14)$$

where the sum in Eq. (2.2.14) is carried out over all values of **k** in the first Brillouin zone and N is the total number of atoms in the crystal, which is equal to the number of allowed values of **k** in the Brillouin zone. For large N the summation in Eq. (2.2.14) can be replaced by integration in the Brillouin zone. It is obvious from Eq. (2.2.14) that $\mathbf{G}(\mathbf{l}, \mathbf{l}'; \omega^2)$ depends on **l** and **l′** only through their difference and therefore, like the force constant matrix, can be labelled by a single index $(\mathbf{l} - \mathbf{l}')$. This is a consequence of the translation symmetry of the lattice. The Green function is also called the response function, because it measures the response of the lattice to an external probe. In the static limit defined by $\omega^2 = 0$, $G_{\alpha\beta}(\mathbf{l}, \mathbf{l}')$ has a simple physical interpretation: it gives the displacement of the atom **l** in the α direction when a unit force is applied on the atom **l′** in the β direction. Thus, physically as well as mathematically, the Green function matrix is the reciprocal of the force constant matrix.

2.3 CONTINUUM THEORY

In the continuum model the discreteness of the lattice structure is ignored and the solid is regarded as a continuum of matter. As mentioned before, this model is a good representation of the solid for calculation of the bulk properties of the solid. The propagation of long elastic waves, which have wavelengths much longer than the interatomic spacing, is also correctly represented by the continuum model. This should be obvious physically because if the wavelength of the passing wave is much longer than the interatomic spacing, it cannot notice the details of the atomic arrangement. Its propagation will therefore depend upon some average properties of an extended region and thus the effect of atoms can be smeared out into a continuum. On account of its relative mathematical simplicity as compared with the lattice model, the continuum model is particularly useful for those engineering applications which do not require a detailed knowledge of the microstructure of the solids.

Let us consider a loaded solid, i.e. a solid upon which some external forces are applied. The effect of the external forces is to deform or 'strain' the solid. The deformation is resisted by the internal forces or 'stresses' arising in the solid. To quantify the stress and the strain in the solid, we assume a system of three orthogonal axes embedded in the solid. Let us consider a point with position vector **r**, given by

$$\mathbf{r} = x\mathbf{f} + y\mathbf{g} + z\mathbf{h} \qquad (2.3.1)$$

where **f**, **g** and **h** are the unit vectors along the positive x, y and z axes, respectively. If **u** denotes the displacement vector of the point originally at **r**, then symmetrized strain tensor $\boldsymbol{\epsilon}$ is defined by

$$\varepsilon_{\alpha\beta} = \tfrac{1}{2}\left(\frac{\partial u_\alpha}{\partial r_\beta} + \frac{\partial u_\beta}{\partial r_\alpha}\right) \tag{2.3.2}$$

where α, β denote the Cartesian components.

The stress is defined as the force per unit area on a given plane. Mathematically the stress tensor $\sigma_{\alpha\beta}$ is the force in the α direction acting on unit area of the surface $\beta = \text{constant}$. If the vector **F** denotes the force per unit volume in a small region surrounding the point at **r**, then we have the relation

$$F_\alpha = -\sum_\beta \frac{\partial \sigma_{\alpha\beta}}{\partial r_\beta} \tag{2.3.3}$$

Let us consider a situation in which the load or the applied stress in a solid is gradually increased from zero to a large value. The response of the solids can be divided into the following three stages.

1. Stage I, or the elastic zone—up to a certain value of the stress, the strain will be proportional to the stress provided cross sectional area remains constant. This stage is treated by the linear elasticity theory and will be described in this section. Most of the work described in this book is based upon the linear elasticity theory, which assumes that the strain (or the stress) is sufficiently small for the solid to be in Stage I.
2. Stage II, or the plastic zone—when the stress or the strain in the solid exceeds a certain value, stress is no longer proportional to strain and the solid enters the plastic zone, or Stage II. According to one theoretical approach for this stage, the stress is written formally as proportional to strain with the constant of proportionality taken as a function of the stress or the strain or both. The value of the stress at the the start of Stage II is sometimes called the yield stress.
3. Stage III, or the fracture zone—when the stress or the strain is increased even further, ultimately the atomic bonds are broken and the solid is cracked or fractured. The solid then enters Stage III. Initially the cracks are localized. They increase in size as the stress is increased further. The cracks, in certain circumstances, can propagate very fast—with the velocity of sound—in a solid. The study of the growth and the propagation of cracks comes under fracture mechanics, which will be briefly reviewed in Section 2.5.

In the linear elasticity theory, which is valid for small deformations, the stress and strain are proportional to each other according to Hooke's law. This is expressed by the following two equations:

$$\sigma_{\alpha\beta} = \sum_{\nu\delta} c_{\alpha\beta\nu\delta}\, \epsilon_{\nu\delta} \tag{2.3.4}$$

and

$$\epsilon_{\alpha\beta} = \sum_{\nu\delta} s_{\alpha\beta\nu\delta}\, \sigma_{\nu\delta} \tag{2.3.4a}$$

c and **s**, which are tensors of the fourth rank, are called elastic stiffness and elastic compliance constants, respectively. The quantities $c_{\alpha\beta\nu\delta}$ are also called moduli of elasticity or simply elastic constants, the last name being more common in solid state science. In engineering applications it is more usual to refer to quantities such as Young's modulus, Bulk modulus, etc., which are derived from the elastic constants as the moduli of elasticity or simply the elastic moduli.

The elastic energy density or the elastic energy per unit volume in a strained solid of volume V is given by

$$\frac{U}{V} = 1/2 \sum_{\alpha\beta} \sigma_{\alpha\beta}\, \epsilon_{\alpha\beta} \tag{2.3.5}$$

which with the help of Eqs. (2.3.4) and (2.3.4a) can be written in the following two equivalent quadratic forms:

$$\frac{U}{V} = 1/2 \sum_{\substack{\alpha\beta \\ \nu\delta}} c_{\alpha\beta\nu\delta}\, \epsilon_{\alpha\beta}\, \epsilon_{\nu\delta} \tag{2.3.5a}$$

or

$$\frac{U}{V} = 1/2 \sum_{\substack{\alpha\beta \\ \nu\delta}} s_{\alpha\beta\nu\delta}\, \sigma_{\alpha\beta}\, \sigma_{\nu\delta} \tag{2.3.5b}$$

The number of independent components of **c** and **s** is greatly reduced by symmetry. In a cubic crystal there are only three independent components. The number of independent components for tetragonal and hexagonal systems is six and five, respectively. In practice the components of **c** and **s** are labelled by two Voigt indices, i and j ($i, j = 1$–6), instead of four Cartesian indices, α, β, etc. Thus, for example, the three independent elastic constants for a cubic crystal are c_{11}, c_{12} and c_{44}. In the case of an isotropic cubic solid only two of them are independent. For a discussion of the symmetry of the elastic constants see Nye (1957) and Fedorov (1968).

In this book we shall be frequently dealing with solids having tetragonal or hexagonal symmetry. In the former case there are six independent elastic constants, namely c_{11}, c_{12}, c_{13}, c_{33}, c_{44} and c_{66}, and of course six elastic compliance constants with the same Voigt indices. (We shall assume that the z or the 3 axis is parallel to the tetragonal axis.) In this case c_{44} can be identified with the longitudinal shear modulus and c_{66} with the transverse shear modulus for a shear strain ϵ_{xy}. Such simple relations cannot be given in general between all the elastic constants and the elastic moduli used in engineering applications, but of course a complicated set of relations can be written, since the two sets of moduli uniquely define each other.

A solid with hexagonal symmetry is transversely isotropic and is characterized by only five independent elastic constants, c_{11}, c_{12}, c_{13}, c_{33} and c_{44}, and the corresponding elastic compliance constants. In fact a solid having tetragonal symmetry will reduce to a transversely isotropic solid or one having hexagonal symmetry if $s_{66} = 2(s_{11} - s_{12})$ or $c_{11} - c_{12} = 2c_{66}$.

For the case of a solid with hexagonal symmetry the relations between the five independent engineering elastic moduli and the elastic constants are given below:

$$E_3 = c_{33} - \frac{2c_{13}^2}{c_{11} + c_{12}}$$

$$K = \tfrac{1}{2}(c_{11} + c_{12})$$

$$M_{12} = \tfrac{1}{2}(c_{11} - c_{12})$$

$$M_{13} = c_{44}$$

$$\nu = \frac{c_{13}}{c_{11} + c_{12}} = \tfrac{1}{2}\left[\frac{c_{33} - E_3}{K}\right]^{1/2} \qquad (2.3.5c)$$

where E_3 is the Young's modulus parallel to the z axis (taken along the hexagonal axis); M_{13} is the shear modulus parallel to the z axis; M_{12} is the shear modulus in the xy plane or normal to the z axis; K is the plane strain bulk modulus for lateral dilatation in the xy plane in the absence of any longitudinal contraction along the z axis; and ν (sometimes also denoted by ν_{31}) is the Poisson ratio connecting strain parallel and normal to the z axis. These moduli, defined in a slightly different way, are also sometimes used in engineering applications. Various other notations are also common, such as G and μ for the shear moduli.

In terms of the elastic compliance constants s_{11} is the inverse of the 'transverse' Young's modulus, i.e. Young's modulus measured normal to the z axis, s_{33} is the inverse of the longitudinal Young's modulus E_3; and s_{44} is the inverse of the rigidity modulus when the shear force is applied in the xy plane. The rigidity modulus when the shear force is applied in a plane containing the z axis depends upon s_{44} and s_{66}. (It should be noted that the shear force in these definitions has been introduced only in order to specify the nature of deformation. The elastic moduli are material properties of the solid and are independent of the applied stress.) Finally, in this scheme s_{12} and s_{13} are sometimes referred to as the Poisson ratio constants.

The use of different definitions of the elastic moduli in engineering applications is dictated by convenience in a particular problem. In this book, accordingly, we have sometimes used slightly different definitions of the elastic moduli. However, this should not create any confu ion, since they can all be related to the tensor elastic constants, i.e. c_{ij}.

The 3 × 3 matrix in the continuum theory, which corresponds to the dynamical matrix in the lattice theory, is called the Green–Christoffel matrix. It will be denoted by Λ and its matrix elements (Fedorov, 1968) are given below:

$$\Lambda_{\alpha\beta}(\mathbf{k}) = \frac{4\pi^2}{\rho} \sum_{\nu\delta} c_{\alpha\nu\beta\delta}\, k_\nu k_\delta \tag{2.3.6}$$

where **k** is the wave vector and ρ is the density of the solid. The factor $4\pi^2$ in Eq. (2.3.6) arises owing to the definition of **k** and is not essential. As in the case of the dynamical matrix in the lattice theory, the eigenvalues of $\Lambda(\mathbf{k})$ are the squared frequencies $\omega_j^2(\mathbf{k})$ and its eigenvectors denote the polarization of the elastic wave. The phase and the group velocities of the wave are given by Eq. (2.2.12) and the Green function is defined as

$$\mathbf{G}(\mathbf{k}) = [-\mathbf{I}\omega^2 + \Lambda(\mathbf{k})]^{-1} \tag{2.3.7}$$

which is similar to Eq. (2.2.13).

As an example, the elements of $\Lambda(\mathbf{k})$ for cubic crystals are given below:

$$\Lambda_{\alpha\alpha}(\mathbf{k}) = \frac{4\pi^2}{\rho}\left[(c_{11} - c_{44})\, k_\alpha^2 + c_{44} k^2\right] \tag{2.3.8}$$

$$\Lambda_{\alpha\beta}(\mathbf{k}) = \frac{4\pi^2}{\rho}(c_{12} + c_{44})\, k_\alpha k_\beta \;(\alpha \neq \beta) \tag{2.3.8a}$$

where ρ is the density of the solid.

An isotropic cubic crystal is characterized by the fact that the velocity of an elastic wave in such a crystal is independent of the direction of the wave. The condition for elastic isotropy is given below:

$$c_{11} - c_{12} = 2c_{44}$$

An effect of the isotropy on the elastic wave propagation is that the three polarizations (sometimes referred to as the branches) of the wave are truely longitudinal (one branch) and transverse (two branches). The two transverse branches are also called the shear waves.

Similarly a uniaxial crystal such as a tetragonal crystal can have transverse isotropy in a plane normal to the tetragonal axis. The velocity of a wave travelling in this plane will be independent of the direction of the wave and it will be truely longitudinal or transverse. The condition for transverse isotropy is the following:

$$c_{11} - c_{12} = 2c_{66}$$

This condition is automatically satisfied by hexagonal crystals, which are therefore always transversely isotropic.

In the case of a general anisotropic crystal, the velocity of the wave depends upon the direction of its propagation. Also, the three branches are not truely longitudinal or transverse except when the wave is propagating in the high-symmetry directions. However, it is usual to refer to the three branches of the wave as the pseudo-longitudinal and the pseudo-transverse waves, according to whether they are nearly longitudinal or nearly transversely polarized.

As mentioned before, the continuum model correctly describes the propagation of long elastic waves, i.e. for small values of k. In this limit, therefore, the two matrices $\mathbf{D}(\mathbf{k})$ and $\Lambda(\mathbf{k})$ should be identical. We see from Eqs. (2.3.8) and (2.3.8a) that $\Lambda(\mathbf{k})$ is quadratic in k, whereas $\mathbf{D}(\mathbf{k})$ is periodic in k (see Eq. 2.2.10). It can be shown by symmetry (Maradudin et al., 1971) that for monatomic Bravais lattices $\mathbf{D}(-\mathbf{k}) = \mathbf{D}(\mathbf{k})$ and therefore the first term in the expansion of $\mathbf{D}(\mathbf{k})$ for low values of k must be quadratic in k, which, according to what has been said above, must be identical with the Green–Christoffel matrix. This is the essence of the method of long waves due to Born (Born and Huang, 1954) and can be expressed mathematically as follows:

$$\lim_{k \to 0} \mathbf{D}(\mathbf{k}) = \Lambda(\mathbf{k}) + \mathbf{O}(k^4) + \ldots \qquad (2.3.9)$$

With the help of Eq. (2.3.9) some useful relations can be derived between the force constants and the elastic constants of a crystal.

2.4 EXAMPLE OF A SIMPLE CUBIC LATTICE

As an illustrative example we shall apply the formalism given in Sections 2.2 and 2.3 to a simple cubic lattice model in which each atom interacts with its nearest neighbours only. In a simple cubic lattice the atoms are located at the vertices of a cube, which is taken to be the unit cell as shown in Figure 2.1. Each atom has 6 nearest neighbours and 12 next-nearest neighbours. Taking a lattice site as the origin, the coordinates of its nearest neighbours are $(\pm 1, 0, 0)$, $(0, \pm 1, 0)$ and $(0, 0, \pm 1)$ in units of a, where a is the lattice constant, i.e. the edge length of the cubic unit cell.

The three real space basis lattice vectors are given by:

$$\left.\begin{aligned} \mathbf{a}_1 &= a\mathbf{f} \\ \mathbf{a}_2 &= a\mathbf{g} \\ \mathbf{a}_3 &= a\mathbf{h} \end{aligned}\right\} \tag{2.4.1}$$

and the reciprocal space basis lattice vectors according to Eq. (2.2.3) are as follows:

$$\left.\begin{aligned} \mathbf{b}_1 &= \frac{1}{a}\mathbf{f} \\ \mathbf{b}_2 &= \frac{1}{a}\mathbf{g} \\ \mathbf{b}_3 &= \frac{1}{a}\mathbf{h} \end{aligned}\right\} \tag{2.4.2}$$

where **f**, **g** and **h** have been defined in Eq. (2.3.1).

We see from the basis vectors $\mathbf{b}_i$ given in Eq. (2.4.2) that the reciprocal lattice also has a simple cubic structure with lattice constant $1/a$. A unit cell of the reciprocal lattice and therefore the Brillouin zone in the present case is a cube. The wave vectors can thus be restricted to the range

$$-\frac{1}{2a} \leqslant k_x, k_y, k_z \leqslant \frac{1}{2a} \tag{2.4.3}$$

The symmetry operators of the cubic group against which the simple cubic lattice is invariant have been given in several texts on group theory (see, for example, Mariot, 1962). With the help of these operators and from Eq. (2.2.7) it can be shown that a nearest-neighbour force constant matrix can be written in the following most general form:

$$\phi(1, 0, 0) = \phi(-1, 0, 0) = -\begin{pmatrix} \lambda & 0 & 0 \\ 0 & \mu & 0 \\ 0 & 0 & \mu \end{pmatrix} \tag{2.4.4}$$

where λ and μ are independent, as yet undetermined parameters. The negative sign for the matrix in Eq. (2.4.4) has been chosen for convenience. The force constant matrices for other nearest neighbours can be written down by symmetry, i.e. by applying appropriate transformations to Eq. (2.4.4). These are given below:

$$\phi(0, \pm 1, 0) = -\begin{pmatrix} \mu & 0 & 0 \\ 0 & \lambda & 0 \\ 0 & 0 & \mu \end{pmatrix} \tag{2.4.4a}$$

and

$$\phi(0, 0, \pm 1) = -\begin{pmatrix} \mu & 0 & 0 \\ 0 & \mu & 0 \\ 0 & 0 & \lambda \end{pmatrix} \tag{2.4.4b}$$

The force constant matrix for an atom with itself, i.e. $\phi(0)$ as obtained with the help of Eqs. (2.2.6) and (2.4.4)–(2.4.4b) is as follows:

$$\phi(0) = 2(\lambda + 2\mu)\begin{pmatrix} 1 & 0 & 0 \\ 0 & 1 & 0 \\ 0 & 0 & 1 \end{pmatrix} \tag{2.4.5}$$

In the approximation in which only nearest-neighbour interactions are allowed, the matrices $\phi(\mathbf{k})$ and $\mathbf{D}(\mathbf{k})$ are diagonal for all values of $\mathbf{k}$. From Eqs. (2.2.10), (2.4.4)–(2.4.4b) and (2.4.5) we obtain

$$\mathbf{D}(\mathbf{q}) = \frac{1}{m}\phi(\mathbf{q}) = \frac{1}{m}$$

$$\times \begin{bmatrix} 2\lambda(1-\cos q_x) + 2\mu(1-\cos q_y) + 2\mu(1-\cos q_z) & 0 & 0 \\ 0 & 2\lambda(1-\cos q_y) + 2\mu(1-\cos q_z) + 2\mu(1-\cos q_x) & 0 \\ 0 & 0 & 2\lambda(1-\cos q_z) + 2\mu(1-\cos q_x) + 2\mu(1-\cos q_y) \end{bmatrix} \tag{2.4.6}$$

where m is the mass of an atom and the dimensionless wave vector $\mathbf{q}$ is defined as

$$q_\alpha = 2\pi k_\alpha a \qquad \text{for } \alpha = x, y \text{ or } z \tag{2.4.7}$$

From Eq. (2.4.3) the range of $\mathbf{q}$ is given by

$$-\pi \leqslant q_x, q_y, q_z \leqslant \pi \tag{2.4.8}$$

In the present case the polarization vectors are obviously independent of $\mathbf{k}$ and are simply given by (1, 0, 0), (0, 1, 0) and (0, 0, 1). The three frequencies for any value of $\mathbf{k}$ are simply the square roots of the diagonal elements of $\mathbf{D}(\mathbf{q})$ as given in Eq. (2.4.6). Some frequencies will of course be degenerate for $\mathbf{q}$ in certain symmetry directions.

In the long-wavelength limit

$$\lim_{q\to 0} \mathbf{D}(\mathbf{q}) = \frac{1}{m}\begin{bmatrix} \lambda q_x^2+\mu q_y^2+\mu q_z^2 & 0 & 0 \\ 0 & \lambda q_y^2+\mu q_z^2+\mu q_x^2 & 0 \\ 0 & 0 & \lambda q_z^2+\mu q_x^2+\mu q_y^2 \end{bmatrix} \tag{2.4.9}$$

The elements of the Green–Christoffel matrix for the cubic lattice have been given in Eqs. (2.3.8) and (2.3.8a). In the present case, since the volume of a unit cell is a^3, $\rho = m/a^3$. From Eq. (2.3.9) and using Eqs. (2.3.8), (2.3.8a) and (2.4.9), we obtain

$$\mu = ac_{44} \tag{2.4.10}$$

$$\lambda = ac_{11} \tag{2.4.10a}$$

and

$$c_{12} = -c_{44} \tag{2.4.11}$$

Thus we see that the elastic constants and the interatomic force constants are related to each other. It should be emphasized that the lattice model described in this section is highly oversimplified and has been given here only for the purpose of illustration. For example, it has the physically unrealistic feature that the atomic displacements in the x, y and z directions are not coupled to one another ($\mathbf{D}(\mathbf{k})$ is diagonal for all values of $\mathbf{k}$), which leads to the relation (2.4.11) among the elastic constants. Of course the procedure given here for the construction of $\mathbf{D}(\mathbf{k})$ and the derivation of some relations between the force constants and the elastic constants is quite general and applicable to real systems. In general, however, there are more force constants than elastic constants in a crystal. The force constants, therefore, cannot be determined from the elastic constants alone without the help of some other measurements (e.g. phonon frequencies) or some restrictive assumptions on the nature and the range of the force constants. In practice, in the field of lattice dynamics, the phonon frequencies are measured by neutron scattering and from these data the force constants can be determined (Maradudin et al., 1971; see, however, Leigh et al., 1971).

2.5 FRACTURE MECHANICS

When the stress in a solid increases beyond a certain value, the atomic bonds are eventually broken and cracks appear in the solid, i e. the solid fractures. The cracks will expand under the effect of any further stress (sometimes even without it). This stage was referred to as Stage III in Section 2.3. The science of fracture mechanics deals with the study of factors governing the stability, growth and propagation of cracks in solids. Whereas it would be highly desirable to have an appropriate lattice theory of fracture in solids, almost all the theories of the fracture properties of solids are based on the continuum model of a solid. A very brief account of some of the basic formulae of fracture mechanics will be given here. For details, a reference should be made to standard textbooks (e.g. Ford and Alexander, 1963) or review articles (such as that by Eshelby, 1971).

We shall first introduce the three fundamental modes of deformation in a solid containing a crack. Let us consider an isotropic solid in the form of a rectangular prism such that its dimensions are much larger than those of the crack itself. We choose a frame of reference in which the origin of the coordinates is at the crack tip, the z axis is normal to the plane of the paper and the crack is taken along the negative x axis (see Figure 2.2) the variable r and θ will denote the polar coordinates with reference to the same origin. Let us also assume that the length of the crack is $2a$.

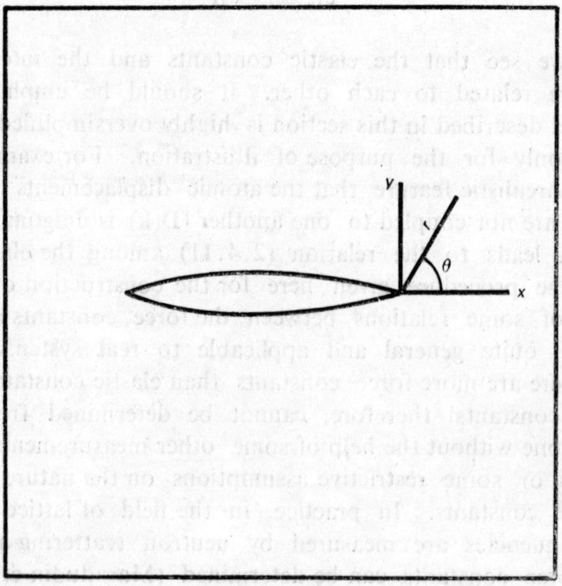

Figure 2.2 The coordinate system used for the study of crack propagation. The origin is at the tip of the crack.

Any deformation of the crack in the solid can be expressed as a combination of the following fundamental modes of deformation.

Plane strain modes

In these modes the displacement field at any point in the solid is confined to the x, y plane, i.e. its z component is zero and it is a function of x and y coordinates only (independent of the z coordinate). There are two types of deformation modes in this class, which are discussed below.

Mode I deformation

This mode is obtained when a transverse plane stress is applied to the crack, which then opens up along the y axis,

If u, v and w denote, respectively, the x, y and z components of the displacement field at any point on the x, y plane, then in the present case $w=0$ and u and v are functions of x and y only. The equations of equilibrium in the present case are simply given by

$$\nabla^4 u = 0 \tag{2.5.1}$$

and

$$\nabla^4 v = 0 \tag{2.5.1a}$$

where

$$\nabla^4 = \nabla^2 \nabla^2$$

and ∇^2 is the two-dimensional Laplacian operator, i.e.

$$\nabla^2 = \frac{\partial^2}{\partial x^2} + \frac{\partial^2}{\partial y^2} \tag{2.5.2}$$

It can be shown that for an isotropic solid the dominating term in the solution of Eqs. (2.5.1) and (2.5.1a) close to the crack tip is given by

$$u = A_2 r^{1/2} \cos\frac{\theta}{2} - \tfrac{1}{2} pxe \tag{2.5.3}$$

$$v = A_3 r^{1/2} \sin\frac{\theta}{2} - \tfrac{1}{2} pye \tag{2.5.3a}$$

where e, the dilatation, is given by

$$e = Tr\,\varepsilon = A_1 r^{-1/2} \cos\frac{\theta}{2} \tag{2.5.4}$$

ε, as defined in Section 2.3, is the strain tensor; Tr denotes trace;

$$p = \frac{1}{1-2\nu} \tag{2.5.5}$$

ν being the Poisson ratio; and A_1, A_2 and A_3 are arbitrary constants which have to be determined with the help of the boundary conditions. Two boundary conditions—(1) the shear stress σ_{xy} is zero on the crack faces and (2) the values of e obtained in terms of strain as $Tr\,\varepsilon$ and from the second part of Eq. (2.5.4) must be consistent with each other—give A_2 and A_3 in terms of A_1. By suitably defining the undetermined constant A_1 in terms of the parameter $K_{\rm I}$ we obtain the following expressions for the elastic field near the crack tip;

$$\sigma_{yy} = \frac{K_{\rm I}}{(2\pi)^{1/2}}\, r^{-1/2} \frac{1}{4}\left[5\cos\frac{\theta}{2} - \cos\frac{5\theta}{2}\right] \tag{2.5.6}$$

$$\sigma_{xx} + \sigma_{yy} = \frac{K_I}{(2\pi)^{1/2}}\, 2r^{-1/2} \cos\frac{\theta}{2} \tag{2.5.6a}$$

$$\sigma_{xy} = \frac{K_I}{(2\pi)^{1/2}}\, r^{-1/2}\, \tfrac{1}{4}\left[-\sin\frac{\theta}{2} + \sin\frac{5\theta}{2}\right] \tag{2.5.6b}$$

$$u = T \cos\frac{\theta}{2} \tag{2.5.7}$$

$$v = T \sin\frac{\theta}{2} \tag{2.5.7a}$$

where

$$T = \frac{K_I}{(2\pi)^{1/2}}\, \frac{4\,(1-\nu^2)}{E}\, r^{1/2}\left[1 - \frac{1}{2\,(1-\nu)} \cos^2\frac{\theta}{2}\right] \tag{2.5.7b}$$

and E is the Young's modulus of the solid.

From Eqs. (2.5.6)–(2.5.7b) we can derive the values of the stress across the plane of the crack just ahead of the tip ($\theta = 0$) and the relative displacement of the crack faces just behind the tip ($\theta = \pm\pi$), which are given below:

$$\sigma^0_{yy} = [\sigma_{yy}]_{\theta=0} = \frac{K_I}{(2\pi)^{1/2}}\, r^{-1/2} \tag{2.5.8}$$

and

$$\Delta\, v^0 = [v]_{\theta=\pi} - [v]_{\theta=-\pi} = \frac{K_I}{(2\pi)^{1/2}}\, \frac{8\,(1-\nu^2)}{E}\, r^{1/2} \tag{2.5.8a}$$

The parameter K_I occurring in the above equations is called the stress intensity factor. The suffix I denotes the mode to which it refers. It is defined in terms of the stress across the plane of the crack just ahead of the crack as the coefficient of the square root singularity at the origin (in units of $\sqrt{2\pi}$). It plays an important role in the field of fracture mechanics. As can be seen from Eqs. (2.5.6)–(2.5.7b), a knowledge of K_I is sufficient to determine the stress and the displacement field near the crack in a solid. The actual calculation of K for a solid is quite involved and depends upon the particular type of loading and the geometry of the crack.

An important property of K is that if at the tip of a certain crack K has the same value as at the tip of a second crack in an identical material, then the elastic fields are identical at the two tips even if the geometry and the loading in the two situations are different. This is because the effect of loading and the geometry of the crack has been absorbed in K.

Mode II deformation

This mode of deformation is obtained by applying a shear stress on the cracked solid which makes the faces of the crack slide over one another parallel to the xy plane. This is also a plane strain mode, so that $w = 0$ and u, v are functions of x and y only. The mathematical treatment of this mode is quite similar to that of mode I. In this case the expression for the stress just ahead of the crack tip and the relative displacement of the crack faces just behind the tip are the following:

$$\sigma_{xy}^{0} = \frac{K_{II}}{(2\pi)^{1/2}} r^{-1/2} \tag{2.5.9}$$

and

$$\Delta u^{\circ} = \frac{K_{II}}{(2\pi)^{1/2}} \frac{8(1-\nu^2)}{E} r^{1/2} \tag{2.5.9a}$$

where K_{II} denotes the stress itensity factor for mode II deformation.

Anti-plane strain mode

Mode III deformation

This mode can be obtained by applying a shear stress so that the crack faces slide over each other perpendicular to the xy plane. This is an anti-plane strain mode, which is characterized by the equations

$$u = v = 0$$

and

$$w = w(x, y) \text{ (i.e. a function of } x \text{ and } y \text{ only)}$$

The equation of equilibrium in this case is the following:

$$\nabla^2 w = 0 \tag{2.5.10}$$

The leading terms in the expressions for the stress and the displacements near the crack tip are given below:

$$\sigma_{zy} = \frac{K_{III}}{(2\pi)^{1/2}} r^{-1/2} \cos\frac{\theta}{2} \tag{2.5.11}$$

$$\sigma_{zx} = -\frac{K_{III}}{(2\pi)^{1/2}} r^{-1/2} \sin\frac{\theta}{2} \tag{2.5.11a}$$

and

$$w = \frac{K_{III}}{(2\pi)^{1/2}} \frac{2}{\mu} r^{1/2} \sin\frac{\theta}{2} \tag{2.5.12}$$

where μ is the shear modulus of the isotropic solid and K_{III} is the stress intensity factor for mode III deformation.

From Eqs. (2.5.11)–(2.5.12) we obtain the following values for the stress just ahead of the crack tip and the relative displacement of the crack faces just behind the crack tip:

$$\sigma^0_{zy} = [\sigma_{zy}]_{\theta=0} = \frac{K_{III}}{(2\pi)^{1/2}} r^{-1/2} \qquad (2.5.13)$$

and

$$\Delta w^0 = [\Delta w]_{\theta=\pi} - [\Delta w]_{\theta=0} = \frac{K_{III}}{(2\pi)^{1/2}} \frac{\Delta}{\mu} r^{1/2} \qquad (2.5.13a)$$

In engineering applications of the theory of cracks in plates, it is usual to define the average of the field across the thickness of the plate, which, for example, in the case of mode I deformation is given by:

$$\overline{\sigma_{yy}} = \frac{K_I}{(2\pi)^{1/2}} r^{-1/2} \qquad (2.5.14)$$

and

$$\overline{\Delta v} = \frac{K_I}{(2\pi)^{1/2}} r^{1/2} \frac{8}{E} \qquad (2.5.14a)$$

Note the absence of the factor $(1 - \nu^2)$ in Eq. (2.5.14a), which is a rather unsatisfactory feature of this averaging process.

Another parameter in fracture mechanics, which is related to K, the stress intensity factor, is the crack extension force, usually denoted by G. It is defined in terms of the work ΔW done during the extension of a crack of length Δa by an amount δa as follows

$$\Delta W = G \, \Delta a \, \delta a \qquad (2.5.15)$$

For equilibrium, ΔW must be equal to the negative of the change in the total energy of the system, i.e.

$$\Delta W = -\Delta U_{tot} \qquad (2.5.16)$$

where

$$\Delta U_{tot} = \Delta U_{el} + \Delta U_{pot} \qquad (2.5.16a)$$

ΔU_{el} and ΔU_{pot} denote the changes in the elastic energy of the solid and the potential energy of the loading mechanism.

The relationships between G and K in the three modes of deformation are given below:

Mode I

$$G_I = \frac{K_I^2}{E}(1 - \nu^2) \qquad (2.5.17)$$

Mode II

$$G_{II} = \frac{K_{II}^2}{E}(1 - \nu^2) \qquad (2.5.17a)$$

Mode III

$$G_{III} = \frac{K_{III}^2}{2\mu} \qquad (2.5.17b)$$

The average of G across a crack in a plate for the plane strain modes is given by:

$$\overline{G}_{\mathrm{I,II}} = \overline{K}_{\mathrm{I,II}}/E \tag{2.5.18}$$

which is based upon an assumption that the average of a product is the product of the averages.

The Griffith–Irwin–Orowan fracture criterion is given by:

$$K_{\mathrm{I}} \geqslant K_{\mathrm{IC}} \tag{2.5.19}$$

or the equivalent condition

$$G_{\mathrm{I}} \geqslant G_{\mathrm{IC}} \tag{2.5.19a}$$

where the subscript C denotes the critical value at which the crack will start to extend. The fracture stress at the critical point can then be written as

$$\sigma_C = \frac{K_C}{(2\pi)^{1/2}} a^{-1/2} \tag{2.5.20}$$

In Griffith's theory, which is applicable to brittle materials, G_{IC} is equal to 2γ, the surface energy of the two faces of the crack. Orowan's extension of the Griffith theory consists of defining G_{IC} as follows:

$$G_{\mathrm{IC}} = 2(\gamma + \gamma_{pl}) \tag{2.5.21}$$

where γ_{pl} is a measure of the work done in plastic deformation near the crack tip. With this modification, Eq. (2.5.21) can be applied to the fracture of ductile materials as well.

It may be remarked that the fracture stress as defined in Eq. (2.5.20) is the stress applied to the plate. It is not a material property of the solid, since it depends on the absolute size of the crack. On the other hand, the stress intensity factor and the crack extension force at the fracture stress are material properties of the solid which express the resistance of the material to fracture. Further, there is a critical size of a crack which will cause fracture at almost any stress level. Some other size effects also exist owing to which two bodies which are geometrically similar but different in size will not fracture at the same level of the applied stress.

2.6 FINITE ELEMENT METHOD

The finite element method is an extremely powerful method for a numerical solution of many types of problems concerning structural and continuum mechanics of solids. In this section only the basic idea of this method will be given. For details of the method and its several applications, a reference may be made to the exhaustive treatise by Zienkiewicz (1971).

Most problems in structural and continuum mechanics of solids can be reduced to the problem of minimization of the integral of a function over a certain region in space. As an example, let us consider the elastic energy of a solid due to a displacement field $\mathbf{u}(\mathbf{r})$ at a point whose position vector is $\mathbf{r}$. For simplicity, let us consider the case in which u is a scalar for any value of $\mathbf{r}$ and $\mathbf{r}$ is a two-dimensional vector. This situation corresponds to, for example, the problem of anti-plane strain when we can identify u as the z component—the only nonvanishing component—of the displacement field on the xy plane. Let us write the elastic energy of such a system in the following simple form:

$$U = \int P(\mathbf{r})\, u(\mathbf{r})\, d\mathbf{r} + \tfrac{1}{2} \int Q(\mathbf{r})\, u^2(\mathbf{r})\, d\mathbf{r} \tag{2.6.1}$$

where P and Q are known and our task is to determine $u(\mathbf{r})$ so that U is minimum.

In the finite element method we divide the whole region into N small, finite regions—called the elements. The choice of geometrical shapes of the elements is arbitrary. It depends on the nature of the particular problem. However, a triangular element appears to be the most popular choice for two-dimensional problems. Having divided the whole space into elements, the integrals in Eq. (2.6.1) are expressed as the sum over the integrals in each element, i.e.

$$U = \sum_{n=1}^{N} \left[\int^{n} P(\mathbf{r})\, u(\mathbf{r})\, d\mathbf{r} + \tfrac{1}{2} \int^{n} Q(\mathbf{r})\, u^2(\mathbf{r})\, d\mathbf{r} \right] \tag{2.6.2}$$

where n over the integral sign indicates that the integration is to be carried out over the nth element.

Let us suppose that we want to obtain the values of $u(\mathbf{r})$ at M grid points in each element. These points can be chosen at the vertices of the element, or some at the vertices and some at the edges of the element. The continuity condition requires that at the grid points which are common to more than one element, the values of u derived from those elements should be the same.

Let the values of $u(\mathbf{r})$ at the M grid points $\mathbf{r}_1, \mathbf{r}_2, \ldots, \mathbf{r}_m$ in a particular element be denoted by $u_1, u_2, \ldots, u_m$. We now choose a set of M functions $F_1(\mathbf{r}), F_2(\mathbf{r}), \ldots, F_m(\mathbf{r})$ and write for any point in that element

$$u(\mathbf{r}) = \sum_{j=1}^{M} A_j F_j(\mathbf{r}) \tag{2.6.3}$$

where A_j are the expansion coefficients, which can be determined by using the following M equations:

$$u_l = u(\mathbf{r}_l) = \sum_{j=1}^{M} F_j(\mathbf{r}_l)\, A_j = \sum_{j=1}^{M} C_{l,j}\, A_j \tag{2.6.4}$$

where the elements of the matrix **C** are defined as

$$C_{i,j} = F_j(\mathbf{r}_i) \tag{2.6 5}$$

Regarding u_i and A_i, respectively, as the elements of column matrices (vectors) **u** and **A** of order M, we can write Eq. (2.6.4) in the following matrix form:

$$\mathbf{u} = \mathbf{CA} \tag{2.6.5a}$$

or

$$\mathbf{A} = \mathbf{Tu} \tag{2.6.5b}$$

where

$$\mathbf{T} = \mathbf{C}^{-1} \tag{2.6.5c}$$

If the functions $F(\mathbf{r})$ are properly chosen, **C** must have an inverse and **T** must be finite.

With the help of (2.6.3), (2.6.2) can be written as:

$$U = \sum_{n=1}^{N} \sum_{i=1}^{M} P_i^n A_i^n + \tfrac{1}{2} \sum_{n=1}^{N} \sum_{i,\,j=1}^{M} Q_{i,j}^n A_i^n A_j^n \tag{2.6.6}$$

where

$$P_i^n = \int^{n} P(\mathbf{r})\, F_i(\mathbf{r})\, d\mathbf{r} \tag{2.6.7}$$

and

$$Q_{i,j}^n = \int^{n} Q(\mathbf{r})\, F_i(\mathbf{r})\, F_j(\mathbf{r})\, d\mathbf{r} \tag{2.6.8}$$

where the superscript n labels the nth element. The quantities P_i^n and $Q_{i,j}^n$ can be regarded, respectively, as the elements of a column matrix **P** of dimension MN and a square matrix **Q** of dimension $MN \times MN$. Similarly A_i^n and u_i^n can be regarded as the elements of the column matrices **A** and **u**, respectively, of dimensions MN. Then Eq. (2.6.6) can be simply written in the matrix notation as follows:

$$U = \tilde{\mathbf{P}}\,\mathbf{A} + \tfrac{1}{2}\tilde{\mathbf{A}}\,\mathbf{Q}\,\mathbf{A} \tag{2.6.9}$$

which with the help of Eq. (2.6.5) is given by:

$$U = -\tilde{\mathbf{R}}\,\mathbf{u} + \tfrac{1}{2}\tilde{\mathbf{u}}\,\mathbf{S}\,\mathbf{u} \tag{2.6.10}$$

where

$$\mathbf{R} = -\tilde{\mathbf{T}}\,\mathbf{P} \tag{2.6.11}$$

and

$$\mathbf{S} = \tilde{\mathbf{T}}\,\mathbf{Q}\,\mathbf{T} \tag{2.6.11a}$$

Now **u** can be determined from the minimization condition on U as follows:

$$\frac{\partial U}{\partial u} = 0 = -\mathbf{R} + \mathbf{S}\mathbf{u}$$

or

$$\mathbf{u} = \mathbf{S}^{-1}\,\mathbf{R} \qquad (2.6.12)$$

which gives the required values of the displacement field.

The characteristics which make the finite element method so powerful are the following:

1. All the triangular elements do not have to be identical. They can be distorted in an adequate manner at the boundaries to take into account the particular boundary conditions. They can also be modified in other regions so that any discontinuity or defects in the solid can be included in the calculation. The method can, for example, be applied to calculate the stress distribution in a plate containing holes or in a fibre composite in which the bonding between some fibres and the matrix is not perfect.

2. The chief numerical advantage of this technique, which arises from the fact that $u(\mathbf{r})$ is expressed in terms of functions separately in each element rather than over the whole region, is that the matrix S is a sparse matrix. Moreover, if the elements of S are properly arranged by suitably numbering the neighbouring elements, S will have a band structure. The inversion of a sparse matrix with a band structure is particularly easy in practice.

2.7 MECHANICAL TESTING OF MATERIALS

The mechanical and of course other physical properties of ideal materials which have a well defined, regular and perfect microscopic structure can be specified once and for all. For such ideal materials no testing would be required and a design engineer could use their known mechanical properties for any desired purpose. Unfortunately, however, no material is ideal or perfect. Even the exact nature of defects, their distribution and concentration in real materials may vary from sample to sample and therefore cannot be universally defined. Since the mechanical properties of materials are quite sensitive to certain types of defects, they will also vary from sample to sample. This is true for all materials and is even more so for fibre composites because of the large variations in their fabrication techniques.

For the purpose of designing a machine or any structure, it is of paramount importance that the design engineer has specific data on the mechanical properties of the particular sample of material which is actually going to be used. Further, engineers have to study the properties of materials in a finished product. This is because, in addition to microscopic defects or discontinuities, the mechanical properties of materials also depend on their shapes, various joints in the machine or the structure, temperatures and other environmental variables and, above all, certain parameters which are time and frequency dependent such as periodic and variable stresses.

These dependences arise because of the stress concentration at bends, joints and other discontinuities which may locally weaken the material. Temperature affects mechanical properties of materials in several ways: a nonuniform temperature may induce significant thermal stresses; some materials like glass lose their strength at high temperatures. The environmental conditions affect the mechanical properties of materials through humidity and gas diffusion which tends to embrittle certain materials. Lastly, the time dependent parameters affect the mechanical properties of materials because of the inability of the molecules of the material to keep in phase with the externally imposed periodic variations. Consequently, a material may be able to withstand a certain value of static stress but may fracture for the same value if applied in a periodic manner at a high frequency. Paradoxically, the importance of such fatigue failures, as they are called, was first emphasized for the case of aircrafts in a fiction (Nevil Shute, *No Highway*) and not in a scientific paper.

The preceding discussion serves to establish the importance of testing of materials. Testing or mechanical testing in the present context, can be defined as the process of determining the usefulness of a material for a specific purpose by a study of its characteristic mechanical properties. Thus, the objectives of testing can be broadly categorized as follows:

1. Measurement of mechanical characteristics of materials for the purpose of studying their fundamental properties. This is useful for exploring new avenues of materials' application and development of new materials.
2. Determination of response of materials to stresses which are expected to be encountered in specific situations.
3. Setting up of limits of reliability for the materials.
4. Characterization and standardization of materials based upon the above studies.

In this section we shall give a brief review of the methods which are generally adopted for mechanical testing of materials. Several books and review articles are available on this subject as for example, Fenner (1965), Beaumont (1954), Tabor (1951), Small (1960), Batson and Hyde (1922).

With appropriate variations, these methods can be adopted for the case of fibre composites. The specific techniques for fibre composites have been referred to in the text at the appropriate places. In order to fully exploit the properties of such materials, it is important that the results obtained by the experimental methods of testing are supplemented by suitable theoretical calculations for which a powerful technique namely the finite element method has been described in the previous section.

Techniques for testing of materials

The techniques for mechanical testing of materials can be mainly classified into the following three classes:

(i) *Measurement of elastic moduli and other mechanical characteristics of the materials.* An excellent review of the techniques for the measurement of elastic moduli has been given by Huntington (1958). Some of these techniques as applicable to fibre composites have been described in Chapter 3.

(ii) *Performance study of materials under controlled conditions.* In this class of methods, the material is subjected to conditions stretching upto its endurance limit and thus its limits of reliability are established.

(iii) *Non-destructive testing techniques.* These methods utilize the propagation of ultrasonic waves in the material, photoelastic response of the material, X-rays, gamma rays and other radio active techniques. Recently electronmicroscopy and holographic techniques are being developed for non-destructive testing of materials. For a review of non-destructive testing techniques, reference may be made to, for example, Krautkramer and Krautkramer (1969) and McClung (1974).

The two classes (i) and (ii) are not mutually exclusive and most of the techniques as described below which are generally used for testing, are common to both these classes. The difference arises in the type of data which are extracted and the limiting stresses to which the material is subjected.

1. *Tensile Testing.* This test is used to determine the Young's modulus, elastic limit, yield point, necking point, ductility and strength of materials. The sample is taken to be in the shape of a rectangular or cylindrical bar (length much greater than lateral dimensions). Usually the cross-section area of the bar is taken to be about 2 cm^2 and the gauge length is about 5 cm.

The test is carried out on a machine called extensometer. The machine provides a mechanism for applying different measurable loads to the sample with a device to measure the change in the length of the sample. The relevant length on the gauge length is identified as the length sample between two specified dots on the sample. The sample is attached to the machine through either a clamp or a screw type grip.

The actual procedure is to apply the stretching (tensile) load to the sample starting from a low value to ultimately a value when the sample fractures by breaking into two pieces. The test results are expressed as a stress strain curve from which various mechanical properties can be read directly. The ductility is determined by measuring the amount of extension which occurred in the sample just before fracture. In certain models of extensometers, this elongation is not properly recorded and in such cases it would be essential to reassemble the broken parts and directly measure the total elongation. The ductility may be expressed in terms of the ratio of the total elongation and the original length of the test piece.

2. *Compression testing.* The compression test is the reverse of the tensile test: the material is compressed rather than stretched. It can be carried on a machine which is just like an extensometer with slight modifications.

The compression test is used to study the deformation of materials such as ceramics which are brittle when stressed by a tensile load. It is also useful for study of mechanical working processes which involve very large plastic strains. However, perhaps the most important application of the compressive test is in determining the crushing strength of building and road materials.

3. *Torsion testing.* In this class of tests, the material is subjected to shear loading, causing a deformation in its shape. The resulting stress strain curve is somewhat similar to that obtained by the tensile testing.

The quantities which are measured from this test are: modulus of rigidity, elastic limit, maximum torque, torsional ductility, modulus of rupture and finally total twist to fracture.

A torsion testing machine, like an extensometer, has a device to twist the sample by applying a measurable load in the plane parallel to the cross-section of the sample. The limiting torque on the sample is roughly proportional to the third power of the lateral dimension of the material. Thus the required size of the torsion testing area of the machines increases much faster than the lateral dimension of the sample. This imposes an operational and economic limit on the torsion testing.

4. *Bending test.* Some of the important types of materials which are tested by bending are given below.

(i) *Plastics* (*mostly thermosetting plastics*). The test is used for measuring the cross-breaking strength of plastics. The test piece is in the form of a rectangular bar (typical approximate dimensions—15 cm × 2 cm × 1 cm). The bar is supported on two V-shaped supports in a symmetrical manner which are approximately 10 cm apart. The load is applied along a sharp line perpendicular to the length of the bar and at a point midway between the two V-supports. The load at the point of fracture is recorded from which the cross breaking strength is determined by using standard formulae.

(ii) *Metals.* The main purpose of the bending test of ductile metals is to determine the ability of the metal to deform in a particular direction without cracking. Essentially the information obtained by this test is of a qualitative nature. Some quantitative information which is available from this test is given in terms of the radius and the angle of bending. The experimental procedure of the test is quite similar to that described earlier. One particular application of this test is in testing of the finished products which are in the form of cylindrical or rectangular bars and also of the samples cut from sheets and plates. The main difficulty with this test is lack of adequate standardization because the response of a material to bending stress depends crucially on the shape of the material.

(iii) *Concrete beams.* The bending test is particularly useful for monitoring the quality of concrete beams. The beam is supported on two points and the load is applied at two points along its length which are separated by a distance $d/3$ where d is the distance between the two supports. The flexural strength is determined in terms of the maximum stress which the beam can withstand just before fracture.

5. *Impact testing.* The impact testing is used to determine the impact strength of a material. The sample is usually taken to be in the form of a rectangular bar on which a V-shape notch is made at its middle point with the axis of the notch perpendicular to the length of the bar. A notch in a bar produces stresses in three different directions—the so called triaxial stress. The technique of the impact tests is to subject the test piece to controllable triaxial stresses. The typical dimension of the sample are 6 cm × 1 cm × 1 cm. The edges of the notch are at 45° to each other and its radius of curvature at the vertex of its V-shape, is 0.025 cm. The following two variants of the impact testing are commonly used.

(i) *Izod test.* In this method the test piece is clamped vertically and a pendulum is made to strike it at the top on the side of the notch. The impact force of the pendulum can be controlled by giving it a calculated swing. A part of the potential energy of the pendulum after it strikes the test piece, is used to fracture the test piece. The result of the test depends crucially not only on the type of the notch but also on the dimension of the specimen. The main disadvantage of this test is therefore, that the absolute impact strength of a material cannot be specified.

(ii) *Charpy test.* This test is similar to Izod test with the following two chief differences: (a) the test piece is clamped across the path of the pendulum (with the notch facing the pendulum) and not vertically, (b) the pendulum is lighter and the speed at the point of striking is higher as compared to the Izod Test. As in the Izod test, the amount of the energy used in fracturing the test piece is recorded. Generally, this test gives a larger area of fracture and a lower value of the fracture energy as compared to the Izod test. For many design applications, the area of fracture is a better guide than the fracture energy. To this extent, Charpy test is more useful than the Izod test but, otherwise, it suffers from the same drawbacks.

6. *Testing of hardness.* Hardness is another mechanical property of materials which is important from the point of view of applications. Hardness is a measure of the resistance of the material to cuts, scratches and other wear and tear of the surface. The testing methods are based on the use of an indentor or a penetrator on the material with a measurable penetrating force. The following methods are commonly used for testing of hardness.

(i) *Rockwell test.* The indentor in this test is a steel ball of about 0.15 cm diameter or a diamond cone with the cone angle 120°. A prescribed load is applied on the indentor and the resistance of the material, i.e., its hardness is directly read from a graduated scale. This is mainly used for making routine tests on items which are produced in large quantities on a production line.

(ii) *Vickers test.* In this method the indentor is a square based pyramid with its vertex made of diamond. The square shaped dent made by the indentor is examined by a microscope and the distance across the diagonals of the dent are measured. The hardness is defined in terms of hardness numbers which can be obtained from standard tables by knowing the dimensions of the dent. Since the size of dent produced in this method is small, it is quite suitable for testing the hardened and polished materials.

(iii) *Brinell test.* This is quite similar to the Vickers test except that the indentor is a hardened steel ball. One advantage of this test is, that it also gives an estimate of the tensile strength. Its utility is limited because it can not be used for very hard materials.

(iv) *Shore scleroscope test.* In this test, a small diamond point hammer (about 2 gm weight) is made to fall on the material from a height of about 25 cm. The hardness is measured in terms of the height to which the hammer rebounds. This method is useful for testing of hardness of the materials on which indenting is not desirable.

7. *Testing of fatigue.* Most material failures in actual situations occur because of fatigue and therefore the importance of fatigue testing cannot

be over emphasized. In view of the very complex processes which are responsible for fatigue failure, it is extremely difficult to obtain reliable information by extrapolating the results of tests carried out in a laboratory. The fatigue tests, therefore, have to be carried out in actual operating conditions or as close to the operating conditions as possible. The laboratory tests are usually made on what is called Wohler machine in which the periodic stress is applied by a variable speed electric motor. This machine gives an estimate of the endurance limit of the material in terms of the stress at which the fracture in the material is initiated after a certain number of stress reversal cycles.

8. *Non-destructive testing*. The importance and the usefulness of testing methods in which, in contrast to previously described methods, the test piece is not destroyed is obvious. The non-destructive testing methods are based on measuring the response of the material (ideal as well as with defects) to an external probe which can interact with the material at the microscopic level.

For the purpose of mechanical testing of materials ultrasonic wave provides a powerful technique. In this method a pulse of ultrasonic wave, produced by a piezoelectric crystal, is introduced in the material and the reflections produced by defects in the material are recorded. It is possible to correlate the nature of the defect in the material with this and other wave propagation characteristics of the material. Such correlation is obtained by a study of the fundamental properties of the materials. The special case of fibre composites will be discussed in detail in Chapters 5 and 6. The details of an experimental technique have also been given in Chapter 5.

Other non-destructive tests for materials are magnetic crack test (suitable for magnetic materials) and fluorescent crack test which detect the presence and size of the cracks using the magnetic property of the material or the optical property of a fluorescent substance. The differential absorption of X-rays and gamma-rays by regions near the cracks in a material as compared to the bulk is also used for non-destructive testing of materials.

CHAPTER **3**

Elastic Constants of Fibre Composites

3.1 INTRODUCTION

The elastic constants of a solid as defined in Section 2.3 are basic parameters in the continuum theory of solids. They play an extremely important role in the characterization of the mechanical properties of solids which are of technological interest, such as the strength and fracture properties of solids, their stress analysis and their response to an external force. It is not surprising, therefore, that a large amount of effort has been devoted to the calculation of elastic constants of fibre composites. Some of this work will be reviewed in this chapter.

Excellent reviews on the calculations of the elastic constants of fibre composites have been written by Chamis and Sendeckyj (1968) who have given a comparative study of the various methods of calculations, Rosen (1970a) and Rosen (1973), the last paper being of a more general nature. The mechanical properties of reinforced thermoplastics from the designing point of view have been reviewed by Ogorkiewicz (1971). Some economic aspects together with the mechanical properties of reinforced thermoplastics have been discussed by Abrahams and Dimmock (1971). The possibility of qualitative predictions of the composite properties from the fibre and matrix properties have been considered by Dimmock and Abrahams (1969).

In this chapter we shall be concerned only with 'static' calculations of the elastic constants, although some work on the stresses in a composite during an impact will be mentioned. In static methods the elastic constants are calculated as the constants of proportionality between an applied stress and the induced strain (or vice versa). The elastic constants can also be determined by a study of the propagation of elastic waves in a composite, since the wave velocities are related to the elastic constants (Section 2.3). The propagation of elastic waves in fibre composites will be discussed in Chapter 5.

The main problem, as one would expect, is to construct a model which is a realistic representation of the composite and at the same time is amenable to not too complicated mathematical techniques. It is unlikely that a mathematically simple model will be applicable to all kinds of composites in all different physical conditions. Several models and many different techniques have, therefore, been developed for a calculation of the elastic constants of composites, some of which may be more suited to the particular problem than others. Although some applications will be mentioned in this chapter, our main emphasis will be on the techniques of calculation rather than their actual applications.

In order to model a fibre composite, a certain degree of idealization is essential. In addition to the advantage of mathematical convenience, such a model can also serve as a reference composite with respect to which the deviation of the physical properties of a real composite be discussed.

The usual assumptions for a model fibre composite are as follows:

1. The fibres are straight, continuous and well-aligned. They extend throughout the length of the composite and are of uniform cross-section and strength.

2. The matrix is homogeneous and is free from voids, cracks and other material defects.

3. The bonding between the fibres and the matrix is perfect: the fibre-matrix interface does not make an independent contribution to the elastic constants of the composite.

4. The composite is macroscopically homogeneous. This implies that if a sample piece is cut out of the composite, it has the same physical properties as the whole composite. The elastic constants of such a sample are sometimes referred to as the 'overall' or 'average' elastic constants. In this book we shall omit the prefix 'overall' or 'average' and it should be understood that any reference to a physical property of the composite will refer to the overall or the average property of the macroscopically homogeneous composite in the sense as defined above.

Almost all the work done on the theory of fibre composites is based upon the assumption of macroscopic homogeneity of the material. A few exceptions are Bolotin (1965a, b; 1966), Herrmann and Achenbach (1967a, b) and Achenbach and Herrmann (1967). Bolotin (1965a, b) has used the displacement equilibrium equations of a Cosserat medium—a medium in which the stress tensor is not symmetric—to take into account the inhomogeneity of the fibre composites. This theory has been applied to a layered medium with randomly distributed defects by Bolotin (1966) and has been extended to the dispersive propagation of elastic waves in fibre composite in the aforementioned papers of Achenbach and Herrmann.

5. The fibres as well as the matrix are elastic, so that the linear theory of elasticity is applicable. In this case, as given in Section 2.3, the stress and the strain are proportional to each other according to Hooke's law, which is valid if the strain is small.

Most of the work which will be reviewed in this chapter deals with a perfect fibre composite which can be modelled as described above. We shall only briefly mention some of the work done on composites which deviate from the perfect model composite, such as composites with short, broken or misaligned fibres or where the strain is so large that one or both of the phases of the composite are in stage II, i.e. in the plastic zone. A considerable amount of work has also been done on more general composites, i.e. composites in which the reinforcing phase can be of arbitrary shape, such as spherical or ellipsoidal, and not necessarily cylindrical, as in the case of fibre composites. We shall not be interested in such calculations in this book. We have, however, referred to some papers on general composites because the techniques used in these papers may be relevant to fibre composites.

In some calculations the fibres have been assumed to form a regular, periodic array. The usual assumptions are a square and a hexagonal array of fibres. In these cases the composite has tetragonal or hexagonal symmetry and the number of independent elastic constants is six and five, respectively. Sometimes the fibres are assumed to be randomly distributed. If the transverse isotropy in a plane normal to the fibre axis can be assumed for the composite, the number of independent elastic constants is five, as for the hexagonal case. Some useful symmetry relations for the elastic stress-strain law in the case of a three-dimensional fibre composite have been given by Rosen and Shu (1971).

It is physically obvious that the elastic constants of the composite will be related to the elastic constants of the fibre and the matrix and their fractional volumes in the material. Such relations based upon the simple laws of mixture are derived in Section 3.2. More rigorous analytical treatments of this problem are reviewed in Section 3.3, and some techiques which are suitable for direct numerical computations of elastic constants of a particular sample are given in Section 3.4. Finally, a brief description of some experimental techniques for the measurement of elastic constants of a fibre composite is given in Section 3.5.

3.2 SIMPLE LAWS OF MIXTURE

Let $\bar{\sigma}_{\alpha\beta}$ and $\bar{\varepsilon}_{\alpha\beta}$ denote the average stress and strain tensors for a composite. If V denotes the volume of the sample, then its total elastic energy as obtained from Eqs. (2.3.5)–(2.3.5c) is given by:

$$U = \frac{V}{2} \sum_{\alpha,\beta} \bar{\sigma}_{\alpha\beta}\, \bar{\varepsilon}_{\alpha\beta} \qquad (3.2.1)$$

In this section, for the sake of simplicity, we shall consider only the traces of $\bar{\sigma}$ and $\bar{\varepsilon}$ or their individual components in a representation in which they are diagonal. In such a simplified case Hooke's law can be written in the form:

$$\sigma_c = K_c \varepsilon_c \qquad (3.2.2)$$

where σ_c, ε_c and the elastic constant K_c are scalars and the subscript C stands for the composite. For example, σ_c and ε_c may denote the volume stress and volume strain, respectively, in which case K_c will be the bulk modulus.

The elastic energy of the composite is then given by either of the following two equations:

$$U_c = \frac{V}{2} K_c \varepsilon_c^2 \tag{3.2.3}$$

or

$$U_c = \frac{V}{2} \frac{1}{K_c} \sigma_c^2 \tag{3.2.4}$$

If $\varepsilon_{f,m}$ and $\sigma_{f,m}$ denote, respectively, the strains and the stress in the fibres or the matrix (to be identified by the subscripts f or m respectively) their elastic energies are as follows:

$$U_{f,m} = \frac{V_{f,m}}{2} K_{f,m} \varepsilon_{f,m}^2 \tag{3.2.5}$$

or

$$U_{f,m} = \frac{V_{f,m}}{2} \frac{1}{K_{f,m}} \sigma_{f,m}^2 \tag{3.2.6}$$

where V_f and V_m denote the volume occupied by the fibres and the matrix, respectively, in the composite, so that

$$V = V_f + V_m \tag{3.2.7}$$

Neglecting the contribution of the fibre-matrix interfaces and fibre-fibre interactions, the total energy of the composite can be taken to be the sum of the the elastic energy of the fibres and the matrix, i.e.

$$U_c = U_f + U_m \tag{3.2.8}$$

We now make the following two alternative assumptions.

1. The strain is uniform throughout the composite, i.e.

$$\varepsilon_f = \varepsilon_m = \varepsilon_c \tag{3.2.9}$$

With the help of Eqs. (3.2.3), (3.2.5) and (3.2.9), we obtain the following result from Eq. (3.2.8)

$$K_c = C_f K_f + C_m K_m \tag{3.2.10}$$

where $C_{f,m} = V_{f,m}/V$, denotes the fractional volume of the fibres or the matrix in the composite.

2. The stress is uniform throughout the composite, i.e.

$$\sigma_f = \sigma_m = \sigma_c \tag{3.2.11}$$

In this ease the use of Eqs. (3.2.4), (3.2.7) and (3.2.11) in Eq. (3.2.8) leads to the result

$$\frac{1}{K_c} = \frac{C_f}{K_f} + \frac{C_m}{C_m} \tag{3.2.12}$$

The values of K_c as predicted by Eqs. (3.2.10) and (3.2.12) are called the Voigt and the Reuss estimates, respectively. From the derivation given above, it is obvious that the two estimates are independent of the geometry of the arrangement of the constituents or their shapes in the composite and are also valid for discontinuous fibres so long as the assumption of uniform stress or strain is valid. However, this assumption may be used to impose some restrictions on the geometrical arrangement of the constituents. By analogy with electric circuits, the Voigt and Reuss estimates are sometimes said to correspond to, respectively, the parallel and the series loadings of the constituents. It may be mentioned in passing that the Voigt and Reuss estimates have been derived above for a two-component composite, but similar expressions can also be obtained for a multiphase composite by a trivial generalization of the above derivation.

The Voigt and Reuss estimates have been obtained with the assumption of uniform strain and uniform stress, respectively. In a real situation neither is uniform. The actual value of K_c is therefore expected to lie between K_V and K_R, which provide the upper and lower bounds of K_c. The usefulness of having upper and lower bounds to the moduli will obviously depend on how far apart they are. The differences between K_V and K_R and their reciprocals are given below:

$$K_V - K_R = \frac{C_f C_m (K_f - K_m)^2}{C_f K_m + C_m K_f} \tag{3.2.13}$$

and

$$\frac{1}{K_R} - \frac{1}{K_V} = \frac{C_f C_m \, K_f K_m}{C_f K_f + C_m K_m} \left(\frac{1}{K_m} - \frac{1}{K_f}\right)^2 \tag{3.2.13a}$$

We notice from the above equations that the difference between K_V and K_R is quite small—of second order in $(K_f - K_m)$—if K_f is nearly equal to K_m and can be quite large if K_f and K_m are very different. Thus, the usefulness of the law of mixtures is limited to composites in which the elastic constants of the components are nearly equal. Subject to this restriction, the formulae for K_V and K_R work quite well for the bulk modulus and the modulus of rigidity. The behaviour of the Young's modulus is somewhat unusual because it does not lie between the two estimates given by Eqs. (3.2.10) and (3.2.12), respectively. In fact Hill (1964a) has proved that the Young's modulus cannot be less than that predicted by Eq. (3.2.10). We shall come back to this point in the next section.

Although both Voigt and Reuss estimates are essentially laws of mixture, it is more usual to refer to the Voigt estimate as the law or rule of mixtures.

3.3 ANALYTICAL CALCULATIONS OF THE ELASTIC CONSTANTS

Important contributions in this field have been made in a series of papers by Hill (1964a, b; 1965a, b) any by Hashin and Rosen (1964). The importance of these papers cannot be overemphasized, as they form the basis for subsequent discussions on this topic. We shall therefore review these papers in some detail. These calculations provide either rigorous bounds for the elastic moduli or approximate expressions for the moduli which are valid subject to certain restrictions on the relative moduli of the fibres and the matrix and the fibre concentration.

Hill (1964a) has derived simple relations for the main elastic constants of a fibre composite based on the assumption of transverse isotropy, i.e. isotropy in a plane normal to the fibre axis. These relations are independent of the geometry of the fibrous arrangement and depend only on the volume concentration of fibres in the composite. The fibres are assumed to be randomly distributed subject to the condition that the composite is macroscopically homogeneous as defined in Section 3.1.

Consider a macroscopically uniform triaxial loading of the composite along and perpendicular to the fibre direction, which is taken to be the z axis. For such a symmetry the stress–strain relations can be written as:

$$\tfrac{1}{2}(\sigma_x + \sigma_y) = K(\varepsilon_x + \varepsilon_y) + L\varepsilon_z \tag{3.3.1}$$

$$\sigma_z = L(\varepsilon_x + \varepsilon_y) + N\varepsilon_z \tag{3.3.1a}$$

$$(\sigma_x - \sigma_y) = 2M(\varepsilon_x - \varepsilon_y) \tag{3.3.2}$$

$$\sigma_{xy} = M\varepsilon_{xy} \tag{3.3.2a}$$

$$\sigma_{xz} = \mu\varepsilon_{xz} \tag{3.3.2b}$$

$$\sigma_{yz} = \mu\varepsilon_{yz} \tag{3.3.2c}$$

where, for the sake of notational brevity, the diagonal components of σ and ϵ are denoted by a single subscript (σ_α for $\sigma_{\alpha\alpha}$) and K, L, M and N are moduli of elasticity. K is the plane strain bulk modulus (sometimes also called area modulus or surface modulus, henceforth to be referred to as bulk modulus) for lateral dilatation without longitudinal extension; M and μ are, respectively, the moduli of transverse and longitudinal rigidity, N and L can be defined through the following equations:

$$E = N - L^2/K \tag{3.3.3}$$

and

$$\nu = L/2K \tag{3.3.3a}$$

E and ν being the Young's modulus and the Poisson ratio, respectively. It can be verified that for complete isotropy $L = K - M$ and $N = K + M$.

With the help of Eq. (2.3.5c) it may easily be verified that the elastic moduli as defined above are related to the elastic constants c_{ij} as follows:

$$\left.\begin{aligned} N &= c_{33} \\ L &= c_{13} \\ K &= \tfrac{1}{2}(c_{11} + c_{12}) \\ M &= \tfrac{1}{2}(c_{11} - c_{12}) \\ \mu &= c_{44} \end{aligned}\right\} \tag{3.3.3b}$$

For notational convenience in this context, E_3, M_{12} and M_{13} as defined in Eq. (2.3.5c) have been replaced by E, M and μ, respectively.

The equations given above are applicable to the composite as well as the fibres and the matrix individually. In what follows, we shall distinguish various quantities for the composite, fibres and the matrix by the subscripts c, f and m, respectively. We also define the following decomposition formula by analogy with Eq. (3.2.10):

$$F_c = C_f F_f + C_m F_m \tag{3.3.4}$$

with

$$C_f + C_m = 1$$

where F denotes any function of coordinates x and y.

The composite is taken to be in the form of a rectangular prism with plane end surfaces which are normal to the fibres. Then by symmetry the xz and the yz components of the shears will be zero and ε_z will be constant in the xy plane.

The relationship between the elastic moduli of the composite and those of its components can be derived by first writing Eqs. (3.3.1) and (3.3.1a) for the composite, i.e. by labelling each quantity by the subscript c. An average lateral strain is defined by:

$$e_c = \varepsilon_{cx} + \varepsilon_{cy}$$

and similarly for fibres and the matrix. Next the stress and strain of the composite are decomposed into the corresponding quantities for the fibres and the matrix, and finally the stresses are written in terms of the strains according to Hooke's law. This procedure leads to the following two equations:

$$C_f\,(K_f\,e_f + L_f\,\varepsilon) + C_m\,(K_m\,e_m + L_m\,\varepsilon) = K_c\,(C_f\,e_f + C_m\,e_m) + L_c\,\varepsilon \qquad (3.3.5)$$

and

$$C_f\,(L_f\,e_f + N_f\,\varepsilon) + C_m\,(L_m\,e_m + N_m\,\varepsilon) = L_c\,(C_f\,e_f + C_m\,e_m) + N_c\,\varepsilon \qquad (3.3.6)$$

where ε simply denotes the constant value of ε_z on the xy plane.

Equations (3.3.5) and (3.3.6) can be written, respectively, in the following form:

$$C_f\,(K_c - K_f)\,x_f + C_m\,(K_c - K_m)\,x_m = C_f\,L_f + C_m\,L_m - L_c \qquad (3.3.7)$$

and

$$C_f\,(L_c - L_f)\,x_f + C_m\,(L_c - L_m)\,x_m = C_f\,N_f + C_m\,N_m - N_c \qquad (3.3.8)$$

where

$$x_{f,\,m} = e_{f,\,m}/\varepsilon$$

If Eqs. (3.3.7) and (3.3.8) were independent, they would determine uniquely x_f and x_m. However, since either of them can be made to vanish separately, we infer that Eqs. (3.3.7) and (3.3.8) cannot be independent and therefore the corresponding coefficients in the two equations must be proportional. This condition yields the following relations:

$$\frac{K_c - K_f}{L_c - L_f} = \frac{K_c - K_m}{L_c - L_m} = \frac{L_c - C_f\,L_f - C_m\,L_m}{N_c - C_f\,N_f - N_m\,C_m} = \frac{K_f - K_m}{L_f - L_m} \qquad (3.3.9)$$

where the last equation is obtained by combining the two sides of the first equation.

It is instructive to write Eq. (3.3.9) in the following form:

$$\Delta L_V = \Delta K_V\,\frac{(L_f - L_m)}{(K_f - K_m)} = \Delta N_V\,\frac{(K_f - K_m)}{(L_f - L_m)} \qquad (3.3.10)$$

where

$$\Delta L_V = L_c - C_f\,L_f - C_m\,L_m = L_c - L_V \qquad (3.3.11)$$

and ΔK_V and ΔN_V are defined by similar equations with L in Eq. (3.3.11) replaced by K and N, respectively. These quantities ΔK_V, ΔL_V and ΔN_V represent the deviation of K_c, L_c and N_c from their corresponding Voigt estimates, denoted by the subscript V.

Using the definitions for E and ν as given in Eqs. (3.3.3) and (3.3.3a), we can also derive the following equation, which is equivalent to Eq. (3.3.10):

$$\Delta\nu_V = \Delta K_R \frac{(\nu_f - \nu_m)}{\left(\dfrac{1}{K_f} - \dfrac{1}{K_m}\right)} = -\frac{\Delta E_V}{4} \frac{\left(\dfrac{1}{K_f} - \dfrac{1}{K_m}\right)}{(\nu_f - \nu_m)} \qquad (3.3.12)$$

where $\Delta\nu_V$ and ΔE_V denote the deviation of ν_c and E_c from their Voigt estimates as in Eq. (3.3.11), and ΔK_R, which gives the deviation of K_c from its Reuss estimate, is defined as:

$$\Delta K_R = \frac{1}{K_c} - \frac{C_f}{K_f} - \frac{C_m}{K_m} = \frac{1}{K_e} - \frac{1}{K_R} \qquad (3.3.13)$$

As mentioned in Section 3.2, we expect the following inequalities to be valid:

$$K_R \leqslant K_c \leqslant K_V \qquad (3.3.14)$$

and

$$\frac{1}{K_V} \leqslant \frac{1}{K_c} \leqslant \frac{1}{K_R} \qquad (3.3.14a)$$

The second equation in Eq. (3.3.12) can be written as:

$$\Delta E_V = -4\, \Delta K_R (\nu_f - \nu_m)^2/(1/K_f - 1/K_m)^2 \qquad (3\ 3.15)$$

We see from Eqs. (3.3.13) and (3.3.14a) that ΔK_R is always negative. Thus we infer from Eq. (3.3.15) that $\Delta E_V \geqslant 0$ or

$$E_c \geqslant E_V = C_f E_f + C_m E_m \qquad (3.3.15a)$$

which proves the result mentioned in the previous section that the Voigt estimate provides a lower bound for the Young's modulus and not the upper bound, as is the case for the bulk modulus. We also note from Eq. (3.3.15) that $\Delta E_V = 0$ for $\nu_f = \nu_m$. Thus, the Young's modulus of the composite is correctly given by the Voigt estimate if the Poisson ratios of the matrix and the fibres are equal.

The discussion so far in this section is valid even if the composite does not have transverse isotropy, provided that it has tetragonal symmetry, i.e. provided that the x and y axes are equivalent symmetry axes. Similar formulae can also be derived by considering a fibre with circular cross-section embedded in a circular cylindrical matrix.

Hill (1964a) has also obtained values for the moduli of elasticity for a composite with transverse isotropy when the rigidity moduli of the fibres and the matrix are equal, i.e.

$$M_f = M_m \equiv M$$

For this derivation a function $F(x, y)$ is defined by the relationships

$$\varepsilon_x = \frac{\partial^2 F}{\partial x^2} \tag{3.3.16}$$

$$\varepsilon_y = \frac{\partial^2 F}{\partial y^2} \tag{3.3.16a}$$

and

$$\varepsilon_{xy} = \frac{2\partial^2 F}{\partial x . \partial y}$$

and

$$\varepsilon_z = \varepsilon \tag{3.3.16b}$$

where F and its gradient are finite and continuous throughout the composite. In terms of $F(x, y)$, Eqs. (3.3.1)–(3.3.2c) can be written as:

$$\tfrac{1}{2}(\sigma_x + \sigma_y) = K_p \left(\frac{\partial^2 F}{\partial x^2} + \frac{\partial^2 F}{\partial y^2} \right) + L_p \varepsilon \tag{3.3.17}$$

$$\sigma_z = L_p \left(\frac{\partial^2 F}{\partial x^2} + \frac{\partial^2 F}{\partial y^2} \right) + N_p \varepsilon \tag{3.3.17a}$$

$$\sigma_x - \sigma_y = 2M \left(\frac{\partial^2 F}{\partial x^2} - \frac{\partial^2 F}{\partial y^2} \right) \tag{3.3.18}$$

$$\sigma_{xy} = 2M \frac{\partial^2 F}{\partial x \partial y} \tag{3.3.18a}$$

where the subscript p stands for f or m.

For self-equilibrium the two-component Laplacian must be piecewise constant, i.e.

$$\frac{\partial^2 F}{\partial x^2} + \frac{\partial^2 F}{\partial y^2} = e_p \tag{3.3.19}$$

The second derivatives of F are therefore discontinuous at the fibre–matrix surface. This discontinuity in the derivatives will be denoted by the subscript m–f and is given by:

$$\cos^{-2}\theta \left(\frac{\partial^2 F}{\partial x^2} \right)_{m-f} = \sin^{-2}\theta \left(\frac{\partial^2 F}{\partial y^2} \right)_{m-f} = (\sin\theta\cos\theta)^{-1} \left(\frac{2\partial^2 F}{\partial x \partial y} \right)_{m-f}$$
$$= e_m - e_f \tag{3.3.20}$$

where θ is the angle measured anti-clockwise from the x axis to the local normal on a fibre–matrix surface.

The condition for the continuity of the traction vector:

$$\cos\theta (\sigma_x)_{m-f} + \sin\theta\, (\sigma_{xy})_{m-f} = \sin\theta\, (\sigma_y)_{m-f} + \cos\theta\, (\sigma_{xy})_{m-f} = 0$$

leads to the equation

$$(K_f + M)\, e_f + L_f \varepsilon = (K_m + M)\, e_m + L_m \varepsilon \tag{3.3.21}$$

In deriving the above equation no assumptions have been made regarding the geometry of the composite. Another equation for e_f and e_m can be derived if we now specify a circular cylindrical composite. For such a sample the fractional change in the area of cross-section, the average stress S over its curved surface and the mean axial tension t are given by

$$e = C_f e_f + C_m e_m \tag{3.3.22}$$

$$S = \tfrac{1}{2}(\sigma_x + \sigma_y) = C_f (K_f e_f + L_f \varepsilon) + C_m (K_m e_m + L_m \varepsilon) \tag{3.3.22 a}$$

and, since the axial stress is piecewise constant,

$$t = \sigma_z = C_f (L_f e_f + N_f \varepsilon) + C_m (L_m e_m + N_m e_m) \tag{3.3.22b}$$

where Eqs. (3.3.1) and (3.3.1 a) have been used with ε_z replaced by ε.

The moduli $L_c = S/\varepsilon$ and $N_c = t/\varepsilon$ can be obtained from Eq. (3.3.22a) or Eq. (3.3.22 b), and following equation:

$$-\frac{e_f}{C_m} = \frac{e_m}{C_f} = \frac{(L_f - L_m)\,\varepsilon}{C_f K_m + C_m K_f + M} \tag{3.3.23}$$

which is derived from Eqs. (3.3.21) and (3.3.22) by putting $e = 0$ and $C_f + C_m = 1$.

These equations yield the following expressions for L and N:

$$L_c = C_f L_f + C_m L_m - \frac{C_f C_m (L_f - L_m)(K_f - K_m)}{C_f K_m + C_m K_f + M} \tag{3.3.24}$$

and

$$N_c = C_f N_f + C_m N_m - \frac{C_f C_m (L_f - L_m)^2}{C_f K_m + C_m K_f + M} \tag{3.3.25}$$

Similarly the bulk modulus $K_c = S/e$ is derived from Eqs. (3.3.22) and (3.3.22 b) by putting $\varepsilon = 0$, which gives:

$$K_c = \frac{S}{e} = \frac{C_f K_f e_f + C_m K_m e_m}{C_f e_f + C_m e_m}$$

and, by using the following equation, obtained from Eq. (3.3.21) with $\varepsilon = 0$:

$$\frac{e_f}{e_m} = \frac{K_m + M}{K_f + M}$$

$$K_c = \frac{C_f K_f (K_m + M) + C_m K_m (K_f + M)}{C_f K_m + C_m K_f + M}$$

$$= C_f K_f + C_m K_m - \frac{C_f C_m (K_f - K_m)^2}{C_f K_m + C_m K_f + M} \tag{3.3.26}$$

The equation for K_c can also be written in the following form, which may also be useful:

$$K_c = \left(\frac{C_f}{K_f + M} + \frac{C_m}{K_m + M}\right)^{-1} - M. \qquad (3.3.26a)$$

It can be observed from Eqs. (3.3.24)–(3.3.26a) that all of these elastic moduli for the composite are less than their corresponding Voigt estimates.

The expressions for Young's modulus and the Poisson ratio can be obtained from the results for L_c, N_c and K_c with the help of Eqs. (3.3.3) and (3.3.3 a) or directly as $E_c = t/\varepsilon$ and $\nu = -e/2\varepsilon$ by putting $S = 0$ in Eqs. (3.3.22)–(3.3.22b) and using the procedure given above for the determination of N_c and L_c. The result is

$$E_c = C_f E_f + C_m E_m + \frac{4\, C_f C_m (\nu_f - \nu_m)^2}{\dfrac{C_f}{K_m} + \dfrac{C_m}{K_f} + \dfrac{1}{M}} \qquad (3.3.26b)$$

$$\nu_c = C_f \nu_f + C_m \nu_m + \frac{C_f C_m (\nu_f - \nu_m)\left(\dfrac{1}{K_m} - \dfrac{1}{K_f}\right)}{\dfrac{C_f}{K_m} + \dfrac{C_m}{K_f} + \dfrac{1}{M}} \qquad (3.3.26c)$$

In the more general case of a composite with $M_f \neq M_m$ Hill (1964 a) has derived upper and lower bounds for the elastic moduli. These are given below ($M_f \geqslant M_m$):

$$\frac{C_f K_f (K_m + M_m) + C_m K_m (K_f + M_m)}{C_f K_m + C_m K_f + M_m} \leqslant K_c$$

$$\leqslant \frac{C_f K_f (K_m + M_f) + C_m K_m (K_f + M_f)}{C_f K_m + C_m K_f + M_f} \qquad (3.3.27)$$

$$C_L \leqslant \frac{E_c - C_f E_f - C_m E_m}{4\,(\nu_f - \nu_m)^2} \leqslant C_u \qquad (3.3.28)$$

and

$$C_L \leqslant \frac{\nu_c - C_f \nu_f - C_m \nu_m}{(\nu_f - \nu_m)\left(\dfrac{1}{K_m} - \dfrac{1}{K_f}\right)} \leqslant C_u \qquad (3.3.29)$$

where

$$C_L = \frac{C_f C_m}{\dfrac{C_f}{K_m} + \dfrac{C_m}{K_f} + \dfrac{1}{M_m}} \qquad (3.3.30)$$

and

$$C_u = \frac{C_f C_m}{\dfrac{C_f}{K_m} + \dfrac{C_m}{K_f} + \dfrac{1}{M_f}} \qquad (3.3.30a)$$

The limits of K_c as given in Eq. (3.3.27) can also be represented by the following two inequalities:

$$\frac{C_f C_m}{C_f K_m + C_m K_f + M_f} \leqslant -\frac{(K_c - C_f K_f - C_m K_m)}{(K_f - K_m)^2} \leqslant \frac{C_f C_m}{C_f K_m + C_m K_f + M_m} \quad (3.3.31)$$

and

$$C_L \leqslant -\frac{\left(\frac{1}{K_c} - \frac{C_f}{K_f} - \frac{C_m}{K_m}\right)}{\left(\frac{1}{K_f} - \frac{1}{K_m}\right)^2} \leqslant C_u \quad (3.3.31a)$$

It can easily be verified from the above set of equations that if we put $M_f = M_m = M$ the upper and the lower bounds of the elastic moduli coincide and become equal to the exact results given by Eq. (3.3.26). Alternatively, if M in Eq. (3.3.2a) is regarded as some kind of average of M_f and M, so that $M_m \leqslant M \leqslant M_f$, it can be observed that the exact results lie between the corresponding upper and lower bounds. Thus, it should be possible to derive self-consistent values of the elastic moduli by evaluating an appropriate average value of M.

Hill (1965a) has calculated the 'self-consistent' values of various elastic moduli on the basis of a model in which a single fibre is assumed to be embedded in a homogeneous medium which has the same physical properties as the composite. For mathematical convenience the system is assumed to be infinitely extended, so that the end corrections can be neglected. This model, therefore, is not valid for a sample of small size.

Under any overall loading of the above-mentioned system consisting of the fibre and the surrounding medium the fibre strain can be taken as uniform if the loading is also uniform at infinity. The uniform fibre strain defined as above is adopted as the average overall fibres in the real composite.

Let us first consider an overall longitudinal shear strain of the composite, which we denote by ε_c, where ε_c can be, for example, ε_{xz} or ε_{yz} as defined in Eqs. (3.3.1) and (3.3.1a). The corresponding stress, which will also be a pure shear, will be denoted by σ_c, which is related to ε_c by Hooke's law:

$$\sigma_c = \mu_c \, \varepsilon_c \quad (3.3.32)$$

where μ_c denotes the longitudinal modulus of rigidity as defined in Eqs. (3.3.2)–(3.3.2c). Similarly, if we write the following equation for the fibre:

$$\sigma_f = \mu_f \, \varepsilon_f \quad (3.3.32a)$$

the stress concentration in the fibre is given by

$$\frac{\sigma_f}{\sigma_c} = \frac{\mu_f\,\varepsilon_f}{\mu_c\,\varepsilon_c} \tag{3.3.33}$$

Using the relation (Hill, 1965a)

$$\sigma_f - \sigma_c = \mu_c\,(\varepsilon_c - \varepsilon_f) \tag{3.3.34}$$

Eq. (3.3.33) can be written as

$$\frac{\sigma_f}{\sigma_c} = \frac{2\,\mu_f}{\mu_c + \mu_f} \tag{3.3.35}$$

In a real composite σ_c and ε_c can be related to the corresponding quantities for the fibres and the matrix as follows (using the decomposition formula as given in Eq. (3.3.4):

$$\sigma_c = C_f\,\sigma_f + C_m\,\sigma_m \tag{3.3.36}$$

and

$$\varepsilon_c = C_f\,\varepsilon_f + C_m\,\varepsilon_m \tag{3.3.36a}$$

From Eqs. (3.3.34), (3.3.36) and (3.3.36a) we obtain

$$\frac{\sigma_m}{\sigma_c} = \frac{2\,\mu_m}{\mu_c + \mu_m} \tag{3.3.37}$$

The modulus μ_c can then be obtained as a solution of the following equation, which is derived from Eqs. (3.3.35), (3.3.36) and (3.3.37):

$$\frac{2\,C_f\,\mu_f}{\mu_c + \mu_f} + \frac{2\,C_m\,\mu_m}{\mu_c + \mu_m} = 1 \tag{3.3.38}$$

Obviously only the positive root of Eq. (3.3.38) will be the physically acceptable solution for μ_c, which then gives the average value of the modulus of longitudinal rigidity.

Proceeding along similar lines, Hill (1965a) has derived the following equation for M:

$$\frac{C_f}{K_f + M} + \frac{C_m}{K_m + M} = \frac{1}{K_c + M} \tag{3.3.39}$$

or

$$\frac{C_f\,K_f}{K_f + M} + \frac{C_m\,K_m}{K_m + M} = 2\left(\frac{C_f\,M_m}{M_m - M} + \frac{C_m\,M_f}{M_f - M}\right) \tag{3.3.39a}$$

The physically acceptable solution of Eqs. (3.3.39) and (3.3.39a), which must be positive, gives M, the average or self-consistent value of the modulus of transverse rigidity. It can be shown (Hill, 1965a) that this value lies between the following bounds provided that $(M_f - M_m)$ $(K_f - K_m)$ is positive:

$$M_m + \frac{C_f M_m (M_f - M_m)}{M_m + C_m b_m (M_f - M_m)} \leqslant M$$

$$\leqslant M_f - \frac{C_m M_f (M_f - M_m)}{M_f - C_f b_f (M_f - M_m)} \quad (3.3.40)$$

where

$$b_p = \frac{K_p + 2M_p}{2(K_p + M_p)} \quad (p = f \text{ or } m)$$

It can also be shown that the physically acceptable value of μ_c as given by Eq. (3.3.38) is confined to the following range:

$$\mu_m + \frac{2 C_f \mu_m (\mu_f - \mu_m)}{2\mu_m + C_m (\mu_f - \mu_m)} \leqslant \mu_c$$

$$\leqslant \mu_f - \frac{2 C_m \mu_f (\mu_f - \mu_m)}{2 \mu_f - C_f (\mu_f - \mu_m)} \quad (3.3.41)$$

provided that $\mu_f \geqslant \mu_m$.

The bounds for M and μ_c as given in Eqs. (3.3.40) and (3.3.41) have been derived by Hashin (1965). The self-consistent values of E_c, K_c and ν_c are then given by Eq. (3.3.26) using the 'average' value of M as predicted by Eq. (3.3.39) and (3.3.39a). The value of μ_c is given by Eq. (3.3.38). Thus, all the five independent elastic moduli for a composite with transverse isotropy can be calculated. It should be remarked, however, that the above derivation may not be valid when $M_f/M_m \gg 1$ or when the fibre concentration is too large.

Some attempts have been made to generalize Hill's self-consistent model as described above. Hermans (1967) assumed the fibres to be randomly distributed with the restriction that the size of the matrix cylinder is large enough to represent the fibre concentration of the whole composite. A model in which a single fibre is taken to be embedded in a matrix has also been assumed by Whitney and Riley (1966), which has been generalized to the case of anisotropic fibres—subject to their being transversely isotropic—by Whitney (1967). In general, the results given by these models fall between the bounds as derived in Section 3.2. The effect of fibre twist on the Young's modulus has been considered by Whitney (1966).

Kilchinskii (1965, 1966) has also given a model for the 'self-consistent' calculation of various thermoelastic constants of fibre composites. He considers a cylindrical composite unit in which the innermost cylinder represents a typical fibre, the intermediate cylindrical shell represents the matrix and the outer cylindrical shell which encloses the first two is assumed to have the same physical properties as the bulk of the composite material. In these calculations the fibres are assumed to form a hexagonal array.

Hashin and Rosen (1964) have calculated the elastic constants of a composite with transverse isotropy by using a variational method. Two cases of the geometrical arrangement of the fibres have been considered: one in which the fibres are uniformly arranged in a hexagonal array and the other when the fibres are randomly distributed with the restriction that the composite is macroscopically homogeneous.

Owing to the transverse isotropy as in the cases considered by Hill, the composite has only five independent elastic constants as given in Section 2.3. It is more convenient, however, to calculate c_{33} and the following four constants as defined below:

$$K_c = \tfrac{1}{2}(c_{11} + c_{12}) \tag{3.3.42}$$

$$M_c = \tfrac{1}{2}(c_{11} - c_{12}) \tag{3.3.42a}$$

$$\mu_c = c_{44} \tag{3.3.42b}$$

and the longitudinal Young's modulus

$$E_l = c_{33} - \frac{2c^2_{13}}{c_{11} + c_{12}} \tag{3.3.42c}$$

We shall refer to these constants as the elastic moduli as in Section 2.3. The notation has been chosen here to show their correspondence with those calculated by Hill, although their definition is slightly different. The more familiar moduli of elasticity are related to the constants defined in (3.3.42)–(3.3.42c) as follows.

Transverse Young's modulus:

$$E_t = \frac{4M_c K_c}{K_c + \psi M_c} \tag{3.3.43}$$

Poisson ratio for uniaxial stress along the fibre direction:

$$\nu_c = \tfrac{1}{2}\left(\frac{C_{33} - E_c}{K_c}\right) \tag{3.3.43a}$$

and

Poisson ratio in the plane normal to the fibres :

$$\nu_t = \frac{K_c - \psi M_c}{K_c + \psi M_c} \tag{3.3.43b}$$

where

$$\psi = 1 + \frac{4K_c \nu_c^2}{E_c} \tag{3.3.43c}$$

The principle of the variational method used by Hashin and Rosen (1964) is similar to that which has been used in Section 3.2, for the derivation of the law of mixtures. The composite is subjected to either of the two following boundary conditions in the displacement field **u** and the stress vector **T** at the surface of the composite :

$$u_\alpha \text{ (surface)} = \sum_\beta \varepsilon^\infty_{\alpha\beta}\, r^\infty_\beta \qquad (3.3.44)$$

and

$$T_\alpha \text{ (surface)} = \sum_\beta \sigma^\infty_{\alpha\beta}\, n^\infty_\beta \qquad (3.3.44a)$$

where $\mathbf{r}^\infty$ denotes the surface coordinate and $\mathbf{n}^\infty$ the normal at the surface. The surface values of ε and σ can be regarded as their average values.

The elastic energy as given in Eqs. (2.3.5)–(2.3.5c) is then minimized subject to one of these boundary conditions, and the bounds on elastic constants are obtained by using the principle of minimum energy. In Section 3.2 the use of Eq. (2.3.5a) in conjunction with the uniform strain condition led to the upper bound of $c_{\alpha\beta\nu\delta}$, whereas Eq. (2.3.5b) together with the uniform stress condition yielded the upper bound of $s_{\alpha\beta\nu\delta}$. Since the tensors **s** and **c** are reciprocal to each other, the upper bound of **s** gives the lower bound of **c** and vice versa. The procedure used by Hashin and Rosen is more rigorous, since only the boundary values of stress and strain tensors are specified and the assumption of their being uniform is not needed.

In practice the geometry of the composite has to be defined for the use of the above variational method. Hashin and Rosen (1974) assume the composite to have a cylindrical shape. They consider each fibre as embedded into a cylinder made out of the matrix material. This system is referred to as the composite cylinder. Each composite cylinder thus contains one fibre surrounded by the matrix. The outer radius of the composite cylinder is chosen to be the maximum possible subject to the condition that the cylinders are non-overlapping. All composite cylinders are identical and are either arranged in a hexagonal array or randomly distributed. In addition, the fibres may be hollow or solid.

Let us consider the plane strain bulk modulus in a plane normal to the fibres. First we shall use the boundary condition (3.3.44), which defines the average value of the strain. For plane strain all the components of the strain are zero except in the plane normal to the fibres. The two diagonal components will be equal, which will be denoted by ε^∞. The elastic energy density (Eq. 2.3.5a) in this case is simply given by :

$$W = 2K_c\, \varepsilon^{\infty 2} \qquad (3.3.45)$$

On account of the transverse isotropy, the composite has cylindrical symmetry. In this case it is more convenient to use the cylindrical coordinates or the polar coordinates in the plane normal to the fibres. Since all composite cylinders, i.e. the unit cells are equivalent, it is enough to consider one such unit cell. The radial and the angular components of the displacement field vector **u** at the surface of the cylinder are then given by:

$$u_r = \varepsilon^{\infty} r_m \tag{3.3.46}$$

$$u_\theta = 0 \tag{3.3.46a}$$

where r_m is the outer radius of the composite cylinder. The former equation is obtained from Eq. (3.3.44), making use of the equivalence of all unit cells, and the latter simply indicates transverse isotropy. For plane strain the displacement field is confined to the plane and therefore its component normal to the plane, i.e. along the fibre direction, is zero.

The general solution for axially symmetric plane strain problems can be written as:

$$u_r = Ar + \frac{B}{r} \tag{3.3.47}$$

$$\sigma_r = 2KA - 2M\frac{B}{r^2} \tag{3.3.47a}$$

where M is the shear modulus, K is the plane strain bulk modulus, and A and B are arbitrary constants to be determined by appropriate boundary conditions.

When the fibres are arranged in an hexagonal pattern, the composite will have translational symmetry in the same sense as described in Section 1.2 (apart from surface effects, which can be neglected). In this case the composite can be divided into unit cells, all unit cells being identical. Each unit cell will consist of the composite cylinder and the remaining volume, say V_2, which can obviously be exactly determined. The elastic energy of the whole composite is given by the elastic energy of a unit cell, which is the sum of the elastic energies of the composite cylinder and the material constituting the volume V_2.

The solution of Eqs. (3.3.47) and (3.3.47a) is obtained inside the composite cylinder by matching the variables at the fibre–matrix surface. The solutions of Eqs. (3.3.47) and (3.3.47a) is thus considered in the following two regions:

$$r_0 \leqslant r \leqslant r_f$$

and

$$r_f \leqslant r \leqslant r_m$$

where r_0 and r_f are the internal and the external radii of the fibre (Figure 3.1) which define its radial thickness δ;

$$\delta = r_f - r_0 \tag{3.3.48}$$

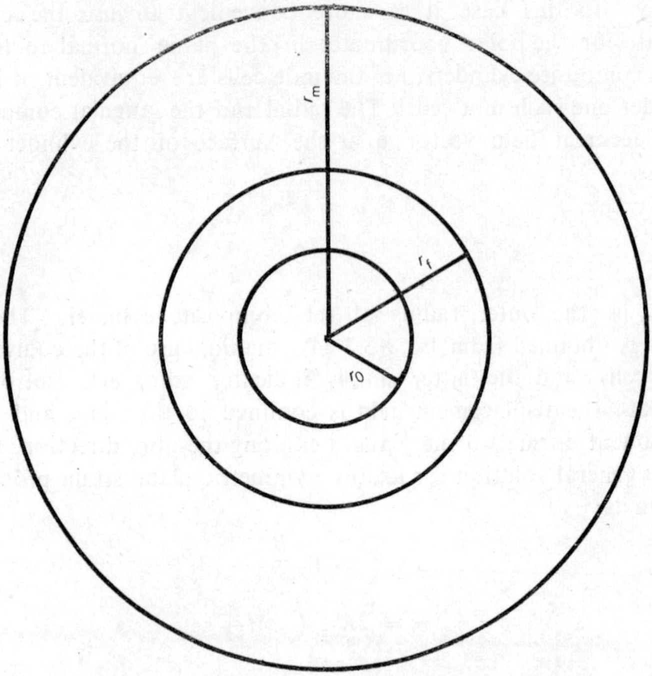

Figure 3.1 Cross-section of the cylindrical composite element studied by Hashin and Rosen. r_0 and r_f are respectively the inner and the outer radii of the fibre and r_m is the outer radius of the matrix material.

As in the previous case, we shall distinguish various quantities for the fibre and the matrix by subscript f and m, respectively. In the present case there are four arbitrary constants which can be determined by four equations: Eq. (3.4.46); and two provided by the condition of continuity of u_r and σ_r in Eqs. (3.3.47) and (3.3.47a) at the interface $r = r_f$, and the fourth by the condition that $\sigma_r = 0$ at $r = r_0$.

For the radial stress at $r = r_m$, Hashin and Rosen have obtained the following expression:

$$\sigma_r\,(r_m) = 2\epsilon^\infty\, K_m \cdot P \tag{3.3.49}$$

where

$$P = \frac{p_3\,(1 - p_1^2)\,[1 + (1 - 2\nu_m)\,p_2^2] + \left[1 + \dfrac{p_1^2}{1 - 2\nu_f}\right](1 - p_2^2)\,(1 - 2\nu_m)}{p_3\,(1 - p_1^2)\,(1 - p_2^2) + \left[1 + \dfrac{p_1^2}{1 - 2\nu_f}\right](p_2^2 + 1 - 2\nu_m]} \tag{3.3.50}$$

$$p_1 = \frac{r_0}{r_f} \tag{3.3.50a}$$

$$p_2 = \frac{r_f}{r_m} \tag{3.3.50b}$$

$$p_3 = \frac{K_f}{K_m} \tag{3.3.50c}$$

and ν_f and ν_m denote the two Poisson ratios.

The elastic energy in the composite cylinder is then given by

$$U_1 = 2K_m P \varepsilon^{\infty 2} V_1 \tag{3.3.51}$$

The total energy of the unit cell is equal to the sum of U_1 and U_2:

$$U = U_1 + U_2 \tag{3.3.52}$$

where U_2 is the elastic energy stored in the volume V_2, i.e.

$$U_2 = 2K_m \varepsilon^{\infty 2} V_2 \tag{3.3.53}$$

According to the principle of minimum energy, the actual value of the elastic energy as given in Eq. (3.3.45) must be the minimum value of U, so that:

$$U_c = W \cdot V_c \leqslant U$$

where $V_c = V_1 + V_2$ is the volume of the unit cell. Thus, we obtain the following upper limit of K_c:

$$K_c \leqslant K_m (P \nu_1 + \nu_2)$$

where ν_1 and ν_2 are the fractional volumes, i.e. $\nu_i = V_i/V_c$ $(i = 1, 2)$, which can easily be determined from geometrical considerations, since ν_1 is proportional to the area of the largest circle, which can be drawn centred at the vertices of a hexagon without overlapping.

It can easily be verified that

$$\nu_1 = \frac{\pi}{2\sqrt{3}}$$

and

$$\nu_2 = 1 - \nu_1$$

The lower bound of K_c is obtained by imposing the boundary condition (3.3.44a) and is given below:

$$K_c \geqslant \frac{P K_m}{\nu_1 + P\nu_2}$$

where P is defined in Eqs. (3.3.50)–(3.3.50c). Thus,

$$\frac{P K_m}{\nu_1 + P\nu_2} \leqslant K_{c\,(h)} \leqslant K_m (P\nu_1 + \nu_2) \tag{3.3.54}$$

where the bracketed suffix for K_c simply identifies the result for the hexagonal geometry. The results for a random distribution of fibres which are discussed below will be identified by the bracketed subscript r.

When the fibres do not form a regular geometrical pattern on a plane normal to their length and are randomly distributed on this plane, it is obviously not possible to define a unit cell. In this case the composite cannot be divided into identical regions in which the stress and strain are simply determined from their values at the outer surface of the composite. However, Hashin and Rosen make the assumption that the procedure as outlined above is valid even for a composite with randomly distributed fibres, the only difference being that in this case the result cannot depend upon v_2. Thus, according to Hashin and Rosen, the results for the hexagonal case reduce to those for a composite with randomly distributed fibres if v_2 is taken to be equal to zero. Then the two bounds of K_c coincide, so that the bulk modulus for the random case is given by:

$$K_{c\,(r)} = K_m P \tag{3.3.55}$$

It should be emphasized that this result crucially depends on the above-mentioned assumption made by Hashin and Rosen, which is not easy to justify. This assumption may be reasonable for a composite with very high fibre concentration, in which case the hexagonal approximation itself may be quite good. On the other hand, if the fibre concentration is too large, this procedure becomes unreliable. It is recommended, therefore, that Eq. (3.3.55) be used with caution and, as in the cases considered by Hill, Eq. (3.3.54) can be used for composites with not too large fibre concentration.

Proceeding along similar lines, Hashin and Rosen have also calculated the bounds for other elastic moduli by specifying appropriate kinds of strains, as, for example, in Eq. (3.3.23). Results have been obtained for both hexagonal and random arrangement of fibres which are quoted below.

1. *Shear modulus* μ_c

$$\frac{P_\mu \mu_m}{v_1 + P_\mu v_2} \leqslant \mu_{c\,(h)} \leqslant \mu_m (P_\mu v_1 + v_2) \tag{3.3.56}$$

and

$$\mu_{c\,(r)} = P_\mu \mu_m \tag{3.3.57}$$

where

$$P_\mu = \frac{p_4 (1 - p_1^2)(1 + p_2^2) + (1 + p_1^2)(1 - p_2^2)}{p_4 (1 - p_1^2)(1 - p_2^2) + (1 + p_1^2)(1 + p_2^2)} \tag{3.3.58}$$

and

$$p_4 = \frac{\mu_f}{\mu_m}\,. \tag{3.3.59}$$

2. *Longitudinal Young's modulus E_c and Poisson ratio ν_c*

$$E_{c\,(r)} = P_E\, E_m \tag{3.3.60}$$

where

$$P_E = \left(\nu_f \frac{E_f}{E_m} + \nu_m\right) \frac{E_m\,(D_1 - D_3 F_1) + E_f\,(D_2 - D_4 F_2)}{E_m\,(D_1 - D_3) + E_f\,(D_2 - D_4)} \tag{3.3.61}$$

$$D_1 = \frac{1 + p_1^2}{1 - p_2^2} - \nu_f \tag{3.3.62}$$

$$D_2 = \frac{1 + \nu_t}{v_m} + \nu_m \tag{3.3.62a}$$

$$D_3 = \frac{2\nu_f^2}{1 - p_1^2} \tag{3.3.62b}$$

$$D_4 = 2\,\nu_m^2\,\frac{v_t}{v_m} \tag{3.3.62c}$$

$$F_1 = \frac{\nu_m\, v_f\, E_f + \nu_f\, v_m\, E_m}{v_f\, \nu_f\, E_f + v_m\, E_m} \tag{3.3.62d}$$

$$F_2 = \frac{\nu_f}{\nu_m}\, F_1 \tag{3.3.62e}$$

v_t is the fractional volume of gross fibres, $v_m = 1 - v_t$ and $v_f = (1 - p_1^2)\; v_t$ is the fractional volume of net fibre (hollow) material. In the limit when $p_1 = 0$, $v_f = C_f$ and $v_m = C_m$ in the notation used earlier in this section.

For the case of random fibre distribution, Hashin and Rosen have suggested that the following law of mixtures is a good approximation to the value of $E_{c\,(r)}$ given above:

$$E_{c\,(r)} \approx v_f\, E_f + v_m\, E_m$$

In this case the approximate value of the Poisson ratio is simply given by

$$\nu_{c\,(r)} = \frac{v_f\, E_f\, L_1 + v_m\, E_m\, L_2\, \nu_m}{v_f\, E_f\, L_3 + v_m\, E_m\, L_2} \tag{3.3.63}$$

where

$$L_1 = 2\,\nu_f\,(1 - \nu_m^2)\; v_t + v_m\,(1 + \nu_m)\; \nu_m$$

$$L_2 = v_t\; [(1 + \nu_f)\; p_1^2 + 1 - \nu_f - 2\nu_f^2]$$

and

$$L_3 = 2\,(1 - \nu_m^2)\; v_t + (1 + \nu_m)\; v_m$$

The evaluation of the Poisson ratio is rather complicated and has not been given by Hashin and Rosen.

For the hexagonal case the result is

$$\frac{P_E' E_m}{v_1 + P_E' v_2} \leqslant E_{c(h)} \leqslant E_m (P_E' v_1 + p v_2) \tag{3.3.64}$$

where

$$p = \frac{1 - \nu_m - 4\nu_m \nu_{c(r)}' + 2\nu_{c(r)}'^2}{1 - \nu_m - 2\nu_m^2} \tag{3.3.65}$$

and v_1 and v_2, as before, are, respectively, the fractional volumes of the composite cylinder and the filling material in a unit cell. The primes on P_E and $\nu_{c(r)}$ indicate that these quantities are given by Eqs. (3.3.61) and (3.3.63) by replacing v_t, v_f and v_m by v_t/v_1, v_f/v_1 and $1-(v_t/v_1)$, respectively.

3. *Elastic constant c_{33}*

$$c_{33c(r)} = E_{c(r)} + 4\nu_{c(r)}^2 K_{c(r)} \tag{3.3.66}$$

where the moduli $E_{c(r)}$, $\nu_{c(r)}$ and $K_{c(r)}$ have been given earlier.

$$\frac{E_m c_{33c(r)}'}{E_m v_1 + L_4 v_2 c_{33c(r)}'} \leqslant E_{c(h)} \leqslant v_1 c_{33c(r)}' + \frac{v_2 E_m (1-\nu_m)}{(1+\nu_m)(1-2\nu_m)} \tag{3.3.67}$$

where

$$L_4 = 1 - 4\nu_m n + 2(1-\nu_m) n^2 \tag{3.3.68}$$

$$n = \frac{2\nu_{c(r)}' + K_{c(r)}'}{c_{33c(r)}'} \tag{3.3.68a}$$

and the prime over a quantity denotes that it has to be evaluated by the corresponding expression with modified v_t, v_f and v_m, as described in connection with Eq. (3.3.64).

4. *Shear modulus M_c.* Even for a random distribution of fibres in a composite, the two bounds of M_c do not coincide. They are given by:

$$M_m \left[1 + \frac{2(1-\nu_m)}{1-2\nu_m} v_t A_4^\sigma\right]^{-1} \leqslant M_{c(r)} \leqslant M_m \left[1 - \frac{2(1-\nu_m)}{1-2\nu_m} v_t A_4^\varepsilon\right] \tag{3.3.69}$$

where v_t has been defined earlier, and A_4^ε and A_4^σ are obtained by solving the following system of equations:

$$A_1^\varepsilon + v_t^{-1} A_2^\varepsilon + v_t^2 A_3^\varepsilon + v_t A_4^\varepsilon = 1 \tag{3.3.70}$$

$$-\frac{3-4\nu_m}{v_t(3-2\nu_m)}A_2^\varepsilon-2v_t^2A_3^\varepsilon+\frac{v_t}{1-2\nu_m}A_4^\varepsilon=0 \qquad (3.3.70a)$$

$$A_1^\varepsilon+A_2^\varepsilon+A_3^\varepsilon+A_4^\varepsilon-B_1^\varepsilon-B_2^\varepsilon-B_3^\varepsilon-B_4^\varepsilon=0 \qquad (3.3.70b)$$

$$-\frac{3-4\nu_m}{3-2\nu_m}A_2^\varepsilon-2A_3^\varepsilon+\frac{1}{1-2\nu_m}A_4^\varepsilon$$
$$+\frac{3-4\nu_f}{3-2\nu_f}B_2^\varepsilon+2B_3^\varepsilon-\frac{1}{1-2\nu_f}B_4^\varepsilon=0 \qquad (3.3.70c)$$

$$A_1^\varepsilon+\frac{3}{3-2\nu_m}A_2^\varepsilon-3A_3^\varepsilon+\frac{1}{1-2\nu_m}A_4^\varepsilon-p_5B_1^\varepsilon$$
$$-\frac{3p_5}{3-2\nu_f}B_2^\varepsilon+3p_5B_3^\varepsilon-\frac{p_5}{1-2\nu_f}B_4^\varepsilon=0 \qquad (3.3.70d)$$

$$-\frac{1}{3-2\nu_m}A_2^\varepsilon+2A_3^\varepsilon-\frac{1}{1-2\nu_m}A_4^\varepsilon+\frac{p_5}{3-2\nu_f}B_2^\varepsilon$$
$$-2p_5B_3^\varepsilon+\frac{p_5}{1-2\nu_f}B_4^\varepsilon=0 \qquad (3.3.70e)$$

$$B_1^\varepsilon+\frac{3p_1^2}{3-2\nu_f}B_2^\varepsilon-\frac{3}{p_1^4}B_3^\varepsilon+\frac{1}{p_1^2(1-2\nu_f)}B_4^\varepsilon=0 \qquad (3.3.70f)$$

$$-\frac{p_1^2}{3-2\nu_f}B_2^\varepsilon+\frac{2}{p_1^4}B_3^\varepsilon-\frac{1}{p_1^2(1-2\nu_f)}B_4^\varepsilon=0 \qquad (3.3.70g)$$

where p_1 has been defined in Eq. (3.3.50a); $p_5=M_f/M_m$; and for A_4^σ Eqs. (3.3.70) and (3.3.70a) have to be replaced by

$$A_1^\sigma+\frac{3}{v_t(3-2\nu_m)}A_2^\sigma-3v_t^2A_3^\sigma+\frac{v_t}{(1-2\nu_m)}A_4=1 \qquad (3.3.70h)$$

$$-\frac{1}{v_t(3-2\nu_m)}A_2^\sigma+2v_t^2A_3^\sigma-\frac{v_t}{(1-2\nu_m)}A_4=0 \qquad (3.3.70i)$$

while the remaining equations (3.3.70b–3.3.70g) remain unchanged except for the superscript ε, which is changed to σ. For solid fibres ($p_1=0$) B_3 and B_4 are zero (Hashin and Rosen, 1964) and Eqs. (3.3.70f) and (3.3.70g) have to be deleted. The form of expression (3.3.69) is unchanged.

When the fibres are arranged in a hexagonal pattern, the bounds of M are as follows:

$$M_m\left[1+\frac{2(1-\nu_m)}{1-2\nu_m}v_tA_4'^\sigma\right]^{-1}\leqslant M_c\leqslant M_m\left[1-\frac{2(1-\nu_m)}{1-2\nu_m}v_tA_4'^\varepsilon\right]$$
(3.3.71)

where $A_4'^{\varepsilon}$ and $A_4'^{\sigma}$ are also given by Eqs. (3.3.70)–(3.3.70i) with v_t replaced by v_t/v_1.

The expressions derived by Hashin and Rosen (1964) for the elastic moduli of a fibre composite and their bounds are rather involved and are not very convenient for numerical applications. Moreover, as discussed earlier, their results for the random arrangement of the fibres in a composite appear to be of questionable validity except in some very special cases. The assumption of elastic isotropy is also not valid for high-strength fibres, which are in fact highly anisotropic. The main advantage of their method is that their results for the hexagonal arrangement of the fibres are exact and their derivation is quite rigorous (apart from the end corrections, which should be small for large samples). These formulae have been applied to a glass fibre composite, for which the reader is referred to Hashin and Rosen (1964).

The upper and the lower bounds of some elastic moduli have also been derived by Tsai (1964), using a variational method. His approach is based on the use of the potential energy theorem. However, the upper and the lower bounds in certain cases are found to be too far apart to be useful in practice. In such cases Tsai has suggested that an appropriate linear combination of the two bounds should be a reasonable representation of the corresponding elastic modulus of the composite. Alternatively, a parameter—the so-called contingency factor (Tsai et al., 1963; Rosen et al., 1964)—could be introduced which can be chosen by fitting the calculated and the measured values of the elastic moduli. Such phenomenological approaches have also been suggested by Wu (1965), Bishop (1966) and Chamis (1967).

Some useful and interesting invariant properties of composite materials have been described by Tsai and Pagano (1968), and the limits on the fibrous composite material properties have been given by Brandmaier (1969).

The effect of fibre anisotropy on the elastic properties of the composite has been discussed by Whitney (1967). He has shown that it is possible to predict the principal elastic moduli of the composite with anisotropic fibres in a simple way by using the analysis developed for a composite with isotropic fibres such as that given by Hashin and Rosen (1964). Whitney's approach is based on qualitative and somewhat intuitive arguments and is quite reasonable for transversely isotropic fibres with their axes as the axis of elastic symmetry. Whitney has also compared his theoretical predictions with the measured values of elastic constants for graphite fibre composites: they are found to be in reasonable agreement. He has shown that the elastic behaviour of a composite is significantly affected by the fibre anisotropy, the most sensitive moduli being the transverse Young's modulus and the rigidity modulus.

As mentioned before, the models given by Hill (1964a) and Hashin and Rosen (1964) become increasingly unreliable as the concentration of fibres increases in a composite. This is due to the fact that each fibre is treated independently of the other fibres and the fibre–fibre interaction is neglected in both the models discussed above. The fibre–fibre interaction arises because the stress field produced by a fibre is modified owing to the presence of neighbouring fibres.

The situation is somewhat analogous to the interatomic interactions in the lattice theory which, as we discussed in Section 2.2, contribute to the elastic constants of the solids. The analogy, however, is quite superficial, because in lattice theory the atoms interact *in vacuo* whereas in a composite the fibres interact in the medium of the matrix. In the case of a crystal lattice the interatomic interaction is solely responsible for the elastic properties of the solid. On the other hand, in a composite the elastic properties of the fibres and the matrix make the major contribution to the elastic properties of the composite and the contribution of the fibre–fibre interaction is likely to be important only at high fibre concentrations. Nevertheless it should be possible to use the techniques of the lattice theory in the present case with considerable advantage. However, no such calculations have been reported in the literature for static problems such as the calculation of elastic constants. The lattice type model of a fibre composite has indeed been used to study the propagation of elastic waves in a fibre composite by Tewary and Bullough (197-) which we shall review in Chapter 5.

The effect of fibre–fibre interaction on the elastic constants of a composite has been included in an approximate manner by Chow and Hermans (1969). They express the elastic moduli of a composite in the form of a series in powers of the fibre concentration, C_f. The first term in this series which is affected by the fibre–fibre concentration is C_f^2. Chow and Hermans (1969) estimate the coefficient of C_f^2 by what they call the 'reflection' theory; the disturbance in the stress field caused by the presence of a fibre creates a reaction around a second fibre and this in turn influences the stress around the first fibre. The authors make the approximation that the reaction of the second fibre can be obtained by treating the above mentioned disturbance locally as a homogeneous field. Although the actual results derived by Chow and Hermans are not rigorous, their technique is quite interesting and can be further generalized. It would be of interest, therefore, to present here a brief review of their technique for the calculation of the elastic constants of fibre composites.

The model of the fibre composite which has been considered by Chow and Hermans is the same as that of Hill (1964a). Each fibre is assumed to be a solid circular cylinder and the composite is also taken to be of cylindrical shape. The z axis is taken to be along the fibre axis and the transverse isotropy in the xy plane is assumed at the outset. The strain ε_z is taken to be zero, by applying a certain stress σ_z if necessary. The problem of a three-dimensional composite is thus reduced to a plane strain problem.

We notice from Eq. (3.3.9) that if one of the three moduli K_c, L_c and N_c is known, the other two can be obtained from that equation. It therefore suffices to obtain K_c, M_c and μ, Chow and Hermans write the solution in powers of C as follows:

$$K_c = K_m [1 + f_1 C + f_2 C^2 + O(C^3)] \tag{3.3.72}$$

and

$$M_c = M_m [1 + g_1 G + O(C^3)] \tag{3.3.72a}$$

where C is the concentration of fibres, the subscript f used in Eq. (3.3.9) being dropped for notational brevity. The problem is to determine the unknown coefficients f_i and g_i, where $i = 1, 2$.

In terms of the Airy stress function χ, defined as

$$\sigma_x = \frac{\partial^2 \chi}{\partial y} \tag{3.3.73}$$

$$\sigma_y = \frac{\partial^2 \chi}{\partial x^2} \tag{3.3.73a}$$

and

$$\sigma_{xy} = -\frac{\partial^3 \chi}{\partial x \, \partial y} \tag{3.3.73b}$$

the compatibility equation reduces to

$$\nabla^4 \chi = \left[\frac{2 L_c M_c}{K_c + M_c}\right] \nabla^2 \varepsilon_z \tag{3.3.74}$$

where

$$\nabla^4 = (\nabla^2)^2 \quad \text{and} \quad \nabla^2 = \frac{\partial^2}{\partial x^2} + \frac{\partial^2}{\partial y^2}$$

is the two-dimensional Laplacian.

In terms of cylindrical coordinates r and θ, we have

$$2K_c = \langle(\sigma_r + \sigma_\theta)\rangle / \langle\theta\rangle \tag{3.3.75}$$

where $\langle\,\rangle$ denotes the volume average and $\theta = \text{div}\,\mathbf{u}$ is the dilatation, $\mathbf{u}$ being the displacement field.

If the correction due to surface effects is neglected, Chow and Hermans showed that

$$K_c = K_0 + (K_f - K_m) N \int \overline{\mathbf{u} \cdot \mathbf{n}} \, dS \tag{3.3.76}$$

where the integration is over the surface S of one cylinder (fibre), N is the number of cylinders of unit length per unit volume, $\mathbf{n}$ is the unit outward normal and the overhead bar denotes averaging over the positions of all other cylinders. If a denotes the radius of a fibre, then, obviously,

$$C = \pi a^2 N \tag{3.3.77}$$

If C is small and the interaction between the fibres is neglected, the solution of Eqs. (3.3.74) and (3.3.76) which is linear in C can be easily obtained by considering a single fibre in an infinite matrix subject to the boundary condition that the radial displacement at infinity is $u_r = r$ (as in Eq. 3.3.46), where r is measured from the fibre axis on the xy plane. As in the case of Hashin and Rosen (1964), it can easily be shown that

$$u_r = r - \alpha a^2/r \qquad (r \geqslant a) \tag{3.3.78}$$

and

$$u_r = (1 - \alpha) r \qquad (r \leqslant a) \tag{3.3.78a}$$

with

$$\alpha = \frac{K_f - K_m}{K_f + K_m} \tag{3.3.78b}$$

to ensure continuity of u_r and σ_r at $r = a$. This solution for u_r when substituted in Eq. (3.3.76) gives the following expression for the bulk modulus which is linear in C:

$$K = K_m + (K_f - K_m)\frac{K_m + M_m}{K_f + M_m}C \tag{3.3.79}$$

To estimate the coefficients f_2 and g_2 in Eqs. (3.3.72) and (3.3.72a) Chow and Hermans considered the interaction between two fibres separated by a distance ρ between their axes (see Figure 3.2). In Eq. (3.3.78) the first term which is linear in r is the displacement imposed on the system, while the second term, which vanishes at $r \to \infty$, denotes the change in the displacement field due to the fibre (fibre I). The first term will obviously be unchanged because of the presence of another fibre. The problem is to calculate the response of the other fibre (fibre II) to the stress field created by the second term in Eq. (3.3.78). Chow and Hermans have calculated this response by assuming that the stress field is homogeneous throughout fibre II and *is equal to its value at the axis of fibre II if fibre I were absent.* Thus,

$$\sigma_r = s \tag{3.3.80}$$

$$\sigma_\theta = -s \tag{3.3.80a}$$

where

$$s = 2M_m \alpha a^2/\rho^2 \tag{3.3.80b}$$

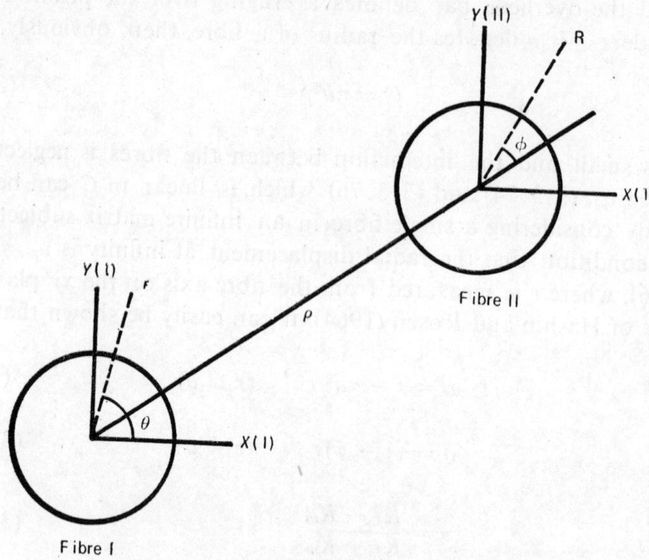

Figure 3.2 The coordinate system used for the calculation of the fibre–fibre interaction by Chow and Hermans.

The response of fibre II to this stress field creates an additional stress field outside fibre II which comes back to fibre I. Thus, there is a reflection of the stress field between fibres I and II.

Taking the origin of coordinates (R, ϕ) in the xy plane at the centre of fibre II (see Figure 3.2 for notation) and assuming a uniform radial stress, Chow and Hermans derive the following expressions for the stress components which satisfy the biharmonic equation $\nabla^4\chi = 0$:

$$\sigma_R = \frac{s}{2} + sD\frac{a^2}{R^2} + s\left(\tfrac{1}{2} - 2A\frac{a^2}{R^2} + 3B\frac{a^4}{R^4}\right)\cos 2\phi \qquad (3.3.81)$$

$$\sigma_\phi = \frac{s}{2} - sD\frac{a^2}{R^2} + s\left(\tfrac{1}{2} + 3B\frac{a^4}{R^4}\right)\cos 2\phi \qquad (3.3.81a)$$

$$\sigma_{R\phi} = s\left(-\tfrac{1}{2} - A\frac{a^2}{R^2} + 3B\frac{a^4}{R^4}\right)\sin 2\phi \qquad (3.3.81b)$$

where A, B and D are constants. The values of these constants are obtained by matching the solutions of the Airy equation at the fibre–matrix interface. We shall require the values of A and D, which are given below:

$$A = -\frac{K_m}{M_m}\beta \qquad (3.3.81c)$$

where

$$\beta = \frac{M_m(M_f - M_m)}{2M_fM_m + K_m(M_f + M_m)} \tag{3.3.81d}$$

and

$$D = -\frac{M_m}{K_m}\alpha = -\frac{M_m}{K_m}\left(\frac{K_f - K_m}{K_f + K_m}\right) \tag{3.3.81e}$$

The other stress field is derived simply by changing s to $-s$ and ϕ to $\phi - \pi/2$ in Eqs. (3.3.81)–(3.3.81e). Superposition of these two stress field gives

$$\sigma_R = 2s\left(\tfrac{1}{2} - 2A\frac{a^2}{R^2} + 3B\frac{a^4}{R^4}\right)\cos 2\phi \tag{3.3.82}$$

$$\sigma_\phi = -2s\left(\tfrac{1}{2} + 3B\frac{a^4}{R^4}\right)\cos 2\phi \tag{3.3.82a}$$

$$\sigma_{R\phi} = 2s\left(-\tfrac{1}{2} - A\frac{a^2}{R^2} + 3B\frac{a^4}{R^4}\right)\sin 2\phi \tag{3.3.82b}$$

It can be verified that the average of the displacement field corresponding to the above stress field over the surface of fibre II is zero. Thus, the stress (3.3.80)–(3.3.80b) does not contribute to the surface integral in Eq. (3.3.76).

Next consider the effect of the stress field (3.3.82)–(3.3.82b) on fibre I. This is the stress fie'd which has been reflected off fibre II in the sense discussed earlier. In the (R, ϕ) coordinate system the coordinates of the centre of fibre I are (ρ, θ) (see Figure 3.2). The stress field at this point obtained from Eqs. (3.3.82)–(3.3.82b) and using Eqs. (3.3.80b) and (3.3.81c) is given by

$$\begin{aligned}\left(\sigma_r + \sigma_\theta\right)_{\mathrm{I}}^{(1)} &= -4As\frac{a^2}{\rho^2}\\ &= 8K_m\alpha\beta\frac{a^4}{\rho^4}\end{aligned} \tag{3.3.83}$$

where the superscript and the subscript denote, respectively, that it is the response of the first reflection at the axis of fibre I. To get the total value of $(\sigma_r + \sigma_\theta)_{\mathrm{I}}^{(1)}$, it is integrated over all possible values of ρ and θ. Thus

$$\sum_r = \text{total } (\sigma_r + \sigma_{\theta\theta})_{\mathrm{I}}^{(1)} = N\int_{2a}^{\infty}\int_{0}^{2\pi}(\sigma_r + \sigma_\theta)_{\mathrm{I}}^{(1)}\,\rho\,d\theta\,d\rho \tag{3.3.84}$$

which, using Eqs. (3.3.83) and (3.3.77), gives

$$\sum_r = 2K_m\alpha\beta C \tag{3.3.85}$$

The original unperturbed stress field at fibre I is $\Sigma_0 = 4K_m$ and therefore the total stress field at fibre I is given by

$$\sum_t = \sum_0 + \sum_r$$
$$= 4K_m + 2K_m\alpha\beta C \tag{3.3.86}$$

which gives the following displacement field:

$$u_r = \frac{K_m + M_m}{K_f + M_m} a(1 + \tfrac{1}{2}\alpha\beta C) \tag{3.3.87}$$

Using Eq. (3.3.87), Chow and Hermans obtained the following value for the bulk modulus of the composite:

$$K_c = K_m + (M_m + K_m)\,\alpha C[1 + \tfrac{1}{2}\alpha\beta C] \tag{3.3.88}$$

Thus, in the approximation that the effect of further reflections can be neglected, the coefficient f_2 in Eq. (3.3.72), which is a measure of the contribution of the second-order term in C, is given by

$$f_2 = \frac{M_m}{2K_m}\left(\frac{K_f - K_m}{K_f + M_m}\right)^2 \frac{(M_f - M_m)(K_m + M_m)}{M_f(K_m + M_m) + M_m(K_m + M_f)} \tag{3.3.89}$$

where Eqs. (3.3.78b) and (3.3.81d) have been used for α and β, respectively.

The coefficient g_2 for rigidity M_c in Eq. (3.3.72a) derived by Chow and Hermans, using the approximation similar to that for K_c, is given below:

$$g_2 = g_2' + g_2'' \tag{3.3.90}$$

where

$$g_2' = -\frac{2\beta^2 K_m (K_m + M_m)}{M_m^2} \tag{3.3.90a}$$

and

$$g_2'' = \frac{K_m + M_m}{2M_m}\beta^2\left(\alpha + \frac{43}{16}\frac{K_m^2}{M_m^2}\beta\right) \tag{3.3.90b}$$

with α and β as defined in Eqs. (3.3.78b) and (3.3.81d), respectively. The value of g_1 can be obtained on the single-fibre model and is as follows:

$$g_1 = \frac{2(M_f - M_m)(K_m + M_m)}{M_f(K_m + M_m) + M_m(K_m + M_f)} \tag{3.3.90c}$$

The contribution of the C^2 term in the rigidity μ_c of the composite as defined by Eqs. (3.3.2b) and (3.3.2c) has been obtained by analogy with the dielectric constant for a composite (Hermans, 1967; Peterson and

Hermans, 1969). Two expressions for μ, one derived by the approximate 'reflection' theory and the other by a rigorous calculation as quoted by Chow and Hermans, are given below.

Reflection theory:

$$\frac{\mu_c}{\mu_m} = 1 + 2pC + (2p^2 + \tfrac{1}{2}p^3 + \tfrac{1}{96}p^5 + \ldots)\, C^2 \tag{3.3.91}$$

Rigorous calculation:

$$\frac{\mu_c}{\mu_m} = 1 + 2pC + (2p^2 + \tfrac{2}{3}p^3 + 0.0555p^5 + \ldots)\, C^2 \tag{3.3.91a}$$

where

$$p = \frac{\mu_f - \mu_m}{\mu_f + \mu_m} \tag{3.3.91b}$$

The calculations of Chow and Hermans (1969) as reviewed above are based on the following three main approximations: (1) the stress field in the region of fibre II is uniform; (2) the second and further reflections between fibres I and II and also multifibre reflections involving other fibres do not make significant contributions; and (3) the distribution of the reflecting fibres (like fibre II) around fibre I is uniform and continuous.

None of these assumptions is strictly valid in a real composite. The stress field, as we can see from Eq. (3.3.83), varies as ρ^{-4}, which is much too fast for the uniform field approximation, particularly for not very large values of ρ. The second and the first approximations are in fact not consistent with each other, as discussed by the authors themselves, because if the stress field were uniform, the successive reflections would make significant contributions. A self-consistent approach like that used in the calculations of dielectric constant and the induced dipole moment in an assembly of polarisable objects may be more useful for treating the successive reflections of the stress field between several fibres. The third approximation which enabled Chow and Hermans to write Σ_r as an integral over the whole region on the xy plane (except that between $r = 0$ and $2a$) is also not valid for small values of $(r - 2a)$ because the fibres have a discrete distribution which cannot be smeared out except in the region where r is very large. A semi-continuum model is one in which the fibres in a region close to fibre I are treated as having a discrete structure and those outside this region can be assumed to be smeared out into a continuum. Such models are used quite often in the lattice theory of crystals (see, for example, Bullough and Tewary, 1972) and should be useful in the theory of fibre composites as well.

Owing to the approximations discussed above, the results derived by Chow and Hermans may not be numerically accurate. Nevertheless their results are very useful for a qualitative understanding of the fibre–fibre interactions in a composite and provide an estimate of the dependence of the elastic properties of a composite on the fibre concentration. An interesting inference which can be derived from their results is that the contribution of the C^2 term vanishes for K_c, all the three moduli K_c, μ_c and M_c if the rigidity moduli of fibres and the matrix are equal, i.e. $\mu_f = \mu_m$ and $M_f = M_m$, and increases with increasing $(\mu_f - \mu_m)$ and $(M_f - M_m)$. This is physically expected because it is this difference which acts as the carrier of fibre-fibre interaction. In addition, the term f_2 in the bulk modulus in Eq. (3.3.89) has a similar though stronger dependence on $(K_f - K_m)$ which can also be physically interpreted on the same lines.

Chow and Hermans (1969) have compared their results for epoxy resin reinforced by glass and boron fibres with those derived by Hill (1965a) and Hermans (1967) on the basis of Kerner's model, in which each fibre is surrounded by a cylinder of matrix material of radius $(\pi N)^{-1/2}$ which itself is surrounded by a material which is indistinguishable from the composite as a whole. The values of f_1 and f_2 in Hill's model are estimated by a Taylor expansion of M in powers of C which is easily obtained by successive differentiations of Eq. (3.3.39) with respect to C for $C = 0$ ($C_f = C$ and $C_m = 1 - C$). Incidentally, it can be verified that g in Eq. (3.3.90c) satisfies Eq. (3.3.39), as would be expected.

The values of f_1 given by all the three models agree with one another, but the value of f_2 derived by Chow and Hermans (1969) is found to be much lower than those estimated from the other two models, namely Hill's (1965a) and Hermans' (1967). Chow and Hermans have given a plausible argument in favour of the low value of f_2 which is based upon the fact that the bulk modulus of the composite for compression in a plane normal to the fibre axis is mainly determined by the compressibility of the matrix, since a load normal to fibres will first attempt to bring the fibres closer, this being resisted only by the matrix for small values of C. The authors have suggested that their results should be reliable for $C < 0.25$. No experimental evidence has, however, been reported. The contribution of fibre–fibre interaction to the Young's modulus is apparently negligible, since it appears to depend only linearly on C.

It may be remarked that the value of f_2 in Hill's (1965a) model does not really correspond to that as derived by Chow and Hermans. This is because the C^2 term as calculated by Chow and Hermans arises only because of the fibre–fibre interaction, whereas in Hill's models it results from the definition of an average value of the rigidity. In fact Hill's treatment is based on a self-consistent approach in which the fibre is assumed to interact with the composite as a whole, and in this sense Hill's model includes some contribution from the fibre–fibre interaction, but this is obviously not the total contribution. The comparison between Hill's results and their own as made by Chow and Hermans is, therefore, not very meaningful.

In addition to the work described above, a considerable amount of work has been done on various other aspects of the elastic behaviour of composites, some of which will be briefly mentioned here. For example, as well as hexagonal, the square or rectangular arrays of fibres have also been considered. In such cases the composite does not have transverse isotropy and it has six independent elastic moduli. More general theories which apply to composites with several phases as well as those with arbitrary phase shapes and geometries, such as spherical or ellipsoidal as well as cylindrical inclusions, have also been developed by several authors: for example, Paul (1960), Hashin and Shtrikman (1961; 1962a, b; 1963), Hashin (1962), Hashin (1965), Hill (1963, 1965b), Walpole (1969) and Budiansky (1965)—a list which is by no means exhaustive. Our interest in this book is of course only in fibre composites. Some papers on general composites have been mentioned here because the techniques used for the calculation of elastic constants of general composites can also be used in the case of fibre composites.

The method of Hashin and Shtrikman (1961; 1962a, b; 1963) has been modified by introducing a normal distribution and applied to calculate the elastic constants of a binary (two-phase) composite by Wang (1970). He assumes a typical reference 'cube' inside the material with rigorously defined boundary conditions regarding the stress and the strain at the surface of the reference cube. This leads to the upper and lower bounds for the elastic constants which can be extrapolated for the whole material. The theory is claimed to be reliable for up to 50% volume concentration of the fibres in the composite.

A general theory for the elastic properties of two-phase composite materials subject to the condition that they are macroscopically homogeneous and elastically isotropic has been given by Davies (1971). His analysis takes into account the geometrical arrangement of the two phases which is specified in terms of joint probability functions. Approximate formulae for the effective elastic constants of the composite have been given with a short discussion of their validity in terms of the general features of the simple geometry.

Returning to the case of fibre composites, a simple though approximate formula has been derived for some of the elastic moduli (Youngs, shear and possibly bulk) of the composite by Halpin and Tsai (1969) with the help of Kerner's (1956) equations. This formula, which is sometimes referred to as the Halpin–Tsai formula, is given below:

$$\frac{Y_c}{Y_m} = \frac{1 + ABC_f}{1 - BC_f} \tag{3.3.92}$$

where

$$B = \frac{Y_f - Y_m}{Y_f + AY_m} \tag{3.3.92a}$$

Y denotes the particular modulus being calculated, with the subscripts

f, m and c labelling the fibres, matrix and the composite; C_f is the fibre concentration; and A is a constant which depends on several factors, such as the geometry of the reinforcing phase, the Poisson ratio of the matrix, etc. In practice A can be treated as a parameter which is to be determined by comparing the theoretical and observed values of Y. The values of A are typically between 1 and 2.

The Halpin–Tsai equation is a reasonably accurate representation of the transverse Young's modulus and the longitudinal shear modulus of the composite, provided that C_f is not very large. On account of its simplicity, it is very useful for crude and quick estimates required for many engineering applications. A modified value of the parameter A which improves the fit between the values calculated from Eqs. (3.3.92) and (3.3.92a) and the corresponding measured values for the Young's modulus has been obtained empirically by Hewitt and de Malherbe (1970) on the basis of the experimental results of Adams et al. (1969). The modified value of A is as follows:

$$A = 1 + 40\, C_f^{10} \qquad (3.3.93)$$

Since this value of A has been obtained empirically, the fact that it works in a particular case, as mentioned above, is obviously no guarantee that it will be reliable for other applications.

A more rigorous generalization of the Halpin–Tsai formula has been derived by Nielson (1970). He has extended Eqs. (3.3.92) and (3.3.92a) to take into account the packing of the filler phase (the reinforcing phase) and has also given the following relation:

$$A = k - 1$$

where k, the so-called Einstein coefficient, is defined by

$$k = \frac{d\,(Y_c/Y_m - 1)}{dc_f} \qquad (3.3.94)$$

as c_f goes to zero and M_f/M_m goes to infinity. These equations are not restricted to fibre composites. As an example, $k = 2.5$ for a suspension of rigid spheres in a matrix whose Poisson ratio is 0.5. The values of k for some other cases have been given by Nielson (1970).

In terms of k as defined in Eq. (3.3.94) the generalized equation suggested by Nielson (1970) for determining the elastic moduli of a composite is as follows:

$$\frac{Y_c}{Y_m} = \frac{1 + (k-1)\,B\,C_f}{1 - B\,\psi\,C_f} \qquad (3.3.95)$$

where the function ψ accounts for the packing of the filler phase. The actual choice of ψ will depend on the particular case to be investigated.

For example, it can be represented by the functions

$$\psi = 1 + C_f\,(1 - C_p)/C_p^2$$

and

$$\psi\, C_f = 1 - \exp\left[\frac{-C_f}{1-(C_f/C_p)}\right]$$

where C_p is the maximum volumetric packing fraction.

Numerical applications of Eq. (3.3.95) for glass spheres in epoxy resin and a model fibre composite have been given by Nielson (1970).

Gillis (1970) has calculated the elastic moduli for plane stress analysis of undirectional fibre composites with anisotropic rectangular alignment using the principles of least work and minimum potential energy. He has derived the upper and the lower moduli for each modulus. The two bounds are found to be quite close to each other and it is suggested that their average is a good approximation for certain types of composites. Some variational principles and bounds for the effective moduli of quasi-isotropic composites have also been derived by Yeh (1970). Numerical applications of these formula and their usefulness for fibre composites have been discussed by Yeh (1971). Pagano (1970) has studied some mechanical properties of rectangular bidirectional composites and sandwich plates for which exact solutions have been given. The elastic constants of layered medium have been obtained by Chou et al. (1972). A simple continuum model of layered composite materials has been given by King (1972).

Various other aspects of elastic properties of composites have been discussed by Amirbayat and Hearle (1969; 1970a, b) and Foye (1972). Foye has calculated the Poisson ratio of a fibre composite with reference to any pair of orthogonal axes which are normal to the fibre axis. The results have been compared with those obtained by the finite element numerical method and also with the bounds predicted by Hashin and Rosen (1964).

The effective shear modulus of a fibre composite subject to a longitudinal shear deformation has been calculated by Sendeckyj (1970). He has considered a fairly general model in which the fibre-fibre spacings and fibre radii are allowed to have a random variation. The shear modulus can also be different for different fibres. An exact though quite involved expression is derived for the shear modulus of the composite. In the case of similar fibres arranged in a regular array the expression is considerably simplified. A comparison with direct numerical calculations shows very good agreement between the two.

In addition to the work described above, many other authors have calculated the elastic constants of composites and have discussed several aspects of their mechanical behaviour. Outwater (1956) considered the mechanics of plastics reinforced in tension and a simplified analysis of filament-reinforced pressure vessels was given by Brown (1961). These authors have calculated only the Young's modulus of the material. A design-oriented analysis and a discussion of the structural synthesis of multilayered filamentary composites have been given by Chamis (1967), who has calculated not only all the elastic constants but also the thermal expansion coefficients of the composites.

Ekvall (1961, 1966) has discussed the elastic properties and the structural behaviour of orthotropic monofilament laminates using a square array model. Greszczuk (1965) has studied the thermoelastic properties of filamentary composite with a similar model but, in addition, has included the effect of voids in the matrix. These authors have obtained similar values for the elastic constants. Berg (1967) has calculated the transverse Young's modulus and the shear modulus of a composite with a rectangular arrangement of fibres. These calculations are also based on the Ekvall model. Shaffer (1964) has calculated the Young's moduli (longitudinal and transverse) of reinforced plastics on the basis of a hexagonal array model.

Rabinovich (1964) has derived equations for the two-dimensional stressed state in oriented glass fibre plastics which give the elastic constants for the composite. The calculated values are found to be in reasonable agreement with the experimental results for the same system obtained by Rabinovich and Verkhovskii (1964). Dow (1966) has calculated the elastic constants of a composite by writing the total elastic energy of the composite as a linear combination of the elastic energies of its phases and then using the principle of virtual displacements to obtain the required coefficients. Abolinsh (1965) has calculated the elastic constants of a composite using the analysis given by Hashin and Rosen (1964), which has been reviewed earlier in this section.

The mechanical properties of composites and the elastic constants have been discussed in several review articles. A comprehensive review of most mechanical aspects of composite materials is available in the series of volumes entitled composite materials: see Metcalfe, ed. (1974), Sendeckyj, ed. (1974), Noton, ed. (1974), Kreider, ed. (1974) and Chamis ed., (1975).

A review of the structural mechanical properties of composite materials has been given by Bert and Francis (1974). A review of carbon fibre reinforced metals has been given by Baker (1975). Baker's review describes mainly the technological aspects of carbon fibre reinforcement of metals.

Some interesting discussion of the concept and significance of effective modulus solution for fibrous composites has been given by Pagano (1974a). The effective moduli of quasi-homogeneous and quasi-isotropic composite materials made of an elastic matrix and of elastic inclusions have been studied by Boucher (1975). A continuum theory for fibre reinforced composites has been given by Hlavacek (1975).

A general discussion of the deformation characteristics of reinforced epoxy resin has been given by Pink and Campbell (1974). The distortional energy of composite materials has been discussed by Pagano (1975). The compression curve of a fibrous composite has been analyzed by Mileiko and Khvostunkov (1973). A method for evaluation of elastic moduli of composite materials by linear programming has been described by Pao and Maheshwari (1974).

Adams (1974) has discussed the longitudinal tensile behaviour of unidirectional carbon-carbon composites. Jones (1974) has studied the stiffness of orthotropic materials and laminated fibre reinforced composites.

The shear modulus of a particulate composite has been calculated by Smith (1974) by using a revised and extended version of Vander Poel's method. The shear moduli of glass reinforced plastics using ring specimens have been determined by Nikolaev (1974). The longitudinal shear of unidirectionally reinforced plastics has also been studied by Bulavs and Birze (1975). The determination of elastic moduli of reinforced plastics during interlayer shear has been described by Protopopov and Piskunov (1975).

Weitsman and Aboudi (1975) have discussed the effects of fibre inextensibility by calculating stress functions for fibre reinforced materials. Finite axisymmetric deformation of ideal fibre reinforced composites has been studied by Pipkin (1975). Pure bending of helically wound ideal fibre reinforced cylinders has been discussed by Spencer et al. (1975). Darlington et al. (1976) have analyzed structure and anisotropy of stiffness in glass fibre reinforced thermoplastics.

The effect of structure of composite materials on their elastic properties has been discussed by Dudukalenko et al. (1974). In this paper a statistical approach is used for estimating upper and lower bounds for bulk modulus and shear modulus of transversely isotropic composite. The structural properties of glass reinforced plastic have been experimentally analyzed by Holmes and Al-Khayatt (1975) using a E-glass/polyster resin sample.

Ko and Sture (1974) have discussed three dimensional mechanical characterization of anisotropic composites. Barnet and Norr (1976) have proposed a three dimensional structural model for a high modulus pan based carbon fibre. A microanalysis of the behaviour of 3–d reinforced carbon-carbon composite material has been carried by Adams (1976). Rhodes and Mikulas (1975) have studied composite lattice structures. In this paper a lattice type structural panel concept is described which exploits the unidirectional character of advanced filamentary composite materials.

The elastic moduli of unidirectional fibre reinforced composite have also been calculated by Mansfield (1976a). The author has used simple analytical technique to obtain lower bounds for the effective moduli of the composite. The technique depends upon a direct method involving stress equilibrium and strain continuity considerations and thus differs from the techniques involving variational methods such as those given by Hashin and Hill as described earlier in this chapter.

The work described so far in this chapter deals mainly with composites having long, continuous fibres. A very crude estimate of the elastic moduli of composites with short, discontinuous fibre distribution can be obtained by the law of mixtures, as given in Section 3.2. A more rigorous analysis for the elastic constants of a composite reinforced by short fibres has been given by Brody and Ward (1971). They have assumed a model in which the whole composite is represented as an assembly of small units of composites. In each unit the fibres are assumed to be continuous and well-aligned and therefore can be treated by usual methods for perfect composites. The authors have remarked that the observed values of the elastic moduli of a composite with short fibres are much lower than those predicted by the law of mixtures. This discrepancy has been attributed by the authors to the continuity of stress rather than strain. This, however, is only an assumption. As discussed in Section 3.2, the conditions regarding the uniformity of stress and strain give the bounds of the elastic moduli. In a real case neither the stress nor the strain will be uniform.

A general discussion of plastics reinforced by fibres has been given by Davis (1971) with special reference to the case of short, randomly oriented fibres and various effects of fibre orientation, fibre lengths and the fibre-matrix interface.

In a real composite, even if the fibres are long and continuous, they may not be properly aligned. In this case most of the formulae derived earlier are not applicable. A crude estimate of the elastic moduli of a composite containing non-parallel arrays of fibres was given by Cox (1952). Both two-dimensional and three-dimensional arrays of fibres were considered. Cox assumed that the fibres are loaded only at their ends and are rigidly joined with one another—the effect of the matrix was totally neglected, which is obviously incorrect. The values obtained by Cox for the Young's modulus, the plane shear modulus and the Poisson ratio are, respectively, $C_f E_f/3$, $C_f E_f/8$ and 1/3 for the two-dimensional case and $C_f E_f/6$, $C_f E_f/15$ and 1/4 for the three-dimensional array. The reliability of these values is obviously limited for a real composite, in view of the approximations mentioned above.

Further discussions on this subject have been given by Gordon (1952) and Cook (1968). Cook's method of calculating the elastic constants of misaligned fibre composites is more general and he has applied his theory to partially aligned ceramic whiskers in resin matrices. He finds that for small misalignment, i.e. when the root mean square angular deviation of the fibres from the reference axis is less than $\pi/30$, the longitudinal modulus of the composite decreases by about 5%. Recently Christensen and Waals (1972) have also studied the case of a composite containing randomly oriented fibres and have discussed the effective stiffness of such a fibre composite.

A laminate approximation for randomly oriented fibrous composites has been suggested by Halpin and Pagano (1969). This model has been further developed by Halpin et al. (1971). Halpin (1969) has obtained some stiffness and expansion estimates for oriented short fibre composites. The problem of layered media has also been considered by Khoroshun (1966a), who has derived some relations between stresses and strains in such media.

The effect of *misalignment of the curvature of the fibres* on the elastic properties of the composites has also been studied theoretically as well as experimentally by Nosarev (1967) and Tarnopolsky et al. (1967). In these calculations the curvature of the fibres is taken to be much larger than their thickness and the misalignment of the fibres is assumed to be in one plane only. The calculations are based on a model in which the curvature of the fibres is represented by several straight line segments with specified inclinations to the axis of reference.

The effect of fibre orientation on the physical properties of composites has been studied by Knibbs and Morris (1974). The elastic moduli of fibrous composites containing misaligned fibres have been analyzed by Swift (1975).

An interesting study has been reported by McLean and Read (1975) on the analysis of storage and loss moduli in discontinuous composites. Fibre orientation distribution in short fibre reinforced plastic has been discussed in Darlington and McGinley (1975).

Some effects of fibre wobbliness in composite plates have been discussed by Benveniste and Weitsman (1974). Similarly, the influence of fibre waviness on the moduli of unidirectional fibre reinforced composites has been considered by Mansfield and Purslow (1976).

The effect of fibre distribution on the modulus of unidirectionally fibre reinforced composites have been discussed by Mansfield (1976b). Some asymptolic modulus results for the effective elastic moduli of composites containing randomly oriented fibres have been obtained by Christensen (1976).

A problem related to the calculation of the elastic constants is the calculation of stress concentration and stress distribution in the fibres, the matrix and fibre–matrix bonds. This problem has been considered by Sendeckyj and Ing-Wu Yu (1971) for the transverse deformation of a fibre composite. The stress distribution for a regular array can be obtained from various analyses given previously. In addition, the treatment of Sendeckyj and Ing-Wu Yu is also applicable to a composite with a random distribution of fibres. An exact solution is presented for a particular composite in which the shear modulus of the fibres is equal to that of the matrix. Some results for uniaxial transverse loading of such a composite with a square and a random arrangement of fibres have been given. Sendeckyj (1971) has studied the stress distribution in a composite which is subject to a longitudinal shear deformation.

Pipes and Pagano (1970) have studied the elastic behaviour of a fibre composite under a uniform axial strain. With the help of a technique involving finite difference equations, they have calculated the stress and the displacement fields in the composite. The theory has been applied to a composite containing high-modulus graphite fibres. An attempt was made to explain the mechanism of shear transfer within a symmetrical laminate. A theoretical analysis of the stress distribution in the fibres as well as the matrix in a composite subjected to uniaxial tensile stress has been given by Umanskii (1969), and a mathematical discussion of the stress distribution in a disc reinforced by fibres has been given by Ismar (1971). The stresses in a unidirectional composite during axial loading have been considered by Bloom and Wilson (1967). Whitney (1971) has described a shell theory for determination of stresses in composite shells. The three-dimensional stress distribution in a fibre composite has been discussed by Haener and Ashbaugh (1967), and a microstress analysis of fibre composites has been described by Kicher and Stevenson (1967), who have also calculated the elastic moduli of the composite.

The stress analysis of a ribbon-reinforced composite has been considered by Chen and Lewis (1970). Iremonger and Wood (1969) have discussed the effects of phase geometry on stress concentration in a plane composite reinforced by unidirectional discontinuous fibres. The stress analysis of thick laminated composites and sandwich plates has been discussed by Whitney (1972). Some work on the stress concentration and stress distribution in composites containing broken fibres will be described in Chapter 4.

The elastic properties of a thin fibre composite, i.e. a thin matrix reinforced by aligned fibres, have also been studied by Conway and Chang (1971). The matrix is assumed to be homogeneous and isotropic. The authors have presented a mathematical analysis of the stress distribution in the fibre–matrix bonds, effective moduli of elasticity and Poisson ratio of a composite in terms of the relative stiffness of the fibres and the matrix, and the fibre–fibre separation which defines the fibre concentration. They find that the bond stresses are larger for composites with higher-modulus fibres. They have also discussed the changes in the distribution of bond stresses associated with an increase in the fibre debonding. These results are of special interest in the study of failure and fatigue properties of composites, which we shall discuss in Chapter 4.

The theory of Conway and Chang (1971) has been further extended by Chang et al. (1972). They have used the point matching technique to analyze the distribution of bond stresses in a composite in two cases: one in which a large single cylindrical fibre is enclosed by an infinite matrix and another in which the composite is made of such overlapping cylinders. The authors find that the gradual failure of bonds can be mathematically represented by a composite with perfect bonds but with fibres of shorter length. This representation is quite useful from a practical point of view, because it enables us to apply the results on stress distribution of various fibre spacings to fibre debonding in the same composite. As mentioned before, this work is more relevant to the discussion in Chapter 4 of this book; it has been given here since the technique used in these calculations is relevant to the calculation of elastic constants.

The effect of temperature on the mechanical properties of fibre composites should be quite important, although not much work in this field has been reported. The somewhat unusual thermal-mechanical behaviour of fibre composites arises from the fact that the fibres and the matrix have quite different thermal expansion coefficients. As a result, even a slight temperature change or temperature gradient in the composite can induce significant thermal stresses. A theoretical treatment of the thermal stresses induced by a temperature change in a fibre composite has been given by Karpinos and Tuchunskii (1968a). They have shown that the stresses depend on the elastic characteristics and the thermal expansion coefficients of the two phases, temperature change and the volume concentration of the fibres. The dimensions of the fibres seem to have no effect. This work has been further extended by Karpinos and Tuchunskii (1968b). Asamoah and Wood (1970) have discussed the thermal self-straining of fibre composites. The thermally induced residual stresses in some eutectic composites have been considered by Koss and Copley (1971). The effect of temperature on mechanical properties of carbon fibres/nickel matrix has been discussed by Sarian (1973).

Another important problem in this context is to calculate the stresses in a composite which has been subjected to an 'impact', in contrast to a steady loading. This problem has been considered by Amirbayat (1971), who has derived differential equations for the stress and the displacement field in a fibre composite during an impact loading. With the help of these equations, the fibre stress, the matrix stress and the interfacial shear can be calculated. A knowledge of these quantities is required for the control of the dynamic modulus and the strength of a composite. Some aspects of the impact behaviour of fibre composites have also been discussed by Morris and Smith (1971). The important toughness of discontinuous born fibres/epoxy composite has been considered by Aldred and Schuster (1973).

Although not directly connected with the present topic, an interesting calculation by Khachaturyan and Shatalov (1977) may be mentioned here. They have calculated the elastic energy of a stack of alternating coherently bonded plates of two phases both in the case of isolated stacks and when the stacks are enclosed in a matrix. The theory takes into account the elastic anisotropy as well as the difference in the elastic moduli of the two phases. The technique given by these authors is also relevant to similar problems with fibre composites.

The problem of composite plates has also been considered by several authors. Whitney (1969a, b) has discussed the effect of transverse shear deformation on the bending of plates and also the cylindrical bending unsymmetrical laminated plates (see also Whitney, 1969c). An exact solution for the composite laminate in cylindrical bending has been given by Pagano (1969). The case of anisotropic plates has been discussed by Wang (1969), Ashton (1969a, b), Ashton and Waddoups (1969), Ashton and Love (1969) and Schumann et al. (1972). The interfacial delamination of a layered composite under antiplane strain has considered by Chen and Sih (1971).

So far in this chapter we have implicitly assumed that the strains imposed on the composite are small, so that Hooke's law is valid for the whole composite. For large strains Hooke's law is not valid any more and, as described in Section 2.3, plasticity or inelasticity, i.e. stage II, sets in. In the case of composites the yield points for various constituents are, in general, quite different, and therefore the composite exhibits mixed behaviour. i.e. elastic–plastic at intermediate strains. For example, in the case of carbon fibre/epoxy resin composites, as the strain is gradually increased, the fibres remain in stage I long after the matrix has entered stage II. Thus, for a two-phase composite, three distinct possibilities exist: namely, fibres and matrix both elastic, fibres elastic–matrix plastic and fibres and matrix both plastic. The first case has already been discussed in detail in the preceding paragraphs. We shall now briefly describe the last two possibilities.

The inelastic behaviour of a fibre composite have been analyzed by Hill (1964b) by generalizing a formalism developed in his previous paper (Hill, 1964a) on the elastic behaviour of fibre composites. The inelastic behaviour of the composite can be analyzed using Eqs. (3.3.1) and (3.3.1a) with σ and ε replaced by $d\sigma$ and $d\varepsilon$, respectively, which denote the differential stress and strain. The coefficients in Eqs. (3.3.1) and (3.3.1a) are now considered to depend on the stress and/or the prestrain. They can, in this case, be identified as instantaneous elastic moduli with their values defined by the state of stress and strain in the composite at a particular instant. Hill (1964b) assumes a model in which the differential stress–strain relations retain their transverse isotropy. The elastic moduli are allowed to depend on tensor invariants which reduce, when the composite is under a longitudinal strain, to the axial strain ε_z, the axial stress σ_z and the hydrostatic part of the lateral stress $\frac{1}{2}(\sigma_x + \sigma_y)$.

The stress–strain relations are written in the following tensorial form:

if $\mu_{\nu\delta}\, d\sigma_{\nu\delta} \leqslant 0$,

$$d\varepsilon_{\alpha\beta} = \sum_{\nu,\delta} s_{\alpha\beta\nu\delta}\, d\sigma_{\nu\delta} \tag{3.3.96}$$

and if $\mu_{\nu\delta}\, d\sigma_{\nu\delta} \geqslant 0$,

$$d\varepsilon_{\alpha\beta} = \sum_{\nu,\delta} s_{\alpha\beta\nu\delta}\, d\sigma_{\nu\delta} \tag{3.3.96a}$$

where

$$s_{\alpha\beta\nu\delta} - s'_{\alpha\beta\nu\delta} = \lambda_{\alpha\beta}\, \mu_{\nu\delta} \tag{3.3.96b}$$

the primed moduli correspond to the pure elastic unloading; $\mu_{\nu\delta}$ specifies the direction of the normal to the yield surface element with its 'sense' defined by the condition that $\mu_{\nu\delta}\, \sigma_{\nu\delta} > 0$; and $\lambda_{\alpha\beta}$ specifies the direction of the plastic part $d\varepsilon^p_{\alpha\beta}$ (p denotes the plastic part) of the strain differential during continued loading. Since, as mentioned in Section 1.3, the tensor s is symmetric with respect to the interchange of the pair of indices $\alpha\beta$ and $\nu\delta$, $\lambda_{\alpha\beta}$ must also be in the normal direction and therefore proportional to $\mu_{\alpha\beta}$. We can therefore write

$$\lambda_{\alpha\beta} = \frac{\mu_{\alpha\beta}}{h} \tag{3.3.97}$$

which gives the following 'flow' rule;

$$d\varepsilon^p_{\alpha\beta} = \mu_{\alpha\beta} \frac{(\mu_{\nu\delta}\, d\sigma_{\nu\delta})}{h} \tag{3.3.98}$$

where h is a positive scalar.

The above equations form the basis for Hill's analysis of the inelastic behaviour of the fibre composites. A particularly simple and interesting result has been derived by Hill for a composite (fibre elastic/matrix plastic) in which there is no work-hardening in the matrix. In this case, considering a cylindrical composite element consisting of a single cylindrical fibre embedded in a cylindrical shell of the matrix, Hill has shown that the value of the instantaneous Young's modulus of the composite jumps from the following elastic value:

$$E_{el} = C_f E_f + C_m E_m + \frac{4 C_f C_m (\nu_f - \nu_m)^2}{\left(\dfrac{C_f}{K_m} + \dfrac{C_m}{K_f} + \dfrac{1}{M_m}\right)} \tag{3.3.99}$$

to the final value given below:

$$E_{\text{fin}} = C_f E_f + \frac{C_f C_m (1 - 2\nu_f)^2}{\left(\dfrac{C_f}{K_{\nu m}} + \dfrac{C_m}{K_f} + \dfrac{1}{M_m}\right)} \tag{3.3.99a}$$

where K_{vm}, is the three-dimensional bulk modulus of the matrix (distinct from K_m, which was defined to be the area bulk modulus or the plane strain bulk modulus. All the moduli on the right-hand side of Eqs. (3.3.99) and (3.3.99a) refer to the elastic phase). The value E_{fin} for the Young's modulus of the composite is then retained up to the fracture of the fibre. This result is, of course, based on the assumption that the fibres remain elastic up to the fracture point.

Hill has also considered the case of a general fibre composite when no restrictions are imposed on the sections and distribution of the fibres. He has assumed that all the phase moduli depend only on the extension and therefore remain piecewise uniform. In this case the bounds on the elastic moduli at each instant are given by the same expressions derived for a pure elastic composite which have been given earlier in this section.

The situation is quite complicated if the condition of non-work-hardening is not satisfied by the matrix. Some interesting observations on this topic for metal matrix composites have been discussed by Stuhrke (1968) and Kelly and Lilholt (1969).

The state of the composite when both fibres and the matrix enter the plastic zone is of interest only in those cases where both the phases of the composite are ductile. This is, therefore, not particularly relevant to carbon fibre composites. Various aspects of this state of a composite when both of its phases are in the plastic zone have been discussed by Kelly and Davies (1965), Piehler (1965), Mileiko (1969), Ahmad and Barranco (1970) and Vennett et al. (1970), and in the review article by Cooper (1971). The elastic/plastic properties of composites have also been discussed by Huang (1971) and Hutchinson (1970).

Sun et al. (1974) have given a theory of physically nonlinear composites reinforced by elastic fibres. The elastic plastic Poisson's ratio of Borsic-aluminium composites has been obtained by Allred et al. (1974). An investigation of non-linear properties of composites consisting of linear elastic components has been reported by Soldatov (1975).

A structural theory of elastic plastic deformation of unidirectionally fibre reinforced materials has been given by Kafka (1974). The overall moduli of non-linear elastic composite materials have been calculated by Ogden (1974). In this work the author has used the second order elasticity theory and carried out detailed calculations for the case of dilute suspension of spherical elastic material embedded in a matrix of different elastic material. The matrix is assumed to be elastically isotropic.

Several other papers on inelastic behaviour of composite materials can be found in the proceedings of A.S.M.E. Meeting at Houston in 1975 (see Herakovich, ed. (1975).

A method involving the use of the principles of virtual work and complementary virtual work in Cosserat media for the analysis of several phenomena such as viscoelasticity, viscoplasticity, etc. in fibre composites has been discussed by Berg (1971). The viscoplasticity non-linear behaviour of the composites has been discussed by Hackett (1971) and Lou and Schapery (1971). The stress analysis of viscoelastic composite materials has been discussed by Schapery (1967). A useful reciprocal theorem for displacement in inelastic bodies has been derived by Lin (1967).

Some work on the elastic/plastic behaviour of the fibre composite has been done using numerical techniques. This will be reviewed in the next section.

3.4 METHODS FOR NUMERICAL CALCULATIONS OF ELASTIC CONSTANTS

In the previous section we reviewed some methods for analytical calculations of the elastic constants of fibrous composites. These methods yield either the upper and the lower bounds of the elastic moduli, which are quite rigorous, or an approximate expression for each modulus, which is based upon a 'self-consistent' model or some other similar assumption. However, if the elastic moduli of the fibres and the matrix are very different, the theoretically predicted bounds become too far apart to be of any practical use and the self-consistent model is not reliable for extreme values of fibre concentration or fibre rigidities. In such cases a direct numerical evaluation of the elastic constants is more convenient. The numerical methods are also useful for the calculation of elastic constants and stress distribution in a non-ideal composite, i.e. one containing broken or misaligned fibres, and for a study of crack propagation, fracture and fatigue behaviour and plasticity of perfect or imperfect composites. Owing to the rapid growth of computer technology and developments in high-speed computing techniques, the numerical methods are quite convenient to use and indeed are preferred by some workers, even in those cases when an analytical solution is available.

We shall first describe the calculations of Heaton (1968) in detail, since the technique given in this paper is quite general and powerful and can be applied to other cases as well. Heaton has assumed the model of a perfect fibre composite as defined in Section 3.1. The fibres are assumed to form a regular square array on a plane normal to their axes, so that the composite has tetragonal symmetry. Any effects arising due to fibre ends or outer surface of the composite are neglected, so that the treatment is valid for a large sample with long continuous fibres and, of course, subject to other restrictions used in defining the perfect composite. In addition, in this paper Heaton has taken the matrix as well as the fibres to be isotropic and therefore this analysis is not valid for high-modulus carbon fibres, which are highly anisotropic. Finally, Heaton's calculations are confined to composites in which the fibre concentration is not too large.

The composite is divided into prismatic unit cells each containing one fibre, which is assumed to be cylindrical. The cross-section of this model composite and the unit cell marked *ABCD* are shown in Figure 3.3.

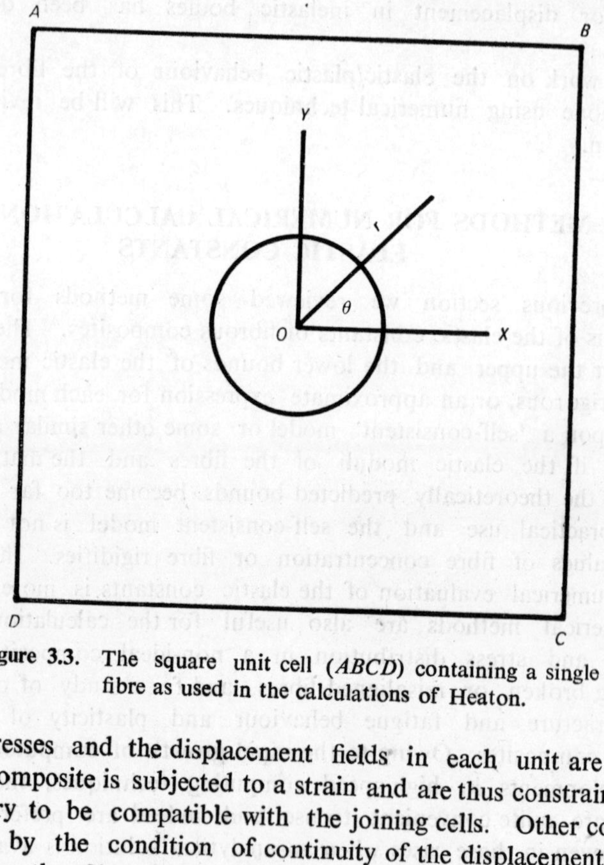

Figure 3.3. The square unit cell (*ABCD*) containing a single fibre as used in the calculations of Heaton.

The stresses and the displacement fields in each unit are the same when the composite is subjected to a strain and are thus constrained at the cell boundary to be compatible with the joining cells. Other constraints are imposed by the condition of continuity of the displacements and the forces across the fibre–matrix interface at the interface. Such problems can be conveniently treated with the help of the Airy stress function.

Heaton has used an infinite series two-dimensional representation of the Airy function. The above-mentioned continuity conditions at the fibre–matrix interface are exactly satisfied but those at the cell boundary are satisfied only approximately. The approximation arises because the infinite series for the Airy function is truncated, which satisfies the necessary conditions only at a finite number of discrete points at the cell boundary. The elastic constants are evaluated by applying appropriate strain on the composite, calculating the stress and the displacement fields with the help of the Airy function and finally using Hooke's law.

As in Section 3.3. the z axis is taken along a fibre axis. The x and y axes are shown in Figure 3.3. The stress-strain relations in terms of six independent elastic constants for the present tetragonal symmetry are given below:

$$\sigma_x = c_{11}\varepsilon_x + c_{12}\varepsilon_y + c_{13}\varepsilon_z \tag{3.4.1}$$

$$\sigma_y = c_{12}\varepsilon_x + c_{11}\varepsilon_y + c_{13}\varepsilon_z \tag{3.4.1a}$$

$$\sigma_z = c_{13}\varepsilon_x + c_{13}\varepsilon_y + c_{33}\varepsilon_z \tag{3.4.1b}$$

$$\sigma_{yz} = c_{44}\varepsilon_{yz} \tag{3.4.1c}$$

$$\sigma_{zx} = c_{44}\varepsilon_{zx} \tag{3.4.1d}$$

$$\sigma_{xy} = c_{66}\varepsilon_{xy} \tag{3.4.1e}$$

where, as in Section 3.2, the diagonal components of the stress and strain tensors are represented by a single Cartesian suffix (σ_α for $\sigma_{\alpha\alpha}$) for the sake of notational brevity.

The elastic constants c_{11}, c_{12}, c_{13}, c_{33} and c_{66} can be obtained by imposing a transverse plane strain ($\varepsilon_z = \varepsilon_{yz} = \varepsilon_{zx} = 0$) or a generalized plane strain on the composite. The generalized plane strain is given by superposing a uniform uniaxial strain in the direction normal to the plane, i.e. the z direction in the present case.

First general expressions for the stress and displacement fields are derived which are applicable to both the matrix and the fibre regions in the unit cell. In polar coordinates (r, θ) on the xy plane the stress components are given by

$$\sigma_r = \frac{1}{r}\frac{\partial \chi}{\partial r} + \frac{1}{r^2}\frac{\partial^2 \chi}{\partial r^2} \tag{3.4.2}$$

$$\sigma_\theta = \frac{\partial^2 \chi}{\partial r^2} \tag{3.4.2a}$$

$$\sigma_{r\theta} = -\frac{\partial}{\partial r}\left(\frac{1}{r}\frac{\partial \chi}{\partial \theta}\right) \tag{3.4.2b}$$

where the Airy function χ is a solution of the equation

$$\nabla^4 \chi = 0 \tag{3.4.3}$$

and can be written in the following general form (Timoshenko and Goodier, 1951):

$$\chi(r,\theta) = a_0\, l_n\, r + b_0\, r^2 + c_0\, r^2\, l_n\, r + d_0\, r^2\,\theta + a_0'\,\theta$$
$$+ \tfrac{1}{2}\, a_1\, r\theta \sin\theta + \left(b_1\, r^3 + \frac{a_1'}{r} + b_1'\, r\, l_n\, r\right)\cos\theta$$
$$- \tfrac{1}{2}\, c_1\, r\theta\cos\theta + \left(d_1\, r^3 + \frac{c_1'}{r} + d_1'\, r\, l_n\, r\right)\sin\theta$$
$$+ \sum_{n=2,3}^{\infty} [a_n\, r^n + b_n\, r^{n+2} + a_n'\, r^{-n} + b_n'\, r^{-n+2}]\cos n\theta$$
$$+ \sum_{n=2,3}^{\infty} [c_n\, r^n + d_n\, r^{n+2} + c_n'\, r^{-n} + d_n'\, r^{-n+2}]\sin n\theta \qquad (3.4.4)$$

where a_n, b_n, etc. are arbitrary constants.

For generalized plane strain the stress–strain relations (3.4.1)–(3.4.1e) can be written in terms of the familiar elastic moduli as follows:

$$\varepsilon_r = \frac{1+\nu}{E}[(1-\nu)\,\sigma_r - \nu\,\sigma_\theta] - \nu\,\varepsilon_z \qquad (3.4.5)$$

$$\varepsilon_\theta = \frac{1+\nu}{E}[(1-\nu)\,\sigma_\theta - \nu\,\sigma_r] - \nu\,\varepsilon_z \qquad (3.4\ 5a)$$

$$\varepsilon_z = \frac{1}{E}[\sigma_z - \nu\,(\sigma_r + \sigma_\theta)] \qquad (3.4.5b)$$

$$\varepsilon_{r\theta} = \frac{1}{M}\,\sigma_{r\theta} \qquad (3.4.5c)$$

where E, ν and M are, respectively, Young's modulus, the Poisson ratio and the shear modulus, which are related as

$$M = \frac{E}{2\,(1+\nu)} \qquad (3.4.5d)$$

We may note in passing that each constituent material but not the composite is characterized by only two independent elastic moduli. This is a consequence of the assumption that each constituent of the composite is linear elastic and isotropic.

The r, θ and z components of the displacement field will be denoted by u_r, v_θ and w_z, respectively, and are given by:

$$u_r = \int \varepsilon_r\, dr \qquad (3.4.6)$$

$$v_\theta = \int r\varepsilon_\theta\, d\theta - \int u_r\, d\theta \qquad (3.4.6a)$$

$$w_z = \varepsilon_z\, z \qquad (3.4.6b)$$

The relations between the polar and the Cartesian components of the stress and the displacement fields are given below:

$$\sigma_x = \sigma_r \cos^2\theta + \sigma_\theta \sin^2\theta - \sigma_{r\theta} \sin 2\theta \tag{3.4.7}$$

$$\sigma_y = \sigma_r \sin^2\theta + \sigma_\theta \cos^2\theta + \sigma_{r\theta} \sin 2\theta \tag{3.4.7a}$$

$$\sigma_{xy} = \tfrac{1}{2}(\sigma_r - \sigma_\theta) \sin 2\theta + \sigma_{r\theta} \cos 2\theta \tag{3.4.7b}$$

$$U_x = u_r \cos\theta - u_\theta \sin\theta \tag{3.4.7c}$$

$$V_y = u_r \sin\theta + u_\theta \cos\theta \tag{3.4.7d}$$

It can be shown from symmetry considerations that the arbitrary constants in Eq. (3.4.4) vanish for odd n. The constants c_0 and d_0 must be zero, since they lead to discontinuous fields, and a_0' can be taken to zero because it corresponds to the uninteresting case of a uniform shear stress applied along the boundary. Thus, the following equations are obtained:

$$\begin{aligned}\sigma_r &= a_0\beta^{-2} - 2b_0 \\ &- \sum_{n=2,4\ldots}^{\infty} \{n(n-1)\,a_n\beta^{n-2} + (n-2)(n+1)\,b_n\beta^n + n(n+1)\,a_n'\beta^{-n-2} \\ &+ (n+2)(n-1)\,b_n'\beta^{-n}\}\cos n\theta - \sum_{n=2,4,\ldots} \{n(n-1)\,c_n\beta^{n-2} \\ &+ (n-2)(n+1)\,d_n\beta^n + n(n+1)\,c_n'\beta^{-n-2} + (n+2)(n-1)\,d_n'\beta^{-n}\}\sin n\theta\end{aligned} \tag{3.4.8}$$

$$\begin{aligned}\sigma_\theta &= -a_0\beta^{-2} + 2b_0 \\ &+ \sum_{n=2,4,\ldots}^{\infty} \{n(n-1)\,a_n\beta^{n-2} + (n+2)(n+1)\,b_n\beta^n + n(n+1)\,a_n'\beta^{-n-2} \\ &+ (n-2)(n-1)\,b_n'\beta^{-n}\}\cos n\theta + \sum_{n=2,4,\ldots}^{\infty} \{n(n-1)\,c_n\beta^{n-2} \\ &+ (n+2)(n+1)\,d_n\beta^n + n(n+1)\,c_n'\beta^{-n-2} + (n-2)(n-1)\,d_n'\beta^{-n}\}\sin n\theta\end{aligned} \tag{3.4.8a}$$

$$\begin{aligned}\tau_{r\theta} &= \sum_{n=2,4,\ldots}^{\infty} n\{(n-1)\,a_n\beta^{n-2} + (n+1)\,b_n\beta^n - (n+1)\,a_n'\beta^{-n-2} \\ &- (n-1)\,b_n\beta^{-n}\}\sin n\theta - \sum_{n=2,4,\ldots}^{\infty} n\{(n-1)\,c_n\beta^{n-2} + (n+1)\,d_n\beta^n \\ &- (n+1)\,c_n'\beta^{-n-2} - (n-1)\,d_n'\beta^{-n}\}\cos n\theta\end{aligned} \tag{3.4.8b}$$

$$\frac{u_r}{a} = -\nu\,\beta\varepsilon_2 + \frac{1+\nu}{E}(-a_0\beta^{-1} + 2(1-2\nu)\,b_0\,\beta$$

$$-\sum_{n=2,4,\ldots}^{\infty}[na_n\beta^{n-1} + \{n-2(1-2\nu)\}\,b_n\beta^{n+1} - na'_n\,\beta^{-n-1}$$

$$-\{n+2(1-2\nu)\}\,b'_n\,\beta^{-n+1}]\cos n\theta - \sum_{n=2,4,\ldots}^{\infty}[nc_n\beta^{n-1}$$

$$+\{n-2(1-2\nu)\}\,d_n\beta^{n+1} - nc'_n\,\beta^{-n-1} - \{n+2(1-2\nu)\}d'_n\,\beta^{-n+1}]\sin n\theta) \qquad (3.4.8c)$$

$$\frac{v_\theta}{a} = \sum_{n=2,4,\ldots}^{\infty}[na_n\beta^{-1} + \{n+4(1-\nu)\}\,b_n\,\beta^{n+1} + n\,a'_n\,\beta^{-n-1}$$

$$+\{n-4(1-\nu)\}\,b'_n\,\beta^{-n+1}]\sin n\theta - \sum_{n=2,4,\ldots}^{\infty}[nc_n\beta^{n-1}$$

$$+\{n+4(1-\nu)\}\,d'_n\,\beta^{n+1} + n\,c'_n\beta^{-n-1} + \{n-4(1-\nu)\}\,d'_n\,\beta^{-n+1}]\cos n\theta \qquad (3.4.8d)$$

$$w_z = \varepsilon_z Z \qquad (3.4.8e)$$

where β is a dimensionless parameter which is proportional to r, i.e,

$$\beta = r/a$$

The stress distributions and the displacement field at any point in the composite—in the fibre region as well as the matrix region—are given by Eqs. (3.4.8)—(3.4.8e). The coefficients in these equations in the fibre and the matrix regions will be labelled by f and m, respectively. These coefficients have to be determined by matching conditions at the boundary.

First we note that the coefficients of the inverse powers of β in the fibre region, i.e. $a'^f_0, a'^f_n, b'^f_n, c'^f_n, d'^f_n$, must be zero, because otherwise they would lead to divergence at $r = 0$. The condition of continuity of the stress and displacement components at the fibre–matrix interface give a^m_n and a'^m_n in terms of b^m_n and b'^m_n and the elastic constants of the two phases and similar relations are obtained for c^m_n and c'^m_n in terms of d^m_n and d'^m_n. Thus, only the coefficients b^m_n, b'^m_n, d^m_n and d'^m_n remain to be determined, which is done by point matching at the boundary of the unit. For this purpose the Cartesian representation of σ and $\mathbf{u}$ as given in Eqs. (3.4.7)—(3.4.7d) may be more convenient to use. The elastic moduli are obtained by applying an appropriate strain on the composite and then relating it to the stress.

For example, let us consider symmetric plane strain defined by

$$\varepsilon_x = \varepsilon_y = \varepsilon \text{ and } \varepsilon_z = 0 \qquad (3.4.9)$$

The only surviving stress components are

$$\sigma_x = \sigma_y = (c_{11} + c_{12})\,\varepsilon \qquad (3.4.10)$$

and

$$\sigma_z = 2c_{13}\,\varepsilon \tag{3.4.10a}$$

where Eqs. (3.4.1)–(3.4.1e) have been used. The conditions to be imposed at the cell boundary are as follows:

$$[U_x]_{x=c} = [V_y]_{y=c} = \varepsilon C \tag{3.4.11}$$

and

$$[\sigma_{xy}]_{x=c} = [\sigma_{xy}]_{y=c} = 0 \tag{3.4.11a}$$

For the displacements to be symmetrical, all d_n, d'_n must vanish and the allowed values of n are 0, 4, 8,. . . . To determine the remaining constants b_n^m, b'^m_n, the infinite series in Eqs. (3.4.8)–(3.4.8e) has to be truncated after a certain number of terms. Then the finite number of unknown coefficients can be obtained by satisfying Eqs. (3.4.11) and (3.4.11a) at the appropriate number of points at the cell boundary. Owing to the symmetry of the unit cell only the points on the boundry which are within the region $0 \leqslant \theta \leqslant \pi/4$ have to be included.

To calculate the effective value of $c_{11} + c_{12}$ in Eq. (3.4.10), the average values of the stress components are required, which can be calculated as follows:

$$\bar{\sigma}_x = \frac{1}{2c}\int_{-c}^{c} [\sigma_x]_{x=c}\, dy$$

$$= \frac{1}{2c}\left\{\left[\frac{d\chi}{dy}\right]_{x=c}\right\}_{y=-c}^{y=c} \tag{3.4.12}$$

where we have used the relation

$$\sigma_x = \frac{\partial^2\chi}{\partial y^2} \tag{3.4.13}$$

Proceeding in a similar fashion, c_{13} can be obtained by calculating the average axial stress, $\bar{\sigma}_z$, which is required to maintain the transverse plane strain. For this purpose the stress has to be averaged over the cross-section of the unit cell.

The above procedure yields the value of $c_{11} + c_{12}$. Heaton has determined $c_{11} - c_{12}$ by considering the problem of anti-plane strain, which is defined by

$$\bar{\varepsilon}_x = -\bar{\varepsilon}_y = \varepsilon$$

and

$$\bar{\varepsilon}_z = 0$$

In this case

$$\bar{\sigma}_x = -\bar{\sigma}_y = (c_{11} - c_{12})\,\varepsilon \tag{3.4.14}$$

and the boundary conditions are given by

$$[U_x]_{x=c} = -[V_y]_{y=c} = \varepsilon C$$

and

$$[\sigma_{xy}]_{x=c} = [\sigma_{xy}]_{y=c} = 0$$

As in the case of symmetric plane strain, d_n, d'_n are zero in the present case as well and the allowed values of n are 2, 6, 10, etc. The procedure for calculating the unknown coefficients and the average value of the stress is the same as in the case of symmetric plane strain which has been described above. Then the use of Eq. (3.4.14) yields the value of $c_{11} - c_{12}$. Thus, the calculations described in this and the previous paragraph give c_{11}, c_{12} and c_{13}.

To determine c_{33} and c_{13} by an alternative method which provides a cross-check to the value obtained in the symmetric plane strain calculation, Heaton considers the case of longitudinal strain. In this case $\varepsilon_x = \varepsilon_y = 0$ and $\varepsilon_z = \varepsilon$, so that the average stresses are given by:

$$\bar{\sigma}_x = \bar{\sigma}_y = c_{13}\,\varepsilon \qquad (3.4.15)$$

$$\bar{\sigma}_z = c_{33}\,\varepsilon \qquad (3.4.15a)$$

subject to the following boundary conditions:

$$[U_x]_{x=c} = [V_y]_{y=c} = 0$$

$$[\sigma_{xy}]_{x=c} = [\sigma_{xy}]_{y=c} = 0$$

By symmetry d_n^m and $d_n'^m$ are zero and the allowed values of n are 0, 4, 8, etc. The procedure for calculating c_{33} and c_{13} is the same as has been given earlier for the calculation of $c_{11} + c_{12}$. By use of the same procedure, the constant c_{66} can be calculated by imposing a transverse shear strain on the composite. In this case

$$[\sigma_x]_{x=c} = [\sigma_y]_{y=c} = 0$$

and

$$[U_x]_{y=c} = [V_y]_{x=c} = \tfrac{1}{2}\varepsilon_{xy} C$$

These equations are sufficient to determine the coefficients $d_n'^m$ and d_n^m by 'point matching', provided, as before, the infinite series for the Airy function is terminated after a finite number of terms. The average applied shear stress can be simply calculated by using the relation

$$\sigma_{xy} = -\frac{\partial^2 x}{\partial x \partial y}$$

Thus, the procedure described above gives the values of five elastic constants c_{11}, c_{12}, c_{13}, c_{33} and c_{66}. It only remains to calculate c_{44}.

The constant c_{44} can be calculated by taking the displacement field within the fibre as well as the matrix regions to be in the z direction but independent of the z coordinate, so that

$$u_r = u_\theta = 0 \tag{3.4.16}$$

and

$$w_z = w_z(r, \theta) \tag{3.4.16a}$$

Considering the medium to be isotropic, the shear stresses are given by

$$\sigma_{rz} = M \frac{\partial w_z}{\partial r} \tag{3.4.17}$$

and

$$\sigma_{\theta z} = M \frac{1}{r} \frac{\partial w_z}{\partial \theta} \tag{3.4.17a}$$

where M is the shear modulus of the medium. The stress components other than those given in Eqs. (3.4.17) and (3.4.17a) are zero. In terms of the stress function χ on the r, θ plane, the stress components are written as:

$$\sigma_{rz} = \frac{1}{r} \frac{\partial \chi}{\partial \theta} \tag{3.4.18}$$

and

$$\sigma_{\theta z} = -\frac{\partial \chi}{\partial r} \tag{3.4.18a}$$

with the compatibility condition

$$\nabla^2 \chi = 0 \tag{3.4.19}$$

The general solution of Eq. (3.4.19) in terms of the dimensionless parameter $\beta = r/a$ is given below:

$$\begin{aligned} \chi = (a_0\theta + b_0)(c_0 \ln \beta + d_0) & \\ + \sum_{n=1,2,\ldots}^{\infty} (a_n\beta^n + b_n\beta^{-n}) \cos n\theta & \\ + \sum_{n=1,2,\ldots}^{\infty} (c_n\beta^n + d_n\beta^{-n}) \sin n\theta & \end{aligned} \tag{3.4.20}$$

The general expressions for the stress and the displacements can be derived as before with the help of Eqs. (3.4.16)–(3.4.20). It may be noted that c_0 must be zero, otherwise the stresses and the displacements will not be single-valued.

Now the composite unit cell is considered under a longitudinal shear strain, so that the displacements satisfy Eqs. (3.4.16) and (3.4.16a). To avoid the singularity $1/\beta = \infty$ in the fibre region at $r = 0$, the coefficients of the inevrse powers of β and $\ln \beta$ must be made zero. The continuity requirements for the displacement w_z and the shear stress σ_{rz} at the fibre–matrix interface are satisfied by taking

$$a_n^f = (1 + f)\, a_n^m \tag{3.4.21}$$

$$c_n^f = (1 + f)\, a_n^m \tag{3.4.21a}$$

$$b_n^m = f a_n^m \tag{3.4.21b}$$

and

$$d_n^m = f c_n^m \tag{3.4.21c}$$

where

$$f = \frac{M^f - M^m}{M^f + M^m} \tag{3.4.22}$$

When a shear strain ε_{yz} is applied, the displacement w_z must be symmetric about the y axis and antisymmetric about the x axis. Thus, only the odd a_n, b_n coefficients will be non-vanishing. Then the stress and the displacement fields in the matrix region can be written as:

$$\sigma_{rz}^m = -\frac{1}{a} \sum_{n=1,3,\ldots}^{\infty} n\, (\beta^{n-1} + f\beta^{-n-1})\, a_n^m \sin n\theta \tag{3.3.23}$$

$$\sigma_{\theta z}^m = -\frac{1}{a} \sum_{n=1,3,\ldots}^{\infty} n\, (\beta^{n-1} - f\beta^{-n-1})\, a_n^m \cos n\theta \tag{3.4.23a}$$

and

$$w_z^m = -\frac{1}{G^m} \sum_{n=1,3,\ldots}^{\infty} (\beta^{n-1} - f\beta^{-n})\, a_n^m \sin n\theta \tag{3.4.23b}$$

Similarly equations are written down for the fibre region but without the terms containing the inverse powers of β, because of considerations of divergence, as mentioned earlier. The boundary conditions at the cell boundary are as follows:

$$[w_z]_{y=c} = \varepsilon_{yz} C$$

and

$$[\sigma_{xz}]_{x=c} = 0$$

The average value of the shear stress is obtained as follows:

$$\bar{\sigma}_{xy} = \frac{1}{2c} \int [\sigma_{xy}]_{y=c}\, dx$$

$$= -\frac{1}{2c} \left\{ [\phi]_{y=c} \right\}_{x=-c}^{x=c} \tag{3.4.24}$$

where we have used the relation

$$\sigma_{xy} = -\frac{\partial \chi}{\partial x}$$

Thus we see that with the help of the Airy stress function and the point matching technique Heaton (1968) has been able to calculate all the six elastic constants of the composite having a square arrangement of fibres and, hence, tetragonal symmetry. Similar calculations can also be performed for a composite with hexagonal symmetry, in which case there will be only five independent elastic constants. The main point in this technique is, as mentioned before, that the continuity conditions at the fibre–matrix interface have been satisfied exactly, which give a_n^m and $a_n'^m$ in terms of b_n^m and $b_n'^m$ and c_n^m and $c_n'^m$ in terms of d_n^m and $d_n'^m$. The coefficients b and d are determined by satisfying the boundary conditions at the unit cell boundary in which the approximation of truncating the infinite Airy series has to be introduced. However, the advantage is that this approximation can be carried to any degree of accuracy desired by keeping more and more terms in the Airy series and satisfying the boundary conditions at the required number of discrete points at the unit cell boundary. Thus, the only limitation on the accuracy in this step is imposed by the available computer.

The square array model limits the fibre volume fraction in the composite to 0.785, which can be improved up to 0.90 for the hexagonal geometry. However, for such large concentrations of fibres a model of the composite based on the regular arrangement of fibres may not be a realistic representation. Moreover, for such large fibre concentrations the effects arising from the multiple fibre–fibre interactions will be very important and their contributions will be very complicated to calculate.

In the calculations as described above Heaton (1968) has assumed the fibres to be elastically isotropic. This is not a valid assumption for most fibres of interest, particulary the carbon fibres. These calculations have been subsequently revised by Heaton (1970) to take into account the effect of fibre anisotropy. The model of the composite and the procedure for the calculation of the elastic constants of the composite used by Heaton (1970) is the same as that by Heaton (1968) which has been reviewed above. The difference arises in the definition of certain moduli—for example, the Young's modulus and the Poisson ratio.

Heaton (1970) has assumed the fibres to be transversely isotropic, so that they can be characterized by five elastic constants. They are taken to define the following moduli; E' and E, the Young's moduli with respect to transverse and longitudinal directions of the fibre, respectively; ν', the Poisson ratio which characterizes the transverse contraction in the transverse plane of the fibre for a stress in the same plane; ν, the Poisson ratio for the contraction in the transverse plane of the fibre for a stress in the longitudinal directions; and finally, M and μ, which denote the transverse and the longitudinal shear moduli, respectively. Out of these six, only five moduli are independent, and E, M and ν are related as follows:

$$M = \frac{E'}{2(1+\nu')}$$

Then the stress–strain relations for the fibre material are given by

$$\varepsilon_x = \frac{1}{E'}(\sigma_x - \nu'\sigma_y) - \frac{\nu}{E}\sigma_z \tag{3.4.25}$$

$$\varepsilon_y = \frac{1}{E'}(\sigma_y - \nu'\sigma_x) - \frac{\nu}{E}\sigma_z \tag{3.4.25a}$$

$$\varepsilon_z = \frac{1}{E}\sigma_z - \frac{\nu}{E}(\sigma_x + \sigma_y) \tag{3.4.25b}$$

$$\varepsilon_{yz} = \frac{1}{\mu}\sigma_{yz} \tag{3.4.25c}$$

$$\varepsilon_{zx} = \frac{1}{\mu}\sigma_{zx} \tag{3.4.25d}$$

and

$$\varepsilon_{xy} = \frac{1}{M}\sigma_{xy} \tag{3.4.25e}$$

The method of calculation of the elastic constants is the same as that used for the isotropic fibres. First a generalized plane strain is imposed for which the equations corresponding to Eqs. (3.4.5)–(3.4.5d) in the present case are as follows:

$$\varepsilon_r = \left[\frac{1}{E'} - \frac{\nu^2}{E}\right]\sigma_r - \left[\frac{\nu'}{E'} + \frac{\nu^2}{E}\right]\sigma_\theta - \nu\varepsilon_z \tag{3.4.26}$$

$$\varepsilon_\theta = -\left[\frac{\nu'}{E'} + \frac{\nu^2}{E}\right]\sigma_r + \left[\frac{1}{E'} - \frac{\nu^2}{E}\right]\sigma_\theta - \nu\varepsilon_z \tag{3.4.26a}$$

$$\sigma_z = E\varepsilon_z + \nu(\sigma_x + \sigma_y) \tag{3.4.26b}$$

$$\varepsilon_{r\theta} = \frac{2(1+\nu')}{E'}\sigma_{r\theta} \tag{3.4.26c}$$

These equations lead to the determination of c_{11}, c_{12}, c_{13}, c_{33} and c_{66}. The remaining constant, c_{44}, is determined by subjecting a unit cell of the composite to a longitudinal shear strain.

It may be noted that Eqs. (3.4.26)–(3.4.26c) contain five independent elastic moduli unlike Eqs. (3.4.5)–(3.4.5d) for the isotropic case, which contain only two independent elastic moduli. It can easily be verified that for the isotropic case, defined by $\nu = \nu'$, $E = E'$, Eqs. (3.4.26)–(3.4.26c) reduce to Eqs. (3.4.5)–(3.4.5d).

Heaton (1970) has applied this method to a calculation of the elastic constants of carbon fibre/epoxy resin composite as a function of the volume fraction of the carbon fibres. Some of his results have been given in Table 5.2.

The main qualitative findings of Heaton (1968, 1970) can be summarized as follows. For not very large concentrations of fibres, the transverse elastic constants (c_{11}, c_{12}, etc.) and the longitudinal shear modulus of the composite are mainly determined from the elastic properties of the matrix. In a direction parallel to the fibres the elastic constant c_{33} is mainly determined by the volume fraction and the elastic constants of the fibre (the predicted longitudinal Young's modulus of the composite in the fibre direction for isotropic fibres is found to be within 1% of the value obtained by the simple law of mixtures). The effect of the anisotropy of the fibres is to increase the basic anisotropy of the fibre composite. Finally, it may also be noted that Heaton's method can also be used to calculate the stress distribution in the fibres.

An approach very similar to that given by Heaton (1968, 1970) has been used by Chen (1970) to calculate the shear modulus of a fibre composite subjected to a longitudinal shear strain. Both the matrix and the fibres have been assumed to be elastically isotropic—an assumption which is similar to that made by Heaton (1968). A Fourier analysis method has been used by Chen (1970) to satisfy the matching conditions at the boundary.

One difficulty with the use of the numerical methods as described above for high fibre volume fraction arises from the lack of convergence of the numerical process. This difficulty has been circumvented by Symm (1970) with the help of a least squares technique which has enabled him to calculate the elastic constants of a fibre composite containing up to 90% fibres by volume.

The unit cell approach is very convenient for treating a regular periodic array of objects (see, for example, Grigolyuk and Fil'shtinskii, 1965; Lomakin and Koltunov, 1965). Numerical methods based upon the unit cell approach have been used by several authors to calculate the elastic constants of composite materials. Using this method, Herrmann and Pister (1963) have calculated the elastic constants of a composite with a square array of fibres (some thermal properties of the composite have also been studied by these authors). A rectangular unit cell was assumed by Wilson and Hill (1965) in studying the mechanical properties of a plate with a rectangular array of rigid inclusions or holes. As in the calculations of Heaton (1968, 1970), the appropriate boundary conditions at the interface of the inclusion and the matrix were satisfied exactly where the conditions at the unit cell boundary were satisfied approximately, using the point matching technique. Unlike Heaton (1968, 1970), however, the calculations of Wilson and Hill are based upon complex variable analysis.

The rectangular unit cell has also been assumed by Adams and Doner (1967a, b), who have calculated some elastic constants of the composite by using a finite difference method (see also Adams et al. 1967). The longitudinal shear and transverse normal loading of the composite has been considered, respectively, in papers referred to as *a* and *b*.

A composite with hexagonal array of fibres has been considered by Pickett (1965), Pickett and Johnson (1966) and Chen and Cheng (1967). A convenient choice of unit cell for a hexagonal array of fibres is a triangular cell. These authors have calculated the elastic constants of the composite by using the polar coordinate representation and a least squares method to satisfy the boundary conditions at the fibre–matrix interface as well as at the boundary of the unit cell. A similar technique has been used by Clausen and Leissa (1967). The hexagonal array of fibres and the triangular unit cell have also been considered by Bloom and Wilson (1967), who have calculated the Young's modulus and the Poisson ratio of the composite. The boundary conditions were satisfied exactly at the fibre–matrix interface and approximately in other regions by using a point matching technique. This method has been applied to the case of fibres with transverse isotropy by Quackenbush and Thomas (1967).

A very powerful method for numerical evaluation of the mechanical properties of solids is the finite element (henceforth referred to as FE) method, which has been briefly described in Section 2.6. This method can be applied to the calculation of elastic constants, stress distribution and several other quantities, and is also applicable to imperfect fibre composite.

The FE method has been used for an analysis of filament-reinforced axisymmetric bodies by Wolson and Parsons (1969). A more detailed application of the FE method has been made by Chew and Lin (1969) in the calculation of transverse stiffness and strength of fibre composites as a function of the fibre volume fraction. Two limiting cases representing perfect fibre–matrix bonding and total debonding of the fibres have been considered. The calculations have been made for both hexagonal and square arrays of fibres. It is shown that the experimental results obtained with some metal/metal and glass/epoxy composites lie between the calculated values in the two limiting cases.

Lin et al. (1972) have applied the FE method to the elastic–plastic analysis of a fibre composite subjected to plastic loading. The analysis has been applied to boron/aluminium and boron/epoxy composites in order to examine the dependence of the composite behaviour on the material properties of the matrix.

Adams (1970) has used the FE method for the inelastic analysis of a fibre composite subjected to transverse normal loading. He finds that extensive yielding and stress redistribution can occur locally in certain regions in the composite without being immediately apparent in the elastic response of the material as a whole. The relevance of this result to fracture behaviour and the fatigue life of the composites has also been discussed by Adams (1970).

The FE method has been used by Adams and Tsai (1969) to calculate the elastic constants of a fibre composite in which the fibres are randomly distributed. Two representations of the composite have been considered, which have been referred to as the square random array and the hexagonal random array model. The square and the hexagon in this context refer to the shape of the finite elements in which the material is divided. A comparison with the experimental results show that the hexagonal random array model gives better agreement with the measured values of the elastic moduli than the square random array model. This is in contrast with some previous inferences, based on regular array models, that the square array model gives better agreement with the measured values of the elastic moduli than the hexagonal array model. This result of Adams and Tsai (1969) is physically plausible, since, judging from the observed cross-sectional geometry of some fibre composites, a random hexagonal array model appears to be a more realistic representation of a composite than the random square model. Other main conclusions of Adams and Tsai may be summarized as follow: (1) the stiffness properties of the composites depend significantly on the distribution of the fibres; (2) there is no particular justification for assuming a regular square array or any other regular array of fibres; and (3) it is possible to predict accurately the composite material behaviour by suitably taking into account the geometric constraints and the non-linearities of the phases forming the composite.

Owen et al. (1969) have applied the FE method to analyze the elastic–plastic behaviour of composites containing discontinuous fibres, taking into account the end effects as well. The authors have found that the matrix yield stress for cylindrical and plane single-fibre models increases as the aspect ratio decreases. Good agreement is found with the experimental results obtained by a photoelastic technique in the case of a carbon fibres/resin composite.

The FE method has been used to calculate the strength of fibre composites containing discontinuous fibres by Iremonger and Wood (1967; also 1970 for plastic flow in the matrix), Carrara and McGarry (1968) and Chen (1971). Asamoah and Wood (1970) have included thermally induced stresses in their FE analysis of fibre composites. For other applications of the FE method, reference may be made to Crisfield (1971), Sahu and Broutman (1972), Isakson and Levy (1971) and Mau et al. (1972). The last two papers have considered, respectively, the problems of interlaminar shear in fibre composites and laminated thick plates.

In addition to the two types of approaches described above, some other methods have also been used which are suitable for a direct numerical evaluation of the mechanical properties of composite. Recently an 'initial strain' method has been suggested by Owen (1972) for the analysis of fibre composites. The author has developed expressions for the stress and the displacement field in the composite, and the results are generated by Fourier methods from the solutions for a periodic distribution of eigenstrains.

Khoroshun (1966b, c; 1968) and Fokin and Shermergor (1967) have used statistical methods for the calculation of the elastic constants of composites in which the fibres may be randomly distributed and randomly oriented. The statistical method consists of writing the total elastic field in the composite in terms of weighted contributions from its phases. The weights of these contributions are allowed to fluctuate. The statistical average is evaluated in terms of the moments of the distribution of fluctuating weights. Some approximations are inevitably introduced in this method, since only a finite number of moments can be calculated.

3.5 MEASUREMENT OF THE ELASTIC MODULI

In the previous sections of this chapter we have reviewed the theoretical calculations of the elastic moduli of fibre composites. In this section we shall describe the principles behind some methods of experimental measurement of the elastic moduli and refer to some experimental work reported in the literature. For a general review of the methods for mechanical measurements, see, for example, Beckwith and Buck (1961) or Turner (1973).

The elastic moduli of a composite, or in fact any solid, can be measured by static methods or dynamic methods.

In the static methods a known static (time-independent) load is applied to the sample according to the type of stress required and the corresponding strain is measured directly. The ratio of the known applied stress and the measured strain gives the corresponding elastic modulus.

The dynamic methods for the measurement of the elastic moduli of a solid can be classified into the following two categories:

1. Methods based on the propagation of elastic waves in the sample. As mentioned before, the velocity of elastic waves in a solid depends on the elastic constants of the solid. Thus, a measurement of the wave velocities in the solid can be used to yield the required information about its elastic constants. The propagation of elastic waves in fibre composites will be discussed in detail in Chapters 5 and 6. Henceforth, in this chapter the dynamical method for the measurement of elastic constants will refer to the second type of method, described below.

2. Methods based upon the vibration of the sample in the form of a

beam fixed at a certain point, e.g. a cantilever beam which is fixed at one end. The vibration frequencies of elastic beams are functions of their elastic moduli. If this functional dependence can be obtained theoretically, then the elastic constants of the beam material can be obtained from the measured values of its vibration frequencies.

The effect of fibre orientation on the vibration modes of cantilever beams of fibre composite materials has been studied both theoretically and experimentally by Abarcar and Cunniff (1972). These authors have also measured the elastic moduli of the composite, which have been used for the theoretical calculation of the vibration frequencies of the beam as a function of the fibre orientation. Good agreement is obtained between the calculated and the observed values of these frequencies. A description of their work will be given here.

Abarcar and Cunniff (1972) consider a solid beam of the composite fixed at one end. The origin of the coordinates is taken at the fixed end, with the x and y axes along the principal axes of inertia of the cross-section and the z axis along the beam axis. To take into account the effect of fibre orientation with respect to the beam axis, a 'composite' coordinate system is also defined in which the axes are labelled by 1, 2 and 3. The z axis is along the fibre axis and the 1 and 2 axes are in the plane normal to the fibres. The angle between the fibres, i.e. the 3 axis and the z axis, is denoted by θ (see Figure 3.4).

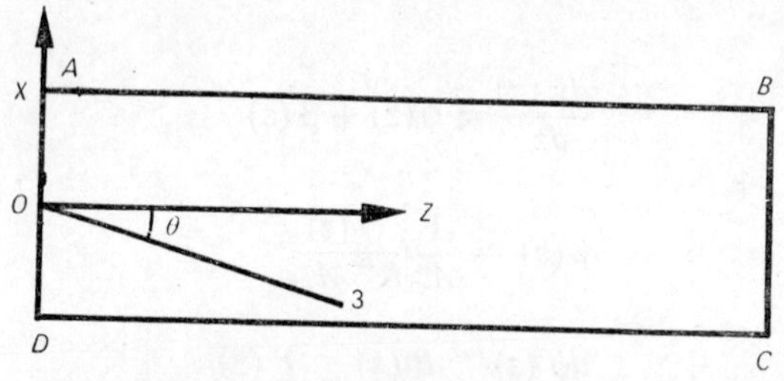

Figure 3.4 The coordinate system used for the analysis of the experimental results by Abarcar and Cunniff. The beam is denoted by *ABCD*. The y axis is normal to the plane of the paper. The distorted 3-axis is at an angle θ from the z axis. The distorted 1- and 2-axes have not been shown.

The deflection of the beam subject to a bending moment B about the x axis plus a torque T about the z axis is given by the following equations (Lekhnitskii, 1963):

$$Y = \frac{1}{2I_1}\left[B\,S_{33} + \frac{S_{35}}{2}\right]z^2 \qquad (3.5.1)$$

$$\frac{d\phi}{dz} = \frac{T}{C_t} - \frac{B}{2I_1}S_{35} \qquad (3.5.1a)$$

where S_{ij} are the coefficients of deformation; I_1 is the second moment of area of the beam cross-section with respect to the x axis; C_t is the generalized torsional rigidity, which depends upon the cross-section of the beam; and ϕ is the angle of twist of the beam cross-section. The two coefficients of deformation in Eqs. (3.5.1) and (3.5.1a) are the follows:

$$S_{35} = 2\left[\frac{\sin^2\theta}{E_1} - \frac{\cos^2\theta}{E_3}\right]\sin\theta\cos\theta + \tfrac{1}{2}\left[\frac{1}{M_{13}} - \frac{2\nu_{31}}{E_3}\right]\sin 2\theta\cos 2\theta \qquad (3.5.2)$$

and

$$S_{33} = \frac{1}{E_z} = \frac{\sin^4\theta}{E_1} + \frac{\cos^4\theta}{E_3} + \left[\frac{1}{M_{13}} - \frac{2\nu_{31}}{E_3}\right]\sin^2\theta\cos^2\theta \qquad (3.5.2a)$$

where ν_{31} is the Poisson ratio; E_1, E_3 and E_z are the Young's moduli along the direction as indicated by the subscripts; and M_{13} is the shear modulus as defined by Eq. (2.3.5c).

In the case of a rectangular bar of orthotropic material of width b and thickness h, the value of C_t is given by:

$$C_t = \frac{bh^3}{3}\,\frac{(b-0.630)\,M_{xz}}{b - 0.630\,h\,S_{35}^2\,E_z\,M_{xz}} \qquad (3.5.3)$$

The free vibrations of the beam are governed by the following set of equations:

$$\frac{dy(z)}{dz} = \psi(z) + \beta(z) \qquad (3.5.4)$$

$$\beta(z) = \frac{1}{A}\,\frac{Q(z)}{K'\,M_{zy}} \qquad (3.5.4a)$$

$$\frac{d\psi(z)}{dz} = \frac{B(z)}{I_1 E_z} - \frac{T(z)}{C_{mt}} \qquad (3.5.4b)$$

$$\frac{dB(z)}{dz} + Q(z) + mk^2\omega^2\,\psi(z) = 0 \qquad (3.5.4c)$$

$$\frac{d\phi(z)}{dz} = -\frac{B(z)}{C_{mt}} + \frac{T(z)}{C_t} \qquad (3.5.4d)$$

where ψ denotes the angle of rotation due to the bending moment; Q is the shear force; K' is a coefficient which depends upon the cross-section of the beam; A is the area of cross-section and m is the mass per unit length of the beam; M_{zy} is the longitudinal–transverse shear modulus; k is the radius of gyration of the beam differential element about an axis through its centre, normal to the yz plane; and C_{mt}, the mutual stiffness coefficient, is given by

$$C_{mt} = \frac{2\, I_1}{S_{35}} \qquad (3.5.5)$$

Abarcar and Cunniff have solved the set of equations (3.5.4)–(3.5.4d) numerically by using the method of transfer matrices. From the solution of these equations the natural frequencies and the 'shapes' of the corresponding normal modes are obtained. The elastic moduli which determine the dynamic response of the cantilever beam are the Young's moduli E_1 and E_3; the shear moduli M_{12}, M_{23} and M_{13}; and the Poisson ratio ν_{31}. For graphite fibre/epoxy and boron fibres/epoxy composites, these moduli have been measured experimentally by Abarcar and Cunniff, using the following procedure.

1. *Determination of* E_1 *and* E_3. These are determined by using the Bernoulli–Euler formula for the fundamental frequency of the transverse vibration of a cantilever beam, as given below:

$$\omega_1 = 3.516\left[\frac{E_z\, I_1}{ml^4}\right]^{1/2} \qquad (3.5.6)$$

Thus, a measurement of ω_1 gives the value of E_z and, hence, E_1 and E_3 when the fibres in the sample are aligned, respectively, as normal and parallel to the z axis.

2. *Determination of* M_{12} *and* M_{23}. These are determined from the vibrations of the Timoshenko beam. For orthotropic materials its natural frequencies depend on the Young's modulus E_z and the ratio of the Young's modulus and the shear modulus, i.e. E_z/M_{zy}. Thus, a measurement of these vibration frequencies together with E_z obtained from Eq. (3.5.6) gives M_{zy}. By using appropriately aligned samples, Abarcar and Cunniff have obtained the values of M_{12} and M_{23}.

3. *Determination of* ν_{31}. The Poisson ratio ν_{31} was determined by taking the average of the ratios of the lateral and the associated longitudinal strains, which were obtained from measurements on specimens in which the fibres were aligned parallel to the z-axis. The value of ν_{31} obtained for graphite fibre/epoxy composites is 0.30, which agrees with the value obtained from static measurements. However, the value $\nu_{31} = 0.1667$ obtained for boron fibre/epoxy composites is obviously incorrect, since it is less than the values of ν_3 for either of the constituents. This has been attributed to the fact that it is difficult to find a smooth surface for gauge placement on the boron fibres/epoxy composite, owing to very hard boron

fibres embedded in soft epoxy matrix. The authors have therefore used the value $\nu_{31} = 0.267$ obtained from the static measurements.

4. *Determination of M_{13}.* First E_z was obtained from Eq. (3.5.6) for two specimens in which the fibres are aligned at angles of 15° and 30° to the z axis, i.e. for $\theta = 15°$ and 30°. Two values of M_{13} corresponding to each value of θ were then obtained from Eq. (3.5.2a) by use of the known values of E_z, E_1, E_3 and ν_{31}. Both the values of M_{13} should of course be equal, and the difference between them can be attributed to the lack of perfect alignment of the fibres in the samples.

Using the values of the elastic moduli as determined by the procedure given above, Abarcar and Cunniff (1972) have calculated the natural frequencies of graphite fibre/epoxy and boron fibre/epoxy beams in several modes. The calculated values are found to agree quite well with the values measured by the same authors. The measurements were carried out on four types of specimens in which the fibres were aligned at angles of 0°, 15°, 30° and 90° to the beam axis. The dimensions of the beams were approximately 1.3×10^{-2} m wide, 0.3×10^{-2} m thick, with a double cantilever length 0.36 m, giving the length-to-thickness ratio as 60 for each cantilever beam.

In the experimental set-up each beam was supported between two fixed cylinders to minimize clamping effects. The exciter was driven by an oscillator through two amplifiers. The beam response was monitored by the strain gauges placed as near as possible to the clamped support, where the internal resisting moment and torque are maximum. The gauges were used either singly or in combination to indicate longitudinal strains, lateral strains or shear strains.

A sinusoidal force was applied to the beam clamp to set the beam into a harmonic motion. When a natural frequency was reached, as indicated by the peaking of the strain in the oscilloscope, the frequency was adjusted to get the maximum increase in the indicated strain. The measurements were made up to a frequency of about 5000 Hz. The modes were observed and photographed by sprinkling chalk powder on top of the beam. The white chalk powder was used to mark the contrast with the black beam.

A problem which arises with the samples containing fibres oriented at 15° and 30° to the beam axis is to distinguish between the bending and the twisting modes of the beam. For this purpose Abarcar and Cunniff used two sets of strain gauges, one indicating longitudinal strain and the other indicating shear strain, observed simultaneously in a dual-beam oscilloscope. The bending modes could then be distinguished from the twisting modes, since the ratio of the shear to longitudinal strains is almost constant for the bending modes, whereas the twisting modes gave an increased value of the ratio, which indicated that the resonance was in the twisting mode. The bending and twisting modes can also be distinguished from each other by an examination of the mode shapes and the direction of the twisting at the free end, shown by the direction of the nodal line nearest to the free end. The direction of the twist of the free end is the same as for bending in the bending modes and opposite to that in a torsional mode.

The dynamic elastic moduli and the damping ratios for a glass fibres/epoxy composite have been measured and analyzed by Schultz and Tsai (1968) in the vibration-frequency range 5000–10000 Hz. They find that the elastic moduli of the composite are not very sensitive to the frequency in this frequency range (5000–10000 Hz) except when the fibres are aligned at an angle of 45° to the beam axis, in which case the modulus is increased by about 15% from its value at the low-frequency end. In general, the low–frequency values of the moduli are 7–27% higher than the values given by the static methods. The damping ratios are about of the order of 10^{-2}, and are sensitive to the frequency as well as the amplitude of the vibrations. The experimental procedure is briefly described below.

The measurements were carried out on beams of composite material which were clamped between two steel cylinders. These cylinders were coupled to the moving element of an electrodynamic vibration exciter. The beams were clamped across their width halfway along their length, thus forming symmetric double-cantilever beams. They were set to vertical oscillations, by sinusoidal driving force applied at their clamped ends. The exciting signal was monitored by an electronic counter, excitation amplitude by accelerometers and the beam response by a foil strain gauge cemented to the top of the beam near the clamped end. The excitation and the response signals were observed with the help of oscilloscopes and voltmeters. The resonance was indicated by the peaking of the response observed as a function of excitation frequencies keeping the excitation amplitude constant. The mode shapes could be observed visually and by stroboscopic lighting in the lower modes. The free vibration decay was studied by exciting the beams at a resonant frequency, switching off the exciter and photographing the response decay trace in the oscilloscope. Decay over 10 and 50 cycles of response was measured from the photographs, and for the higher modes the damping ratios were derived from the measurements of the bandwidth of the resonance peak. The authors also measured the static elastic moduli by hanging small weights on the free end of the beam and measuring directly the induced strain from the strain gauge.

A resonant beam method was also used by Dudek (1970) to measure the Young's and the shear moduli of a graphite fibre/epoxy composite. He found that the resonant frequencies of such composite beams are quite different from those predicted by classical isotropic beam theories at higher vibration modes. The experimental results were analyzed by using the Timoshenko beam theory. Good agreement was obtained between the dynamic and the static values of the longitudinal Young's modulus. Adams et al. (1969) have studied the dynamic properties of carbon and glass fibre composites in torsion and flexure. They have measured the damping, longitudinal Young's modulus and longitudinal shear modulus of the two fibre composites as a function of the fibre volume fraction. They found that the damping decreases with increasing fibre volume fraction in both cases. Good agreement with the theoretical values was obtained for the shear modulus but the Young's modulus was found to be consistently lower than that predicted by the law of mixtures for both carbon and glass fibre composites.

Recently a detailed experimental analysis of the moduli of carbon fibre composites, using static as well as dynamic methods, has been carried out by Goggin (1973). The samples were taken in the form of bars as well as blocks. Goggin's measurements of the elastic moduli are described below.

1. *Dynamic Young' modulus.* Measured by a Foster resonant bar technique in which only the transverse modes were excited. Up to six vibration modes could be excited in the long bars.

2. *Static Young's modulus.* Goggin used a three-point bending technique. He used the following expression for the observed deflection δ_o of the beam supported at its ends and loaded at its centre:

$$\delta_o = PL^3 + QL + R \tag{3.5.7}$$

where L is the gauge length of samples, and P, Q and R are constants. In addition to the applied load and the cross-section of the sample, the values of P and Q depend upon the Young's modulus and the shear modulus, respectively. The constant R depends upon the machine softness, depression at the loading points and the applied load. The procedure used by Goggin is to measure δ_o as a function of L for the same maximum load and from this curve estimate P and, hence, the modulus by using Eq. (3.5.7).

3. *Tensile Young's modulus and Poisson ratio.* Obtained by measuring the tensile and the lateral strains.

4. *Torsional rigidity.* The samples were mounted in specially constructed apparatus which enabled a torque to be applied on the samples without bending them. Mirrors were attached to the face of the specimens and their deflection was measured by the standard lamp and scale technique. The

deflection was measured as a function of torque applied clockwise as well as anticlockwise. The torsional rigidity was obtained by taking the ratio of the maximum stress to maximum strain.

Goggin (1973) has measured the elastic moduli for composites in different shapes—long and short bars, rectangular blocks—with two types of carbon fibres, referred to as Types I and II, and with different resin systems. The experimental values have also been compared with some theoretical predictions. For the purpose of illustration we have reproduced some of the results obtained by Goggin (1973) in Figures 3.5–3.9.

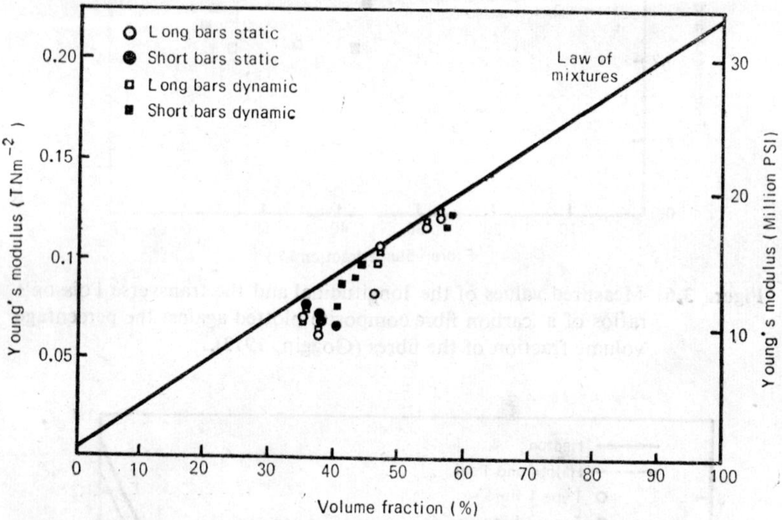

Figure 3.5 Measured values of the Young's modulus (parallel to the fibres) of a carbon fibre composite plotted against the percentage volume fraction of the fibres (Goggin, 1973). The straight line shows the values as predicted by the law of mixture.

It will be noticed that the measured values of the longitudinal Young's modulus agree quite well with those predicted by the law of mixtures. In this case the theories of Heaton (1968, 1970) and Halpin and Tsai (1969) predict almost the same values as those given by the law of mixtures. The values of the longitudinal shear modulus also agree with the theoretical values predicted from the theories of Heaton and Halpin and Tsai. The measured values of Poisson ratio show a large scatter (Figure 3.6). The agreement between the calculated and the experimental values is quite bad in the cases of transverse Young's and shear moduli, as can be seen from Figures 3.8 and 3.9. However, Goggin has shown that the theoretical values obtained from the Halpin and Tsai formula can be made to agree with the experimental results by taking a modified value of the ratio between the elastic moduli of the fibres and the matrix. This procedure

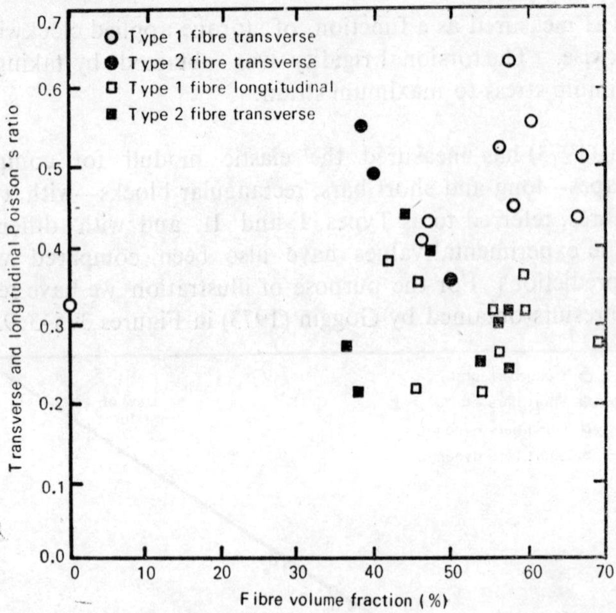

Figure 3.6 Measured values of the longitudinal and the transverse Poisson's ratios of a carbon fibre composite plotted against the percentage volume fraction of the fibres (Goggin, 1973).

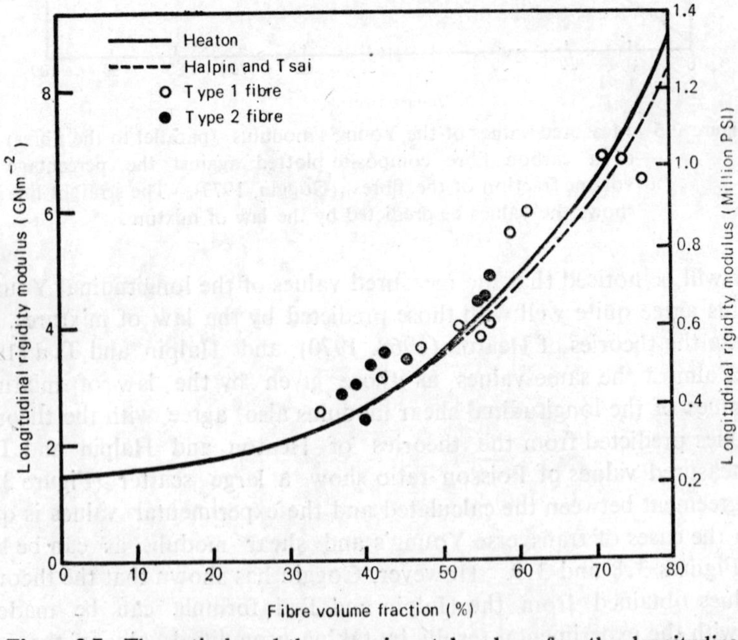

Figure 3.7 Theoretical and the experimental values for the longitudinal rigidity modulus of a carbon fibre composite (Goggin, 1973).

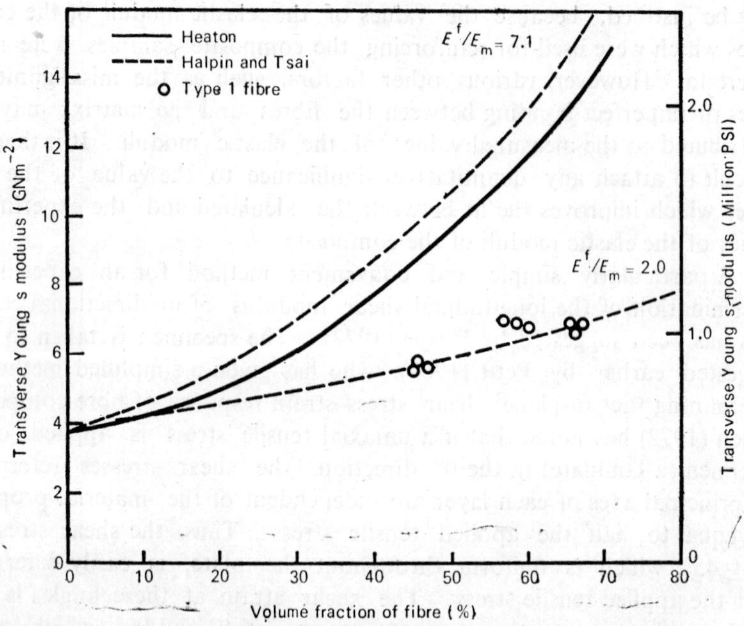

Figure 3.8 Theoretical and the experimental values of the transverse Young's modulus of a carbon fibre composite (Goggin, 1973).

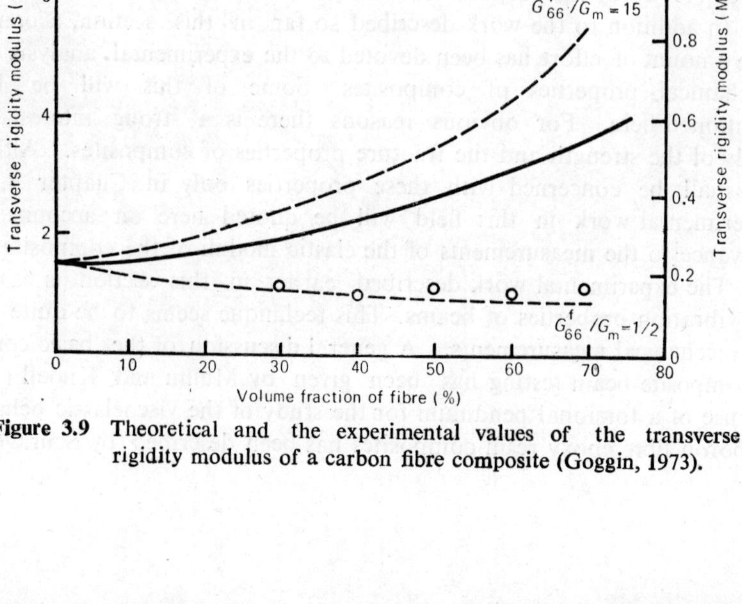

Figure 3.9 Theoretical and the experimental values of the transverse rigidity modulus of a carbon fibre composite (Goggin, 1973).

may be justified, because the values of the elastic moduli of the carbon fibres which were used for reinforcing the composite samples were rather uncertain. However, various other factors, such as the misalignment of fibres or imperfect bonding between the fibres and the matrix, may have contributed to the measured values of the elastic moduli. It is therefore difficult to attach any quantitative significance to the value of the parameter which improves the fit between the calculated and the experimental values of the elastic moduli of the composite.

A particularly simple and convenient method for an experimental determination of the longitudinal shear modulus of unidirectional composites has been suggested by Rosen (1972). The specimen is taken to be as suggested earlier by Petit (1969), who has given a simplified method for determining the in-plane shear stress–strain response of fibre composites. Rosen (1972) has noted that if a uniaxial tensile stress is applied on the specimen (a laminate) in the 0° direction, the shear stresses referred to the principal axes of each layer are independent of the material properties and equal to half the applied tensile stress. Thus, the shear stress acts at $\pm 45°$, which is uniform throughout the plate, is easily determined from the applied tensile stress. The shear strain at these angles is equal to the difference of the laminate strains at 0° and 90°, which can be easily measured. Thus, the longitudinal shear modulus M_{12} is simply given by

$$M_{12} = \frac{\sigma_x}{2\,(\varepsilon_x - \varepsilon_y)}$$

where σ_x is the applied uniaxial stress at 0°; and ε_x and ε_y, sign different from ε_x, are, respectively, the longitudinal and the transverse laminate strains.

The above formula has been applied to a boron/epoxy laminate by Rosen (1972) for the purpose of illustration.

In addition to the work described so far in this section, a considerable amount of effort has been devoted to the experimental analysis of the mechanical properties of composites. Some of this will be briefly mentioned here. For obvious reasons there is a strong interest in the study of the strength and the fracture properties of composites. Although we shall be concerned with these properties only in Chapter 4, some experimental work in this field will be quoted here on account of its relevance to the measurements of the elastic moduli of the composite.

The experimental work described earlier in this section is based on the vibration properties of beams. This technique seems to be quite useful for mechanical measurements. A general discussion of the basic concepts in composite beam testing has been given by Mullin and Knoell (1970); the use of a torsional pendulum for the study of the viscoelastic behaviour of boron fibre/epoxy resin composites has been described by Schrager and

Carey (1970). Daniels et al. (1971) have discussed short-beam shear tests for graphite fibre composites. Some damping properties of aluminium-particle-filled epoxy composites have been described by Sierakowski et al. (1971). The applications and the limitations of rail shear tests have been discussed by Whitney et al. (1971).

Jackson et al. (1972) have examined the tensile and the flexural properties of carbon fibres/aluminium composites, and the transverse tensile properties of boron fibre/aluminium composites have been investigated by Prewo and Kreider (1972). The transverse properties of unidirectional aluminium matrix composites have also been measured by Lin et al. (1971). The stress–strain curves and the fracture properties of fibre-reinforced aluminium alloys have been discussed by Nixdorf et al. (1971). Some elastic moduli of several metal-matrix composite systems have been measured by Zecca and Hay (1970).

Chiao and Moore (1971) have measured the effect of strain rate on the ultimate tensile stress of fibre/epoxy strands. The mechanical performance of cross-plied composites has been examined by Lavengood and Ishai (1971). The stress–strain behaviour of glass composites at various strain rates have been investigated by Nicolai and Narkis (1971). Moehlenpath et al. (1971) have examined the effect of time and temperature on the mechanical behaviour of epoxy composites. The tensile behaviour of fibrous materials at high rate of strain and sub-ambient temperature has been investigated by Skelton (1970). The use of instrumented Hounsfield tensometers and instrumented Avery Izod pendulums for the measurements of the effect of strain rate on the mechanical properties of composites has been described by Hazell (1970, 1971). Hardy (1969) has described the use of the Instron tensile testing machine.

The high temperature strength of boron, silicon-carbide-coated boron, silicon carbides, stainless steel and tungsten fibres has been examined by Veltri and Galasso (1971). Kreider and Marciano (1969) have reported their measurements on the mechanical properties of borsic (silicon-carbide-coated boron filaments)/aluminium composites and compared the results with those predicted by the law of mixtures.

The tensile properties of graphite fibres have been measured by Ezekiel (1971). The metallic matrices have been analyzed by Bhattacharyya and Parikh (1970, 1971). These authors have also investigated the influence of filament orientation and fabrication variables on the properties of fibre-reinforced metals. Garmong and Shephard (1971) have investigated the effect of fibre diameter and fibre concentration by volume on the mechanical properties of iron fibre/copper matrix composites. Some general principles of mechanical testing of metal-matrix composites have been described by Adams (1969).

Strain measurement techniques for determination of fibre modulus have been described by Moore and Lepper (1974). De Vekey (1974) has described the use of an LVDT (linear variable differential transformer) extensometer for tensile testing of composite materials. A 10° off axis tensile test for interlaminar shear characterization of fibre composites has been described by Chamis and Sinclair (1976).

The plasticity of unidirectionally solidified silver–germanium eutectic alloys has been analyzed by Krummheuer and Alexander (1971). These alloys behave like fibre composites in some ways. An elastoplastic analysis of residual stresses in axial loading in composite cylinders has been given by Hecker et al. (1970). Wall and Card (1971) have measured the shear strength of filament-wound glass epoxy tubes, and the torsional stiffness of plastic tubes reinforced with glass fibres has been measured by Ogorkiewicz and Sayigh (1971).

Jones (1971) has investigated the effect of fibre diameter on the fracture strength and the Young's modulus of carbon fibre composites. These fibres were made from polyacrylonitrite.

Optical methods for testing the mechanical properties of fibres and fibre composites have also been used. Middleton (1968–70) has used this method for testing graphite fibres. He measured the microphotometric spectral reflectivities for 50 individual fibres ranging in modulus from about 70 to 350 GN/m². The mean maximum reflectivity R_0 of these samples was found to be linearly correlated with their statically determined Young's modulus through the following relation:

$$E=(11.09 \text{ Max } R_0-65.93)\times 6.9 \text{ GN/m}^2$$

The author has claimed that such a method can be used with a fair degree of accuracy.

Stone (1969) has suggested that photoelastic methods may be quite useful in the determination of microstresses in systems involving complex materials. He has discussed destructive and non-destructive testing techniques for the determination of the stress distribution in a three dimensional fibre array.

An X-ray diffraction technique has been used by Cheskis and Heckel (1970) to study the deformation behaviour of continuous fibre/metal matrix composites. The mechanical properties of carbon fibres have been studied by optical methods by Perry et al. (1971) (see also Perry et al., 1970; Phillips et al. 1971).

Another useful technique for the analysis of stresses and strains is based on the use of Moiré fringes. Daniel and Rowlands (1971) have determined the strain concentration in composites by Moiré techniques, and a Moiré analysis of the interlaminar shear edge affect in laminate composites has been given by Pipes and Daniel (1971).

A review of various optical techniques, including Moiré fringes, together with the usual mechanical methods for an experimental analysis of strains in fibre composites has been given by Kedward and Hindle (1970).

Recently an optical method for determination of the cross-sectional shape of fibres by using a fibre morphometer has been described by Mann and Campbell (1975). The authors have also discussed the theory of the method in some detail. Leach and Ashbee (1974) have described development of a light pipe technique for investigation of interfacial phenomena in fibre reinforced composites.

A X-ray method for monitoring of fatigue in boron/epoxy composite has been used by Roderick and Whitcomb (1975). The method shows fibre failure during fatigue of the composite. The method is based on the ability of the X-ray techniques to recognize the fibre breaks.

Acoustic emission which refers to emission of audible sound from materials under certain conditions of mechanical loading provides a useful technique for mechanical testing. This technique has been used for determination of the load time history of fibre composites by Rotem and Baruch (1974). Acoustic emission from aluminium oxide-molybdenum fibre composite has been observed by Lloyd and Tangri (1974).

CHAPTER **4**

Fracture and Failure of Fibre Composites

4.1 INTRODUCTION

In the previous chapter we described the elastic properties of a fibre composite and some methods for the calculation of its elastic constants. The elastic constants give the response of the solid to an applied stress as long as the strain is small, i.e. the solid is in stage I (see Section 2.3). As the applied stress is increased, the solid gradually travels through stage II to stage III, when the failure occurs.

The importance of the study of fracture and failure properties for engineering applications should be obvious. If all the factors contributing to the failure of a solid were known, it would be possible in principle to avoid it. A less ambitious but nevertheless useful achievement would be to be able to predict the occurrence of failure at a certain value of the applied load and/or after a certain duration due to material fatigue. Thus, there is a strong case for calculating the critical stress or stress intensity factors as defined in Section 2.5.

The characteristic problem in the case of a composite which makes it different from an ordinary solid arises from the fact that a composite has more than one component. The different components have, in general, different elastic constants and different ranges pertaining to the three stages on the stress–strain diagram. As a result, the three stages of the composite are, in general, not so distinct as in the case of an ordinary solid. For example, at a certain stress the matrix may enter stage II while the fibres may still be in stage I, or vice versa. The behaviour of the composite in this region will therefore be neither purely elastic nor plastic. An appropriate theory for the fracture properties of the composite must therefore account for the mixed response of its components.

4.2 GENERAL CONSIDERATIONS

Most of the calculations on the stress–strain relations and the critical stress in a composite are based upon the law of mixtures as originally given by Dietz (1954) for plastics reinforced by glass. Although the law of mixtures has only a limited validity, as discussed in Section 3.2, it is very useful for a rough estimate of the overall failure properties of a composite on account of its simplicity. In this section we shall consider the two cases separately—failure under compression and failure under tension.

Failure under compression

Let us consider an unstressed fibre composite as shown in Figure 4.1a. If a compressive load is applied to the composite along the axes of the fibres, the fibres will 'buckle' as shown in Figure 4.1b, c. The buckling

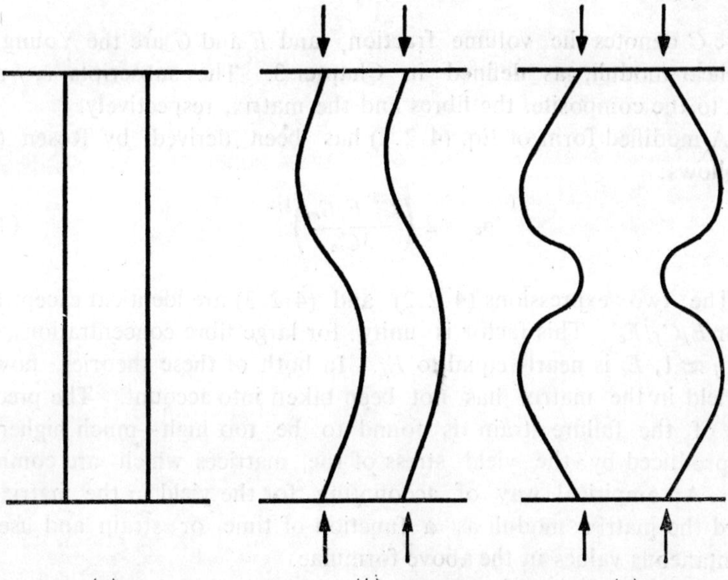

Figure 4.1 Fibre buckling due to a compressive load: (a) no buckling (b) in-phase buckling (c) out of phase buckling.

of two neighbouring fibres can be in phase or out of phase and also in the same plane or in different planes. The in-phase and out-of-phase buckling modes for the same plane buckling have been illustrated in Figure 4.1b, c, respectively. Theoretically the main difference between these two modes arises from the fact that in the case of out-of-phase buckling the matrix is subjected to a compression in some regions and a tension in other regions. There is no such reversal in the case of the in-phase modes.

A rigorous theoretical analysis of the fibre buckling in a real composite is obviously a very complex problem, in view of the fact that in a real case there are always mixed modes of buckling—the calculation of the out-of-plane modes is particularly complicated. Fortunately, however, it appears that the strength of a composite is reasonably well predicted by the law of mixtures (see Kelly and Tyson, 1965a, b; De Ferran and Harris, 1970).

Schuerch (1966) has given a theory for the failure of a laminated composite in compression in which only in-plane buckling is possible. He has obtained the following expressions for the critical stress in the two modes:

$$\sigma_c = 2G_m/C_m \quad \text{(in-phase mode)} \quad (4.2.1)$$

$$\sigma_c/E_c = 2\left(\frac{C_f E_m}{3C_m E_f}\right)^{1/2} \quad \text{(out-of-phase mode)} \quad (4.2.2)$$

where C denotes the volume fraction, and E and G are the Young's and the shear moduli, as defined in Chapter 3. The subscripts, c, f and m refer to the composite, the fibres and the matrix, respectively.

A modified form of Eq. (4.2.2) has been derived by Rosen (1965) as follows:

$$\sigma_c = 2\left(\frac{C_f^3 E_f E_m}{3C_m}\right)^{1/2} \quad (4.2.3)$$

The two expressions (4.2.2) and (4.2.3) are identical except for a factor $E_f C_f/E_c$. This factor is unity for large fibre concentrations, since, for $C_f \approx 1$, E_c is nearly equal to E_f. In both of these theories, however, the yield in the matrix has not been taken into account. The predicted value of the failure strain is found to be too high—much higher than that produced by the yield stress of the matrices which are commonly used. An empirical way of accounting for the yield in the matrix is to regard the matrix moduli as a function of time or strain and use their instantaneous values in the above formulae.

The effect of plasticity in the yielded matrix has also been considered by Dow et al. (1966), who have derived the following expression for the critical stress;

$$\sigma_c = \left(\frac{C_f E_f \sigma_m'}{3C_m}\right)^{1/2} \quad (4.2.4)$$

where σ_m' is the yield stress of the matrix.

The above formulae provide a reasonable estimate of the critical stress and are particularly convenient to use because of their simplicity. The experimental aspects of this problem have been discussed by Schuerch (1966), Yue et al. (1968), Lager and June (1969), De Ferran and Harris (1970) and Pinnel and Lawley (1970).

Failure under tension

According to Cooper (1971), the fracture of a composite under tension can be classified into two categories—single fracture and multiple fracture. A composite fails by a single fracture if both the components are fractured at the same point along their axes, and in this case the composite will immediately break into two pieces. Multiple fracture occurs if only one of the components is fractured while the other is intact. In this case, as the applied stress is increased, more and more cracks will appear in the damaged component until finally the other component is also fractured. In this case it is physically obvious that during the interval between the appearance of the first crack in the weaker component and the final fracture, the applied stress will produce a larger strain than that produced in the perfect composite. In other words, during this interval the composite will behave in a partially plastic fashion. Thus, the behaviour of the composite will be similar to that of a ductile metal (Cooper, 1971) and the stress at which the first crack appears in the weaker component can be considered to be the yield stress of the composite.

Following Cooper (1971), we can derive simple criteria for the occurrence of single and multiple fractures in a two-component composite. Let us consider a two-component cylindrical composite under a tension along its axis. We shall assume that both the components are parallel to each other and to the axis of the cylindrical composite. The two components will be identified by the subscripts 1 and 2. In the case of a fibre composite, for example, one of the components will be the fibres and the other will be the matrix.

Let $C_{1,2}$ and $E_{1,2}$ denote the concentration (volume fraction) and the Young's modulus, respectively, of the components. Suppose that a longitudinal stress σ produces a uniform strain ε in the composite. So long as ε is small, so that both the components and, hence, the composite are in stage I, we can write the following equation with the help of the law of mixtures:

$$\sigma = C_1E_1\varepsilon + C_2E_2\varepsilon$$

Let ε_2 denote the strain at which component 2 fails. Assuming elastic behaviour throughout the region $0 \leqslant \varepsilon \leqslant \varepsilon_2$, the value of σ at $\varepsilon = \varepsilon_2$ is given by

$$\sigma_c' = C_1E_1\varepsilon_2 + C_2\sigma_2 \tag{4.2.5}$$

where $\sigma_2 = E_2\varepsilon_2$ and we have also assumed that ε_1, the failure strain of component 1, is larger than ε_2.

σ_c', as given by Eq. (4.2.5), is the stress at which component 2 will fail. Thus, the conditions for single and multiple fractures can be written as

Single fracture:

$$C_1\sigma_1 < \sigma_c' \tag{4.2.6}$$

Multiple fracture:

$$C_1\sigma_1 > \sigma_c' \tag{4.2.6a}$$

where $\sigma_1 = E_1\varepsilon_1$ is the stress at which component 1 will fail.

In the case of a multiple fracture, i.e. when Eq. (4.2.6a) is satisfied, the critical stress at which the whole composite will fail is given by

$\sigma_c = C_1\sigma_1$. In this case, as discussed earlier, the composite will behave like a ductile metal even if both the components are brittle. The yield stress of the composite will be equal to σ_c' and it will exhibit a plastic-type behaviour for stresses between σ_c' and σ_c. This apparent ductility is a quite useful feature of the multiple fracture, an advantage which is not available with the composites which fail by single fracture.

Several other theoretical and experimental aspects of single and multiple fracture in various composites have been given in the following papers: Jech et al. (1959), Cratchley (1963), Kelly (1964), Tyson (1964), Kelly and Tyson (1965a, b), Kelly and Davies (1965), Piehler (1965), Cline (1966), Pinnel and Lawley (1970), Cooper and Silwood (1972) and Aveston and Kelly (1973).

In the derivation of σ_c as given above it was assumed that the tension on the composite was applied in a direction parallel to the axis of the cylindrical components. If the tension is applied at an angle ϕ to the axis of the cylinder, then, according to Stowell and Liu (1961), the composite may fail by any of the three mechanisms given below along with the corresponding values of the stress.

1. Transverse failure ($\phi \approx \pi/2$):

$$\sigma_t(\phi) = \sigma_{\pi/2} \operatorname{cosec}^2 \phi \tag{4.2.7}$$

2. Longitudinal failure ($\phi \approx 0$):

$$\sigma_l(\phi) = \sigma_0 \sec^2 \phi \tag{4.2.7a}$$

3. Shear failure:

$$\sigma_s(\phi) = 2\tau \operatorname{cosec} 2\phi \tag{4.2.7b}$$

where the subscripts, t, l and s denote the various failure modes; σ_0 and $\sigma_{\pi/2}$ denote the failure stresses for $\phi = 0$ and $\pi/2$, respectively; and τ is the failure stress in shear parallel to the axis of the cylinder.

A unified treatment of the above-mentioned three failure mechanisms has been given by Tsai (1965), who has derived the following dependence of σ on ϕ:

$$\sigma(\phi) = \sigma_0 \left[\cos^4 \phi + \left(\frac{\sigma_0^2}{\tau^2} - 1\right) \cos^2 \phi \sin^2 \phi + \frac{\sigma_0^2}{\sigma_{\pi/2}^2} \sin^4 \phi\right]^{-1/2} \tag{4.2.8}$$

It can easily be verified that the functional dependence of σ on ϕ as predicted by Eq. (4.2.8) in the three limiting cases $\phi \approx \pi/2$, $\phi \approx 0$ and $\phi \approx \pi/4$ is similar to σ_t, σ_l and σ_s, respectively, as given in Eq. (4.2.7).

The above equations are obeyed quite well in the case of glass fibre composites (Tsai, 1965) and also for several other systems (Cooper, 1971).

For a further discussion of these failure mechanisms see Kelly and Davies (1965), Jackson and Cratchley (1966), Cooper (1966), George et al. (1968), Prager (1969), Kreider and Marciano (1969) and Bhattacharyya and Parikh (1970).

Before closing this section we shall give a brief review of some of the recent papers dealing with the general failure properties of fibre composites.

A quite detailed analysis of interfibre failure of composite materials has been given by Koeneman (1970). He has suggested a micromechanical failure criterion based upon certain conditions which are satisfied at the failure of fibre composites; the suggested failure criterion is a critical creep rate (see Section 4.5) normal to the fracture. From an application of this criterion, the author infers that the transverse strength of unidirectional composites should increase as the strain hardening of the matrix increases. The effect of residual stresses on failure strength is predicted to be small but the shear stresses existing around fibre ends or where a fibre passes through a void are found to affect significantly the matrix yield and the failure of the composite.

An explanation has been given for the observed difference in the temperature dependence of cross-plied strengths between glass fibre and graphite fibre composites. The width effect of cross-plied strength and the strengthening effect of elastomer inclusions on epoxies are also explained. The theory predicts a width effect and a layer effect on the cross-plied creep rate which was also experimentally verified.

A study of some mechanical and fracture properties of carbon fibre/nickel composites has been carried out by Braddick et al. (1971). Jackson et al. (1971), from the same school, have also published their results on the fracture of boron/aluminium composites. The fracture properties of some fibre-reinforced aluminium alloys have been given by Nixdorf et al. (1971). The flow and fracture of aluminium/boron composite sheet under biaxial strain has been discussed by Wright and Ebert (1972).

Various other aspects of the fracture of metal-matrix composites have also been discussed in the literature. Hietman et al. (1973) have studied the effect of brittle interfacial compounds on the deformation and fracture of molybdenum–aluminium fibre composites. Some observations on the tensile fracture in an aluminium sheet reinforced by boron fibres have been reported by Herring et al. (1973). Kellerer et al. (1972) have attempted to derive a relationship between strength and ductility in some dispersion-hardened aluminium alloys.

Multiple fracture in a steel-reinforced epoxy resin composite has been discussed by Cooper and Sillwood (1972). The mixed mode fracture of graphite fibres/epoxy composites has been analyzed by McKinney (1972).

An experimental investigation of fracture in advanced fibre composites has been reported by Konish et al. (1972). An off-axis tensile coupon test for the evaluation of the strength of a laminar composite has been given by Schneider (1972). Harrison (1972) has discussed the release rates of strain energy for turning cracks. The longitudinal tensile feature of unidirectional fibre composites has been analyzed by Lifshitz and Rotern (1972). The fracture properties of polyphenylene oxide composites have been discussed by Trachte and Dibenedetto (1971).

Mullin and Mazzio (1972b) have studied the effects of matrix and interface modification on local fractures of carbon fibres in epoxy resin under axial tensile and compressive loading for composites having low volume fractions of fibres. The effects of shear damage on the torsional behaviour of carbon fibre composites have been examined by Adams et al. (1973). The effect of vibrations on the strength of glass fibres has been studied by Bartenev and Motorina (1971). The effect of specimen and testing variables on the fracture of some fibre-reinforced epoxy resins has been discussed by Ellis and Harris (1973).

Some observations on the effect of diameter on the fracture strength and the Young's modulus of carbon and graphite fibres have been reported by Jones (1971). The effects of impurities in carbon fibres on its properties has been studied by Liberman and Noles (1972). Crane and Tressler (1971) have discussed the effect of surface damage on the strength of *c* axis sapphire filaments, and the effect of surface softening on the tensile strength of fibre glass rods has been discussed by Reifsnider and Kelly (1972). The effect of temperature on the strength of a polymer impregnated ceramic has been studied by Gebauer et al. (1972). Berg et al. (1971) have examined fracture in the grip section of reduced cross-section specimens of fibre-reinforced materials.

A discussion of the off-axis strength test for anisotropic materials—in particular, boron/epoxy composites—has been given by Pipes and Cole (1973). Cruse and Stout (1973) have given a fractographic study of graphite/epoxy laminated fracture specimens. A non-destructive technique using vibration tests for a determination of fatigue crack damage in composites has been described by Dibenedetto et al. (1972).

The non-linear behaviour of unidirectional composite laminate has been described by Hahn and Tsai (1973). The yield criterion of laminated media has been discussed by Chou et al. (1973). Dvorak et al. (1973) have discussed yielding in fibre composites under external loads and temperature changes. Foye (1973a, b) has given a detailed analysis of the post-yielding behaviour of composite laminates.

Bert (1974) has described some static testing techniques for filament wound composite materials. Craig and Courtney (1975) have described tension test for studying the fibre composite failure mode. A graphical method for determination of stiffness and strength of composite laminates has been proposed by Hahn and Tsai (1974). The fracture characteristics of a metal matrix composite have been analyzed by Baldwin and

Sierakowski (1975). The improved impact resistant boron-aluminium composites for use as turbine engine fan blades has been described by McDanels and Signorelli (1976).

A review of mainly the Russian work on strength of fibrous composites has been given by Skudra (1974a). Skudra (1974b) has also discussed properties of fibre glass plastics reinforced with high modulus fibres. Skudra and Bulavs (1973) have discussed the effect of the sign of shear stresses on the shear strength of reinforced plastics. The edge effect studies in fibre reinforced laminates has been studied by Oplinger et al. (1974). Methods for determination of strength and plasticity characteristics of multilayer composite materials have been described by Gurev et al. (1975).

The effect of certain structural and technological parameters on the tensile strength and deformation of unidirectionally reinforced plastics has been analyzed by Medvedev (1975). In this paper a model has been proposed for such a study which has been shown to be in qualitative agreement with the experimental results.

Majumdar and McLaughlin (1975) have discussed effects of phase geometry and volume fraction on the plane stress limit analysis of a unidirectional fibre reinforced composite. The interlaminar shear strength of fibre reinforced composites has been studied by Markham and Dawson (1975).

Some aspects of fracture in brittle fibre composites have been discussed by Chaplin (1974). Hanasaki and Hasegawa (1974) have studied compressive strength of unidirectional fibrous composites. The transverse compressive behaviour of unidirectional carbon fibre reinforced plastics has been analyzed by Collings (1974).

The compression strength of a unidirectional carbon fibre reinforced plastic has been studied by Hancox (1975). The deformation of a fibre composite under hydrostatic pressure has been discussed by Weaver and Williams (1975). Greszczuk (1975) has analyzed microbuckling failure of circular fibre-reinforced composites.

The intermittent bonding for high toughness/high strength composites has been discussed by Atkins (1975a). The related problem of strength of mortar has been considered by Romualdi (1974). He has presented a theory for fibre reinforcement for strengthening of brittle materials and given some experimental data for mortar in support of the theory.

The problem of stress distribution in layered composites has been attempted by Partsevskii (1973). Pagano (1974b) has calculated interlaminar normal stress in composite laminates and Pipes and Pagano (1974) have given an approximate elasticity solution to estimate the interlaminar stresses in composite laminates.

A problem of stress concentration in elastic composites with limiting shear properties has been analyzed by Theocaris and Paipetis (1975).

These authors have shown that the composite can be fabricated with a high E-modulus by choosing fibre and matrix which have equal shear moduli but a greater poisson ratio for the fibre.

A method has been suggested by Shorshorov (1976) for determining the residual stresses in fibre composites. In this method, which is based on a consideration of the longitudinal deformation, the amount of pre-straining and the volume fraction of the fibre are determined so that the compressive residual stress in the matrix is equal to the compressive yield stress. A similar problem concerning longitudinal residual stresses in boron fibres has been discussed by Behrendt (1976). One interesting problem is the strength retention and the stability of the mechanical properties over a period of time in case of the fibre composites. This has been considered by Chiao et al. (1976). Kolevatov et al. (1976) have analyzed combined effect of external factors on the stability of the strength and strain characteristics of fibre glass reinforced plastics. Matrix strengthening in continuous fibre reinforced composites under monotonic and cyclic loading conditions has been discussed by Lee and Harris (1974).

There has been some attempt to develop macrofibres which are expected to have better mechanical properties. Cooper et al. (1974) have discussed the development of such fibres and the macrofibres composite of large diameter.

An interesting development in the fibre composites is in terms of what is referred to as a hybrid composite. Such a composite contains two types of reinforcing fibres with different stiffness which are embedded in a common matrix. An example would be carbon and glass fibres in vinyl ester matrix. There has been some controversy whether there is really a hybrid effect, i.e. whether the toughness of the composite is improved by hybridization. Phillips (1976) has carried out some investigations on this subject. His experimental results from a tensile test show that a load sharing takes place between the carbon and the glass fibres from which the author concludes that the hybrid composite has a much greater toughness than the corresponding carbon fibre composite. The hybrid composite also has a much greater fatigue resistance. The impact properties of a glass/carbon fibre hybrid composite have also been studied by Harris and Bunsell (1975).

A design of an advanced composite Aileron has been discussed by James and Vanghn (1976) for commercial aircraft. They have considered tha application of graphite and Kevlar 49/epoxy composite material for long term service. The combined load stress strain relationship for advanced fibre composites has recently been analyzed by Chamis and Sullivan (1976).

An investigation of the compressive strengths of Kevlar 49/epoxy composite has also been reported by Kulkarni et al. (1975). Chamis et al. (1975) has described boron/aluminium and graphite/resin advanced fibre composite hybrids. Some Charpy impact experiments on graphite/epoxy hybrid composites have been reported by Perry and Adams (1975).

4.3 STATISTICAL THEORIES

If the strength of a composite is expressed in terms of the strengths of its components according to the simple law of mixture, it is implicitly assumed that the strength of all the fibres is the same. In fact the fracture properties of a material depend also upon its size. The above assumption therefore also implies that the length and the cross-section of all the fibres is the same. In a real composite this assumption is not valid. Although the gross elastic constants of the composite may not be so sensitive to this assumption as mentioned in Section 3.2, it may lead to very misleading estimates of the strength of the composite.

In case the strength, bonding and length of the fibres are different, we can attempt to use the law of mixture for the strength of the composite by defining the mean values of all the relevant quantities. This will obviously give an overestimate of the strength, because the failure will start at the weakest fibre and not at the fibre with 'average' strength. Even if the number of weak fibres is small, which contribute very little to the 'average', they may substantially lower the strength of the composite by initiating an early fracture. Moreover, the strength of the fibres will also depend upon their local environments in the matrix, which may have a substantial, regional fluctuation.

To account for such random variations, various statistical theories have been developed for the calculation of the mechanical properties of a composite. The basic technique consists of assuming a certain probability distribution for the defects in fibres and then calculating their integrated effects. A Gaussian distribution was assumed by Pierce (1926) for determining the strength of yarns as a function of their length. However, the distribution function as given by Weibull (1939, 1951), now referred to as the Weibull distribution, has been very successful in predicting the strength of the composite materials and is now widely used. The Weibull distribution $W(x)$ of a quantity x is given by

$$W(x) = 1 - \exp(-\alpha L x^{\beta}) \qquad (4.3.1)$$

where α and β are two parameters which define the strength and the spread of the Weibull distribution, and L is a geometrical characteristic of the specimen, which may be, for example, the length of the specimen.

Early work on the statistical theories was carried out by among others, Daniels (1945), Coleman (1958), etc. These theories have been analyzed and reformulated by Gücer and Gurland (1962). More recent developments in the statistical theories can be found in, for example, Rosen (1964), Arridge (1965), Zweben (1968, 1969), Zweben and Rosen (1969, 1970) and Rosen (1970b).

A recent review of the statistical theories of the fracture of composite materials has been given by Argon (1972). He has also described a fairly simple model for the calculation of the strength of composites. The efficiency of fibre reinforcement of brittle matrices has been discussed by Laws (1971) on the basis of the statistical theory. She has proposed some 'efficiency' factors which can serve as a measure of the effectiveness of utilization of fibre stiffness and strength when a random distribution of extensible fibres is incorporated into a brittle matrix.

Recently a very useful bounding approach to the strength of composite materials has been given by Zweben (1972) which we shall review later in this section.

First, however, we shall consider the effects of broken and discontinuous fibres on various mechanical properties of a composite which are relevant to discussion of the strength of a composite.

Effect of broken or discontinuous fibres on the mechanical properties of composites

First we shall calculate the effective modulus of a composite due to a discontinuous fibre. This problem has been studied by Cox (1952), Dow (1963) and Rosen (1965). The treatment given here is due to Cox (1952). The results obtained by Dow (1963) and Rosen (1965) are similar to those obtained by Cox (1952).

As a representative element of the composite we shall choose a composite system of two concentric cylinders (see Figure 4.2). The inner cylinder will represent the fibre and the outer cylinder will represent the

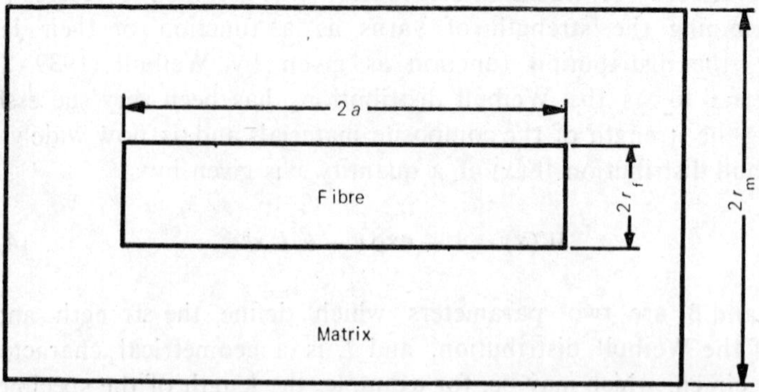

Figure 4.2 Model used for the calculation of the effective modulus of a composite containing discontinuous fibres (Cox 1952).

matrix. We shall assume that $2a$, the length of the fibre, is less than the length of the outer cylinder. Let r_f and r_m be the radii of the fibre and the matrix, respectively.

Consider the case when a longitudinal stress σ is applied on the system, which causes an overall strain ε. Let u_f and u_m denote the displacement fields in the fibre and the matrix regions, respectively. Then we have from the definition of the overall strain.

$$\varepsilon = \frac{du_m}{dx} \tag{4.3.2}$$

and from Hooke's law, applied to fibres,

$$\sigma = E_f \frac{du_f}{dx} \tag{4.3.3}$$

where x is the distance measured from one end of the fibre chosen as origin and E_f is the Young's modulus of the fibre.

It is reasonable to assume that the change in the stress along the length of the cylinder is proportional to the difference between the displacement fields in the fibre and the matrix regions (Cox, 1952). If this constant of proportionality is denoted by A, we can write

$$\frac{d\sigma}{dx} = A\,(u_f - u_m) \tag{4.3.4}$$

With the help of Eqs. (4.3.2) and (4.3.3), Eq. (4.3.4) can be written in the form

$$\frac{d^2\sigma}{dx^2} = \frac{A}{E_f}\sigma - A\varepsilon \tag{4.3.5}$$

This is a linear differential equation of the second degree which can be solved by taking an exponential dependence of σ and x. It is easy to verify that

$$\sigma = E_f\,\varepsilon + P\,e^{-\alpha x} + Q\,e^{\alpha x} \tag{4.3.6}$$

where P and Q are arbitrary constants, will be a solution of Eq. (4.3.5) provided that

$$\alpha^2 = \frac{A}{E_f} \tag{4.3.7}$$

The arbitrary constants P and Q can be determined from the boundary conditions that σ must vanish at $x = 0$ and $x = 2a$. After a little manipulation we obtain

$$\sigma\,(x) = E_f\,\varepsilon \left[1 - \frac{\cosh \alpha\,(a - x)}{\cosh \alpha x}\right] \tag{4.3.8}$$

The effective modulus can be obtained in terms of the average stress σ as follows:

$$(E_f)_{\text{eff}} = \frac{\bar{\sigma}}{\varepsilon}$$

where

$$\bar{\sigma} = \frac{1}{2a} \int_0^{2a} \sigma(x)\, dx$$

$$= E_f\, \varepsilon \left[1 - \frac{\tanh \alpha a}{\alpha a} \right]$$

This gives the following result for the effective modulus:

$$\eta = \frac{(E_f)_{\text{eff}}}{E_f} = 1 - \frac{\tanh \alpha a}{\alpha a} \tag{4.3.9}$$

The value of α or the constant of proportionality A in Eq. (4.3.4) depends upon the elastic constants of the two components and their geometrical arrangement. For the model given above, α is given below (Cox, 1952):

$$\alpha^2 = \frac{2\pi M_m}{\ln (r_m/r_f)} \quad (r_m > r_f) \tag{4.3.10}$$

where M_m is the shear modulus of the matrix.

We note from Eq. (4.3.9) that η, the ratio of the effective to the real modulus, goes to zero in the limit of very small fibres ($a \approx 0$) and approaches unity in the limit of long fibres ($a \to \infty$). This behaviour of η in the two limiting cases could, of course, be derived intuitively. The more important fact, however, is that η depends on αa and not on a alone. Thus, if we choose a fibre with large α, η can be brought closer to unity for any finite value of a. We infer, therefore, that the effective strength of a composite containing discontinuous fibres can be increased by choosing a fibre-matrix system for which M_m/E_f (see Eq. 4.3.10) is large.

The strength of composites reinforced by discontinuous fibres has been discussed by Kelly (1964), Kelly and Tyson (1965b) and Spencer (1965), on the assumption that the law of mixtures is valid for the strength of the composite. They write the following for the critical stress in a composite:

$$\sigma_c = \bar{\sigma}_f C_f + \bar{\sigma}_m C_m \tag{4.3.11}$$

where $\bar{\sigma}_f$ is the average failure stress in the fibres and $\bar{\sigma}_m$ is the stress in the matrix when the strain in the composite reaches the failure point. The limitations of Eq. (4.3.11) have been discussed earlier in this section.

In the present analysis the concept of a critical length a_c is introduced which is defined as the length of the fibre for which the peak stress in the fibre is just equal to the failure stress. If the fibres are shorter than a_c, then the failure will occur by plastic flow of the matrix round the fibres. In this case Kelly (1964) has obtained the following expression for the strength of the composite ($a < a_c$):

$$\sigma_c = \frac{a\,\sigma_f}{2a_c} C_f + \sigma_m C_m \tag{4.3.12}$$

where σ_m in Eq. (4.3.12) refers to the ultimate strength of the matrix.

For $a \triangleright a_c$, the strength of the composite is given by

$$\sigma_c = C_f \sigma_f \left(1 - \frac{a_c}{2a}\right) + \bar{\sigma}_m C_m$$

where $\bar{\sigma}_m$ has been defined in Eq. (4.3.11).

For further work in this field, reference may be made to Petrasek et al. (1967), Riley (1968) and Hancock and Cuthbertson (1970). More recently, the strength of a composite has been discussed by Lavengood (1972). The strengthening effects of discontinuous coatings on metal filaments has been examined by Thornton and Thomas (1972). Helfet and Harris (1972) have discussed the fracture toughness of composites reinforced by discontinuous fibres. The impact toughness of epoxy composites reinforced by discontinuous boron fibres has been discussed by Allred and Schuster (1973). The failure of plastics reinforced by short randomly distributed fibres has been examined by Owen and Rose (1972).

A technique for the statistical analysis of fracture strength data for carbon fibres has been suggested by Jones and Wilkins (1972). They have shown that the distribution of the measured fracture strength values as a function of the cumulative probability of failure can be compared with the standard unimodal distributions. Any deviations from the standard distribution which can be revealed in one plot can be taken as the characteristics of the measured set of values. This technique has been applied to examine the effects of fast neutron irradiation on the strength of the fibres.

In this sub-section we shall also quote the formula derived by Hedgepeth (1961) for the stress concentration K_n in a composite due to n broken fibres. It is as follows:

$$K_n = \frac{2^{2n+1}\,[(n+1)!]^2}{(2n+2)!}. \tag{4.3.13}$$

Further work on the stress concentration due to broken fibres can be found in the following papers: Hedgepeth and Van Dyke (1967), Fichter (1969), Greszczuk (1969) and Franklin (1970). For the experimental work in this field reference may be made to Schuster and Scala (1963, 1964), Tyson and Davies (1965), Allison and Holloway (1967), Durelli et al. (1970) and Daniel (1970).

Among the more recent papers, Lavendel and Kalinka (1974) have discussed the interaction of a polymer matrix and short reinforcing fibres. An experimental study of strength and elastic modulus of a randomly distributed short fibre composite has been reported by Knight and Hahn (1975).

Motavkin et al. (1975) have attempted to model the mechanical characteristics of a glass reinforced polymer composite with randomly oriented fibres. The results of a experimental study of such composite have been analyzed on the basis of the model. The samples were tested in compression as well as tension under static loading. In the model a friction mechanism of interaction between the matrix and the fibres has been assumed. Good agreement with the experimental results has been obtained. An interesting result was that the fracture strain of the composite was observed to be only 65% of the fracture strain of the matrix. The model predicts embrittlement of the composite. Some approximations for the strength of random fibre composites have been obtained by Hahn (1975). Reinforcing effectiveness of discreet unidirectional fibres have been studied by Khoroshun and Shevchenko (1975). The cracking of composites consisting of discontinuous ductile fibres in a brittle matrix has been studied by Morton and Groves (1974).

Statistical bounds for the strength of a composite

A statistical theory for the strength of a fibre composite in which the fibre strengths follow the Weibull distribution has recently been developed by Zweben (1972). This theory gives upper and lower bounds for the strength of a composite which appear to be very useful from the practical point of view. In this section we shall describe briefly the results derived by Zweben (1972).

Let us consider a composite in which the fibres have a random distribution of strengths. If a gradually increasing stress is applied to the composite, first the weakest fibres will fracture. For the assumed random distribution of the fibre strengths, the damaged fibres will also be randomly distributed in the composite. Now two distinct failure modes can be defined.

1. *Weakest link mode.* The failure of a composite is said to occur in this mode if the weakest fibres which have fractured induce cracks in their neighbourhood in the matrix which quickly propagate through the

composite and trigger off failure. The stress at which the first fibre breaks are expected to occur is denoted by σ_1 and is given by (for tensile loading)

$$\sigma_1 = \left(\frac{\beta - 1}{NL\,\alpha\beta}\right)^{1/\beta} \qquad (4.3.14)$$

where α and β are the parameters of the Weibull distribution, as defined in Eq. (4.3.1), and L is the length of the composite. The strength σ of the fibres is assumed to follow the Weibull distribution, i.e. the number of fibres between strengths σ and $\sigma + d\sigma$ is $W(\sigma)\,d\sigma$, where $W(\sigma)$ has been defined in Eq. (4.3.1), with L identified as the length of the sample. The stress σ_1 in this case gives a lower bound for the strength of the composite.

2. *Fibre break propagation mode.* Some composites, however, may not fail in the weakest link mode, which has been described above. This is because the isolated, initially damaged fibres may be surrounded by strong fibres which may arrest the cracks induced in the matrix due to the damaged fibres and thus delay the failure of the composite. For such composites the lower bound given by Eq. (4.3.14) will be too conservative.

A more reasonable lower bound for the strength of such composites can be obtained by noting that the damaged fibres, even if isolated, will nevertheless form the nucleus for the propagation of fibre breaks and thus for the eventual failure of the composite. Suppose that some fibres have been broken but the composite has not yet failed. As the stress is increased, more fibres will break. These will be randomly distributed throughout the composite. The stress distribution in a composite with broken fibres will not be uniform. There will be a large local concentration of stress around the broken fibres. Thus, the fibres surrounding the broken fibres will be overstressed.

As the applied stress is increased further, the overstressed fibres will also fracture. Since the local stress concentration increases with the number of broken fibres, the fracture of the overstressed fibres marks the next stage for the fracture propagation in the composite and its failure. This mode of failure is referred to as the fibre break propagation mode (Zweben 1972).

The stress at which the first overstressed fibres are expected to break is denoted by σ_2 and provides the next lower bound for the strength of the composite. Proceeding along a similar chain of arguments one can define the higher modes of failure of the composite and the corresponding minimum stress for each mode,

The strength σ_2 in the present mode is obtained as a solution of the following equation (Zweben, 1972):

$$Mn\,W(\sigma)\,(1 - [1 - W(k_{1n}\sigma) + W(\sigma)]^n) = 0 \qquad (4.3.15)$$

where M is the effective length ratio for the composite, n is the number of fibres surrounding a broken one and k_{1n} is the stress concentration factor. The factor k_{1n} gives the stress in each of the n fibres which surrounded the broken fibre.

The values of n and k_{1n} depend upon the geometrical arrangement of the fibres in the composite, and k_{1n} also depends upon the behaviour of the matrix. For a square array of fibres n and k_{1n} are, respectively, 4 and 1.146, whereas for the hexagonal case their values are 6 and 1.104, respectively. The quoted values of k_{1n} are applicable to an elastic matrix. For a matrix with low yield stress when plastic effects are important or when the fibre debonding occurs, the value of k_{1n} is reduced.

The quantity M in Eq. (4.3.15) which was referred to as the effective length ratio is actually defined as $M = L/\delta$, where δ is the 'ineffective' length and refers to a region on both sides of the break in which the contribution of the broken fibre to the strength can be assumed to be zero. The value of δ depends upon the strength of the fibres and the shear strength of the inrerface.

In the case of whisker reinforced composites the distribution function $W(\sigma)$ has to be modified to account for the discontinuities. Zweben (1972) has given the following expression for the modified distribution function:

$$W'(\sigma) = \frac{\delta}{a} + \left(1 - \frac{\delta}{a}\right) W(\sigma) \tag{4.3.16}$$

where a is the whisker length. The value of σ_2 in this case can also be obtained from Eq. (4.3.15), with $W(\sigma)$ in that equation replaced by $W'(\sigma)$.

For σ_u, the upper bound for the tensile strength of a composite, Zweben (1972) has suggested the following expression:

$$\sigma_u = (\alpha\beta\delta e)^{-1/\beta} \tag{4.3.17}$$

This expression was derived by Rosen (1965), assuming a Weibull distribution for the fibre strengths. It gives an upper bound for σ, because it does not include the effect of stress concentrations which effectively weaken the composite.

Zweben (1972) has compared the theoretical results with the observed values of stresses at failure for glass, graphite and boron fibres in epoxy matrix and boron fibres in aluminium matrix. The theoretical values are found to be in reasonable agreement with the observed values in all the systems mentioned.

Zweben (1971) has also studied the strength of notched fibre mocposites by analyzing their failure mechanism when subjected to a tensile load along the fibre axis. This paper includes a study of the effects of fibre debonding, matrix and fibre plasticity and scatter in the fibre strength and also a discussion of the applicability of Griffith–Irwin–Orowan theory to the fracture of composites. Very useful formulae for statistical lower bounds on the fracture propagation stress in a composite are given. The tensile strength of notched composites has also been discussed by Beaumont and Phillips (1972a) and Cruse (1973). The failure of glass-fibre-reinforced notched beams in flexure has been discussed by Brown (1973).

The effect of notches and specimen geometry on the pendulum impact strength of uniaxial carbon fibre reinforced composite has been studied by Bader and Ellis (1974). Bullock (1974) has discussed the strength ratios of composite materials in flexure and in tension using the Weibull method.

The gauge length and surface damage effects on the strength distributions of silicon carbide and sapphire filaments has been considered by Kotchick et al. (1975). In this paper the assumption that the Weibull distribution parameters are constant with respect to deviations in gauge lengths has also been analyzed. Some factors affecting the dynamic notch toughness of fibre composites have been discussed by Butcher (1976a, b).

4.4 FRACTURE MECHANICS OF FIBRE COMPOSITES

In this section we shall consider crack propagation in a fibre composite from the fracture mechanical point of view. The basic formalism of fracture mechanics for an ordinary isotropic solid has already been described in Section 2.5. It should be emphasized that the formalism given in that section applies only to isotropic solids. Its rigorous extension to general anisotropic solids is quite complicated.

The complications arise from the tensorial nature of the elastic moduli. The propagation of cracks in anisotropic solids depends upon the direction of its propagation as well as the orientation of the flow, which may change during propagation. In such cases, in addition to the critical intensity factors, the condition of crack propagation must also include the critical orientation of the flow.

Fortunately, considerable simplification in the mathematical analysis of crack propagation can be achieved by exploiting the rotational (point group) symmetry of the elastic tensors. For example, it is possible to write a tensor of the second rank as a diagonal matrix in the representation of its principal axes. For many purposes, then, the tensor can be regarded as a vector with its components equal to the elements of the tensor in the diagonal representation. If all the diagonal elements are equal, the tensor can be regarded as a scalar, which corresponds to the isotropic case.

Such simplifications can also be achieved for tensors of higher rank.

These simplified formulae can then be applied to practical cases, provided that certain conditions involving the relative orientation of various quantities with respect to the symmetry axes are satisfied. These conditions have been discussed by Wu (1967) for the applicability of the isotropic theory of crack propagation in anisotropic solids and are summarised thus: (a) the relative orientation of the flow with respect to a principal axis should not change during propagation; (b) the critical orientation should be along a symmetry axis.

These conditions are generally satisfied by most cases of interest in fibre composites. An additional difficulty, however, arises in the definition of the critical intensity factors or the crack extension force. In Section 2.5 these quantities were defined in terms of the Young's modulus and the Poisson ratio for the solid which can be uniquely defined only for the isotropic case. It is important, therefore, to define the critical intensity factors in such a way that it is consistent with the isotropic case in stress distribution and crack displacement modes. For this purpose it is useful to define an 'effective' modulus in terms of the tensorial components of the elastic moduli.

An approximate form of the Griffith criterion in terms of the effective elastic moduli can be written for anisotropic materials which may be applicable to the fibre composites. Such approaches have been used by several authors, such as, for example, Beaumont and Harris (1972). (See also Sih et al. 1965; Wu, 1967.) The relation between K_I, the stress intensity factor, and G_I, the crack extension force—see Section 2.5 for their definitions—for an orthotropic solid is given below (Beaumont and Harris, 1972):

$$G_{Ic} = K_{Ic}^2\left(\frac{s_{11}s_{22}}{2}\right)^{1/2}\left[\left(\frac{s_{22}}{s_{11}}\right)+\left(\frac{2s_{12}+s_{66}}{2s_{11}}\right)\right]^{1/2} \qquad (4.4.1)$$

where s_{ij} are the elastic compliance constants, which have been defined in Section 2.3. Their relationships with the engineering moduli have also been given in that section. Equation (4.4.1) can easily be generalized to composites with other symmetries.

By analogy with Eq. (2.5.17) we can write Eq. (4.4.1) in the form

$$G_{Ic} = \frac{K_{Ic}^2}{E_{eff}} \qquad (4.4.2)$$

where E_{eff}, the effective elastic moduli of the composite, is given by

$$E_{eff} = \left(\frac{2}{s_{11}s_{22}}\right)^{1/2}\Big/\left[\left(\frac{s_{22}}{s_{11}}\right)+\left(\frac{2s_{12}+s_{66}}{2s_{11}}\right)\right]^{1/2} \qquad (4.4.3)$$

and ν in Eq. (2.5.17) has been neglected.

Thus, we see that the crack extension force in Griffith theory can be defined for anisotropic composites in terms of their effective moduli, which can be calculated if the elastic constants c_{ij} or s_{ij} of the composite are known. The determination of the elastic constants of composites has been discussed in Chapter 3. It is therefore possible to analyze the fracture or crack propagation properties of fibre composites with the help of the Griffith–Irwin–Orowan criterion which has been given in Eq. (2.5.19), provided that the moduli entering the fracture equations are replaced by the effective moduli as given by Eq. (4.4.3). In what follows we shall omit the subscript eff in the moduli for reasons of notational brevity. We shall also omit the subscripts I, II and III referring to plane or anti-plane strain modes unless a distinction between different strain modes is essential for the particular argument.

According to the Griffith–Irwin–Orowan criterion, a crack will extend if the local value of G or K exceeds the critical value (Eq. 2.5.19). The problem is therefore to calculate G_c (or K_c), which is given in Eq. (2.5.21), in terms of the surface energy of the material, including the plasticity contribution, if any. In the case of a fibre composite the relevant surface energy is not found to be simply the sum of the fibre and the matrix surface energies. It is, therefore, obvious that certain other processes which are characteristic of the fibrous nature of the composite must contribute to the absorption of the fracture energy.

The exact nature of various energy-absorbing processes and their contributions to the fracture of composites are matters of some controversy. However, some of the more important processes will be briefly discussed here. For details, reference may be made to, for example, Cooper (1971) and Phillips and Tetelman (1972).

Let us consider a wedge-shaped crack trying to move in a direction normal to the fibres in a composite and visualise the various physical processes which may absorb energy. The fibre at the crack tip will be broken (it may also break at other weak points along its length away from the crack). As the crack moves further and its gap increases at the break point of the fibre, the broken ends of the fibre will be pulled out. The pulling out of the fibre will obviously absorb some energy owing to fibre–matrix frictional forces which contribute to the shear stress at the fibre–matrix interface.

It is of course not essential for a fibre to break in the cracked region. The matrix crack may propagate past a fibre, leaving it unbroken. The unbroken fibre will thus act as a bridge across the crack. As the crack in the matrix tends to expand, it has to do some work against the above-mentioned interfacial friction. Thus, the unbroken fibre across the crack will also contribute to the absorption of energy. This is called the crack bridging process.

If both fibres and the matrix are ductile, we also have to account for their plastic deformation, which contributes to the absorption of energy.

Most high-strength fibres commonly used in modern composites are mainly brittle and their contribution to the plastic deformation energy is usually small. Another process which contributes to the absorption of energy is the fibre stress relaxation. This energy is equal to the stored elastic energy lost from a fibre when it snaps without breaking its bond with the matrix. Finally, the fibre–matrix bond may break, which will also absorb some energy. This process is referred to as the debonding process.

We shall now describe the contribution of these processes to the total fracture energy of a fibre composite.

1. *Fibre pull out*. The energy absorbed in this process has been calculated by Kelly (1964, 1966) and Cottrell (1964) (see also Kelly and Davies, 1965). The energy depends on whether the length a of the fibres is less than or more than the critical length a_c. For $a \leqslant a_c$, it can be assumed that no fibres are broken during the fracture and all pull out, whereas if $a > a_c$, then a fraction a_c/a of the total number of fibres will be pulled out the rest will be broken. (For experimental verification see Kelly and Tyson, 1965a, b.) For these two cases the value of G, the crack extension force, is given below:

$$G = \frac{C_f \sigma_f a^2}{6a_c} \qquad (a < a_c) \tag{4.4.4}$$

and

$$G = \frac{C_f \sigma_f a_c^2}{6a} \qquad (a > a_c) \tag{4.4.5}$$

where C_f is the volume fraction of the fibres and σ_f is the ultimate tensile strength (UTS) of the fibre.

These calculations have been extended by Cooper (1970) to a composite with continuous but defective fibres.

2. *Crack-bridging process*. This process has been studied in detail by Piggott (1970). The contribution of the crack-bridging process to the crack extension force is given below (Phillips and Tetelman, 1972):

$$G = \frac{C_f r_f \sigma_f^3 (1 + \nu_f)(1 - 2\nu_f)}{3\tau_{fm} E_f (1 - \nu_f)} \tag{4.4.6}$$

where r_f is the radius of the fibre, ν_f and E_f are the Poisson ratio and the Young's modulus of the fibre, respectively, and τ_{fm} denotes the shear strength of the fibre–matrix interface.

The contribution of this process increases with the volume fraction of the fibres (C_f) as well as with the fibre size (r_f). However, if the matrix is ductile, the contribution of this process is much smaller than the energy absorbed in the fracturing of matrix bridges during the plastic deformation of the matrix.

3. *Plastic deformation of matrix.* The contribution due to the plastic deformation of the matrix has been calculated by Cooper and Kelly (1967) and is given below:

$$G = \frac{C_m^2}{C_f} \frac{\sigma_m r_f U_m}{\tau} \quad (4.4.7)$$

where U_m is the work required to fracture a unit volume of matrix (ductile) and σ_m is the UTS of the matrix (ductile).

4. *Plastic deformation of fibres.* As mentioned earlier, this process will make a significant contribution only in the case of ductile fibres. In this case Cooper and Kelly (1969) have given the following formula:

$$G = C_f \frac{\sigma_f r_f U_f}{\tau} \quad (4.4.8)$$

where U_f is the work required to fracture a unit volume of the fibres.

5. *Fibre stress relaxation.* The approximate contribution of this process is as follows (Phillips and Tetelman, 1972):

$$G \lesssim \frac{C_f \sigma_f^2 a_c}{3E_f} \quad (4.4.9)$$

6. *Fibre debonding.* The contribution from the fibre debonding process to the fracture energy of the composite has been calculated by Outwater and Carnes (1967) and Outwater and Murphy (1969). They assume a model in which a long single fibre is embedded in a block of matrix and debonding occurs over a certain distance x from the free surface. The contribution of the fibre debonding can be written as follows (Phillips and Tetelman, 1972):

$$G = \frac{C_f \sigma_f^2 y}{2E_f} \quad (4.4.10)$$

where y denotes the debonded length of the fibre.

In the calculation of the contribution of the fibre debonding process as quoted above no account has been taken of the effect of fibre–fibre interaction. It is important to realize that in a composite the fibres inevitably affect their neighbours through the matrix. In other words, it can be said that each fibre exerts a force on its neighbours through the matrix. In a perfect composite at equilibrium (in the absence of any fracture) the net force on each fibre due to all surrounding fibres is zero. If, however, one of the fibres is debonded, the net force on its neighbour will be non-zero and they will tend to move, which requires some energy.

This energy can be called the relaxation energy and must be added to the debonding energy given in Eq. (4.4.10).

An important point in this connection is that, even if the fibre–fibre interaction has a short range, i.e. if the force due to a fibre extends only to its few near neighbours, the relaxation effect is very long-ranged. This is because each relaxing fibre will force its own neighbours to relax, which in turn will affect their neighbours and so on. Thus, the effect of a debonded fibre propagates throughout the composite and each relaxed fibre contributes to the relaxation energy. In other words, the whole composite will respond to the debonding of a single fibre.

The situation is analogous to a vacancy in an ordinary crystal lattice. A vacancy is created by breaking the atomic bonds which displaces all the atoms in the lattice (the displacement of an atom at a distance r from the vacancy in a three-dimensional lattice decreases as $1/r^2$; see, for example, Tewary, 1973). These atomic displacements contribute to the formation energy of the vacancy.

It is obvious from the preceding discussion that the response of the whole composite has to be considered in order to calculate the relaxation energy due to a debonded fibre. This can be done with the help of the Green function method, which has been developed by Tewary (1969), Bullough and Tewary (1972) and Tewary (1973) for normal solids. An extension of this technique to fibre composites and its application to the calculation of the contribution of the fibre–fibre interaction to the debonding energy will be given in Chapter 6.

It may be mentioned that not all of the energy-absorbing processes which have been described earlier in this section contribute at all the stress levels in all systems. The two processes which have been most widely used in the interpretation of the experimental data are the fibre pull-out and the fibre debonding processes. There is some controversy in the literature regarding the relative importance of these two processes, but it appears that the fibre pull-out process is more important for carbon fibre composites whereas the fibre debonding process is more important for boron and glass fibre composites (Phillips and Tetelman, 1972). Further experimental and theoretical work which could resolve this uncertainty should be of great interest. A reliable estimate of fracture energy of fibre composites is very important for their technological applications, since the fracture energy of a material is a measure of its resistance to crack propagation and therefore can be regarded as a measure of its toughness.

Many papers have been published over the past few years dealing with the fracture mechanics of fibre composites. Some of the recent papers will be briefly reviewed here.

Earlier in this section we have commented on the applicability of the Griffith theory to fibre composites. This point has also been examined by Whitney and Kimmel (1972), who have directly measured work to fracture and flow size in carbon fibres. The calculated flow size was compared with the observed values, which are found to be in reasonable agreement.

Waddoups et al. (1971) have discussed macroscopic fracture mechanics of advanced composite materials. They obtained static and fatigue data for a graphite–epoxy laminate with various stress concentrations. They find an apparently paradoxical result that although the material is statically brittle, it shows no propensity for the nucleation and growth of a through crack. The authors have shown that this result together with some others of their observations can be explained with the help of classical fracture mechanics.

Several different experimental as well as theoretical studies of fracture energy, bond strength, toughness, etc. for various systems have been reported in the literature. Lange and Radford (1971) deal with the fracture energy of an epoxy composite system. Owen and Lyness (1972) have investigated bond failure for fibre composites, using the finite element method (see Section 2.6 for a description of the finite element method). The toughness of a glass fibre/epoxy resin composite has been discussed by Harrison (1971). Some theoretical considerations of fibre pull-out from an elastic matrix have been given by Lawrence (1972). A general discussion of the potential of metastable high-strength fibres for strong and tough composites has been given by Gerberich and Zackay (1972).

Plastic flow and fracture in fibre composites have been discussed by Thomason (1972a). The plastic behaviour and the intrinsic strength of carbon fibres have been discussed by Jones and Johnson (1971). Fitz-Randolph et al. (1972) have studied the fracture energy and acoustic emission in the case of a boron/epoxy composite. The acoustic emission produced during burst tests of carbon fibre filament-wound bottles has also been studied by Hamstad and Chiao (1973). A comparative study of tensile fracture mechanisms has been given by Mullin and Mazzio (1972a).

The fracture energy of a glass fibre composite has been discussed by Beaumont and Phillips (1972b). Chan and Patterson (1972) have given theoretical values for the cracking stress of glass-fibre-reinforced inorganic cement. Dibenedetto and Wambach (1972) have studied the fracture of glass/epoxy composites. The fracture energy of carbon-fibre-reinforced glass has been studied by Phillips (1972). Kendall (1972) has examined crack growth resistance in a laminated glass epoxy sheet. Marom and White (1972) have studied the fracture surface energy and the mechanism of transverse fracture in glass fibre composites. The fracture mechanics of fibre–glass laminates has also been discussed by Hamilton and Berg (1973). Owen and Bishop (1973) have discussed the critical intensity factors for glass-reinforced resins.

Another type of crack propagation which results in splitting of fibres in glass fibre/epoxy as well as carbon fibre/epoxy composites has been considered by Harrison (1973). A crack in the plane normal to the axis of the fibres can propagate parallel to the fibres by splitting the fibres provided the strain energy release rate for splitting is equal to that required to create a unit area of the split.

Sih and Chen (1973) have also given a fracture analysis of fibre composites. The fracture toughness of carbon–carbon composites has been examined by Guess and Hoover (1973). The same authors have also discussed the dynamic fracture toughness of carbon–carbon composites. The interlaminar shear strength of a carbon fibre composite under impact conditions has been measured by Sayers and Harris (1973). Chiang and Slepetz (1973) have reported their measurements on cracks in composite materials.

If there are any geometrical disorders in the material, such as bends, sharp corners or holes, then its response to crack propagation will be modified. This occurs as a result of a non-uniform distribution of applied load or stress inside the materials. It is, therefore, important to calculate the stress concentration in structural materials around various specified geometrical perturbations. Such a calculation has been carried out by Kulkarni et al. (1973), who have obtained the load concentration factors for circular holes in composite laminates.

A considerable amount of work has been done on the fracture mechanics of metal composites. Hing and Groves (1972) have discussed the strength and the fracture toughness of polycrystalline magnesium-oxide-containing metallic fibres. The surface energy of fracture in a metal matrix reinforced by continuous fibres has been studied by Thomason (1972b). Garmong and Thompson (1973) have given a theory for the mechanical properties of metal-matrix composites at ultimate loading. The calculations of strain, ultimate strength and the work done to ultimate loading are presented in this paper. The fracture mechanics of a brittle matrix reinforced by ductile fibre composites has been discussed by Tardiff (1973). The multiple yield phenomenon in iron fibre/copper composites has been discussed by Garmong (1972).

The strength of materials is usually quite sensitive to the distribution of defects such as dislocations in the lattice. With this in view, Chawla and Metzger (1972) have studied the initial distribution of dislocations in tungsten fibre/copper composites.

Before closing this section, we shall briefly mention some more recent work. Strelyaev and Sachkovskaya (1974) have discussed determination of shear fracture toughness of fibre glass reinforced plastics using an edge sliding model. The effect of interfaces on the behaviour of in situ metallic composites has been studied by Copley (1974). Marston et al. (1974) have discussed the interfacial fracture energy and the toughness of composites. The strength of fibrous metallic composites as influenced by brittle boundary layers has been considered by Friedrich et al. (1974). The effect of aspect ratio on toughness in composites has been discussed by Piggott (1974). The influence of the interface on the strength and elastic properties of low aspect ratio fibre composites has been considered by Mallick and Broutman (1974). Interfacial bonding and toughness of carbon fibre reinforced glass and glass ceramics has been analyzed by Phillips (1974).

Surface energy analysis of carbon fibres and films has been given by Dynes and Kaelble (1974). Atkins (1975b) has considered large fracture toughness of boron epoxy composites. Kendall (1976) has used the energy balance theory of brittle fracture to study the interfacial cracking of a composite. Morton and Groves (1975) have discussed large work of fracture values in wire reinforced, brittle matrix composites. Energy absorption at high rates of deformation in fibrous composites with non-fracturing reinforcing elements has been analyzed by Millman and Morley (1976). Some model experiments illustrating fibre pull out have been designed by Kendall (1975).

Stochastic finite element simulation of parallel fibre composite has been carried out by Larder and Beadle (1976). In this analysis the fibres are given randomly varying strength properties. Iterative solutions are performed following individual fibre break.

4.5 FATIGUE AND CREEP PROPERTIES OF FIBRE COMPOSITES

In previous sections we have discussed the failure of composites in terms of the critical stress. However, materials are known to fail or deform at much lower stresses with continued usage. Such phenomena are called the fatigue and the creep of the materials, which we shall briefly describe in this section.

Fatigue failure occurs, as the name suggests, when the materials get 'tired' under a stress with the passage of time. Two types of fatigue can be distinguished—static fatigue, which occurs even when the applied stress does not vary with time; and cyclic fatigue, when the material is exposed to an alternating stress. An example of static fatigue is provided by glass, which fractures without any apparent plastic deformation under a relatively small load provided the load has been applied for a long enough period. Cyclic fatigue is commonly observed in structures which are exposed to mechanical oscillations, such as aircraft wings, which are affected by turbulences in air, fuselage stresses due to air compression and decompression, etc. A usually quoted example is the fatigue failure of British Comet planes in the 1950s.

A material can tolerate a much higher stress under static loading than under cyclic loading. The static strength of a material, therefore, is a poor estimate of its strength if it were to be subjected to cyclic loading. This is the reason why dynamic failure tests are essential for materials which have to be used in moving parts of machinery.

The inhomogeneity of materials is mainly responsible for their fatigue, at least under cyclic loading. Each stress cycle produces small irreversible strains. Repeated stressing causes local strain-hardening, which results in the formation of sub-microscopic cracks. Finally, the stress concentration near these cracks causes complete fracture and the failure of the material. Thus, it is non-uniform plastic deformation which leads to fatigue failure.

In the case of fibre composites, most of the commonly used fibres are normally elastic until fairly high stress levels are reached, and the amount of plastic deformation is small. Such fibres should, therefore, be quite resistant to fatigue failure. This, of course, does not apply to the matrix. However, in composites with a high volume fraction of fibres, a sufficient number of fibres may survive to carry the load even after the matrix is locally cracked. Then if the matrix does not have to carry the major part of the load, its properties will be irrelevant so far as the load-carrying capacity of the composite is concerned. This will be the case when a composite is reinforced with well-aligned continuous fibres and is subjected to a tensile load. For such composites the fatigue effect is almost negligible (Owen and Morris, 1970a, b; Beaumont and Harris, 1971). This, however, will not be true if the matrix has to carry any significant load, which will be the case if the fibres are discontinuous or not properly aligned (Lavengood and Gulbransen, 1969; Smith and Owen, 1969a, b). In such cases, therefore, the fatigue effects will be quite important.

Another possibility is that the fatigue cracks in the matrix may also fracture the neighbouring fibres. This is more likely to happen if the fibre–matrix bond is strong, as in the case of metal-matrix composites (Ham and Place, 1966; Morris and Steigerwald, 1967; Bomford, 1968; see also Baker and Cratchley, 1964, and Baker et al., 1966). More details regarding the fatigue behaviour of composites can be found in the following papers: Forsyth et al. (1964); Boller (1964); Courtney and Wulff (1966); Dally and Broutman (1967); Baker (1968) and Colclough and Russel (1972).

A review of the fatigue properties of fibre-reinforced plastics has been given by Hughes and Way (1973). The fatigue behaviour of the matrix in some fibre composites subjected to repeated tensile loads has been examined by Varschavsky (1972). Berg and Salama (1972) have discussed the fatigue of fibre composites when subjected to compression. The cumulative fatigue damage of glass-fibre-reinforced plastics when subjected to repeated tensile impact load has been examined by Fuji et al. (1972). Owen and Morris (1972) have discussed some interlaminar shear fatigue properties of carbon fibre composites. Anderson (1973) has described the fatigue behaviour of a ribbon-reinforced composite. The fatigue failure of boron-fibre–aluminium and carbon-fibre–aluminium composites has been discussed by Baker et al. (1972).

Berg and Salama (1973) have presented some rather interesting results on the 'coaxing' effect in the fatigue of composites. The coaxing effect refers to the increase in the fatigue life of a material at high stress levels after it has been 'prefatigued' at a low level of stress. This effect is not very pronounced in metals but, according to Berg and Salama (1973), is very strong in the case of fibre composites. In some cases Berg and Salama (1973) found that their fatigue life can increase by as much as a factor of 5.

Agarwal and Dally (1975) have described an experimental approach for prediction of low cycle fatigue behaviour of glass fibre reinforced plastic. A method for early fatigue damage detection in composite materials by non-destructive inspection for glass/epoxy and graphite/glass/epoxy composite has been suggested by Nevadunsky et al. (1975). Sumsion (1976) has analyzed environmental effects on graphite epoxy fatigue properties.

Interesting work on the fatigue behaviour of biological composites can also be mentioned here. Freeman et al. (1971) have carried out a study of fatigue fracture in the subchondral bone. They have also measured the strains produced in the subchondral bone plate around the margins of the articular surface of the femoral head subjected to the normal physiological load.

Although the fatigue behaviour of fibre composites is not yet fully understood, in general, it can be said that carbon fibre composites have a good resistance to fatigue failure (see, however, Chapter 5 and Tewary and Bullough, 1971). They are therefore very attractive from this point of view. Their specific fatigue strength—defined as the ratio of the maximum permissible stress to the density—can be up to five times that of steel. Although this figure itself may be misleading owing to the contribution of various other flaws in a realistic application, it does give some idea of the usefulness of carbon fibre composites.

The phenomenon of creep refers to the slow deformation with time of a solid subjected to a constant stress which is less than its yield stress. If the strain of a stressed solid is examined as a function of time, it is found to increase steadily (except for an initial short duration when it may increase quite fast) up to a certain time. This increase of strain is called creep, and the slope of the strain–time curve is called the creep rate. As the deformation increases beyond a certain limit, the effective stress on the solid increases owing to a decrease in its area of cross-section (the so-called necking effect). The increased stress results in a further deformation, which in turn increases the stress even more. Thus, the deformation suddenly accelerates, which leads the solid to rupture.

In an ordinary elastic solid creep occurs as a result of the presence of certain kinds of mobile defects, such as dislocations in the solid and their interaction with other defects. At low temperatures the dislocations are not very mobile and their movement is arrested by the presence of grain boundaries or impurities in the lattice. At high temperatures, however, a dislocation can easily move out of its initial slip plane. Thus, we expect the creep rate to increase with temperature, which is indeed the case. In fact at low enough temperatures the creep rate is almost zero. With the help of a similar qualitative argument, it should be easy to visualize that the creep rate will have a similar dependence on stress—it increases with increasing stress. Thus, the creep rate and the creep-induced rupture are determined by the applied stress as well as the temperature of the solid.

Mathematically, the creep rate can be written in the form of a differential equation, as follows:

$$\frac{d\varepsilon}{dt} = F(\sigma, T) \tag{4.5.1}$$

where ε is the strain; t is the time variable; and $F(\sigma, T)$ is a function of the stress σ and temperature T, which for intermediate values of ε or t is independent of ε as well as t. For a constant value of T, it is usually possible to take $F(\sigma, T)$ to have a power dependence on σ, viz.

$$F(\sigma, T) = A\,\sigma^p \tag{4.5.2}$$

where the exponent p and the constant A are a measure of the resistance of the material to creep. In general, both A and p will be a function of T.

The discussion given above applies to an ordinary single-phase solid. In the case of a composite, F will be a function of the stresses produced in all the components, since the net creep resistance of the composite will depend on the creep resistance of all the components. It should be physically obvious that if a composite is made up of two components with different creep resistances, the creep of the low-resistance component will be checked by the high-resistance material owing to adhesion between them. This tendency will obviously increase with the strength of the adhesive bond between the two components. The creep resistance of the composite should therefore be, in general, greater than that of its components.

High creep resistance is a major advantage of composite materials over ordinary solids. It is thus possible to fabricate materials with very useful high-temperature properties by choosing the fibres of strong creep resistant materials which themselves may be otherwise unsuitable for structural applications owing to some other defects, such as brittleness or poor oxidation-resistance.

The creep properties of composites have been studied by several authors, including Jech and Weber (1959), Cratchley (1965), Kellv and Tyson (1966), Kreider and Leverant (1966), Ellison and Harris (1966), McDanels et al. (1967). Dean (1967), de Silva (1968), Wilcox and Clauer (1969) and Mileiko (1970). More recently, the mechanism of creep in glass-reinforced plastics has been studied by Diggwa and Norman (1972). The rupture of nickel-based reinforced composites has been discussed by Karpinos and Bespyatyi (1971).

Tunik and Tomashevskii (1974) have discussed creep and long time strength of oriented glass reinforced plastics in interlaminar shear. Tensile creep of a unidirectional glass fibre-epoxy laminate has been studied by Weidmann and Ogorkiewicz (1974). Auzukains et al. (1974) have analyzed deformation and strength properties of carbon reinforced plastics in compression. In this paper the elastic constants, fracture and creep up to 3 months have been studied. Creep strength of discontinuous fibre composite has also been discussed by Bocker-Pedersen (1974).

A lot of work—theoretical as well as experimental–still remains to be done on the creep properties of fibre composites. At present the general inference seems to be that the dominating contribution to the creep resistance of a composite with continuous fibres comes from the fibres, whereas the matrix makes a significant contribution in the case of a composite with discontinuous fibres.

4.6 POSSIBILITY OF DUCTILE FIBRE COMPOSITES

A major problem in the application of fibre composites arises from the fact that, unlike most metals, they are brittle. In the language of fracture mechanics, their fracture energy is relatively small. A solution of this problem using what is called a duplex fibre has been suggested by Morley (1970, 1971). The effect of the duplex fibres is to introduce multiple and variable shear strength interfaces between the fibres and the matrix which would increase the work of fracture and, hence, the toughness of the composites.

The problem of brittleness in an ordinary fibre composite arises from the conflicting requirements on the fibre–matrix bond. To see this, first let us consider the critical length (also called the transfer length), which was introduced in Section 4.3. It is given by

$$a_c = \frac{r_f \sigma_{uf}}{\tau} \tag{4.6.1}$$

where r_f is the radius of the fibre, σ_{uf} is its ultimate tensile strength and τ

is the shear strength of the interface. The quantity a_c is a measure of the ineffective length of the fibre, since it is the length of the fibre over which the stress transfer takes place and is therefore relatively ineffective in reinforcing the composite.

We note from Eq. (4.6.1) that a_c is inversely proportional to τ. Thus, to increase the efficiency of the reinforcemet, we should have τ as large as possible. A large τ requires a strong fibre–matrix bond, which, as we have seen in the previous section, is also desirable for better creep resistance. However, if the fibre–matrix bond is made stronger, the interaction between neighbouring fibres also increases. If the fibre–fibre interaction is large, then the fibres will not fail independently of each other—failure of one fibre will trigger the failure of the whole composite.

Thus, we see that a high fibre–matrix bond strength is required for better reinforcement efficiency of the composite, whereas a weak fibre–matrix bond is desirable for a better transverse crack resistance of the composite. In ordinary composites a compromise is reached by taking an intermediate value of τ.

The solution of this problem, according to Morley, is to use a duplex fibre, which introduces two interfaces. A duplex fibre consists of a core and an outer sheath. The outer sheath should be strongly bonded to the matrix whereas the bond between the core and the outer sheath of the fibre should be weak. The strong outer bond will give a high shear strength and thus increase the reinforcement efficiency according to Eq. (4.6.1), whereas the crack propagation across the fibres will be checked by the core of the fibre, since it will easily debond from the sheath owing to its low bonding strength.

In order to increase the work of fracture, it is desirable that the inner core should pull out rather than break during crack propagation. This can be done by choosing a bond with decreasing shear strength as the tensile strain in the fibre increases. This is accomplished in Morley's model fibres which consist of tubes with kinked wires instead of straight wires in the core (Livesey, 1971). When such a system is subjected to a strain, the inner element contracts away from the outer tube, which reduces the shear strength of the bond between the core and the sheath. Thus, the inner element will have the tendency to pull out of the sheath without breaking, which will increase the work of fracture and the toughness of the composite. Some practical suggestions on how to make such complex fibres have also been given by Morley.

Although the practical feasibility of duplex fibres is yet to be fully demonstrated, in principle they seem to be very interesting. So far as the general elastic response of a duplex fibre composite is concerned, it would be quite similar to a metal. They will have an initial elastic region corresponding to stage I (see Section 2.3), in which the stress and strain follow Hooke's law. Stage II will start after the failure of the outer sheath of the fibres, because then the deformation of the composite, caused by the pulling out of their inner elements, will take place at an approximately constant stress. This will continue until the composite finally fails in stage III. Such a composite will therefore have all the useful features of fibre reinforcement as well as the ductility of metals.

Some more recent work on duplex fibres has been reported by Chappell and Millman (1974) who have described fabrication of ceramic coated carbon fibre duplex elements. The characteristics of duplex reinforcing elements with non-fracturing cores and simple theoretical analysis of core sheath debonding have been given by Chappell et al. (1975).

CHAPTER 5

Propagation of Elastic Waves in Fibre Composites

5.1 INTRODUCTION

In Chapters 3 and 4 of this book we have discussed the static elastic properties of fibre composites. In this and the following chapter we shall consider a dynamic elastic property of fibre composites—namely, the propagation of elastic waves. The dynamic models which describe the wave propagation in a solid can also be used for the calculation of various thermodynamic functions of the solid, such as specific heat, free energy, etc. (see, for example, Kittel, 1976, for such calculations for ordinary solids). In this book, however, our interest is only in the mechanics of fibre composites. No attempt will, therefore, be made to discuss the calculation of thermodynamic functions of the composites.

Wave propagation through matter provides a powerful method for a study of several properties of matter which may be inaccessible to other probes. The power of this method arises from the fact that the parameters which describe the chief characteristics of wave propagation—its direction, velocity and frequency—can easily be measured. These parameters, in general, change when the wave passes through matter on account of their interaction. Thus, a measurement of the change in the wave parameters can yield useful information about the properties of matter.

A large number of examples of the usefulness of wave propagation techniques can be found in the physical and chemical sciences. For example, the change in the wavelength of light on scattering from molecules gives the energy levels of the molecule (Raman effect). Neutron and X-ray scattering experiments are now widely used for a determination of various properties of solids and liquids. The scattering of γ-rays from the atomic nuclei is used to study the nature of nuclear forces.

The techniques quoted above are the examples of the use of electromagnetic waves. The mechanical properties of a solid can be studied by using ultrasonic beams, which are mechanical or elastic waves. The velocity of an ultrasonic wave in a solid depends on the elastic constants of the solid. It also depends on various types of imperfections and inhomogeneities in the solid. Thus, it is possible to study the mechanical properties of a perfect solid as well as an imperfect solid with the help of ultrasonics.

In this chapter we shall discuss the propagation of elastic waves in 'perfect' fibre composites and show how useful information can be derived regarding the strength of the fibre-matrix bond. In the following chapter we shall discuss the effect of imperfections in the composites on the propagation of elastic waves. Since the passage of an ordinary elastic wave causes no damage in the solid, these results are useful for the development of non-destructive testing techniques for fibre composites.

Two limiting cases concerning wave* propagation in fibre composites can be distinguished at the outset—long waves and short waves, according to their wavelengths, being much longer than or comparable to the inter-fibre spacing. It should be intuitively obvious that the propagation of long waves will not be very sensitive to those features of the composite which arise from the discrete arrangements of the fibres. Thus, the propagation of long waves in a composite can be studied on the basis of the continuum model of a solid which has been described in Section 2.3. The propagation of long waves will of course depend on the elastic symmetry of the composite. They can, therefore, be used to study the alignment of fibres in a composite. This will be discussed in the next chapter.

Those properties of a composite which can be directly attributed to the discrete arrangement of the fibres on a plane normal to their axis can be studied with the help of short waves. The propagation of short waves in a fibre composite is 'dispersive', in the same sense as the propagation of electromagnetic waves (neutron beams, X-rays) in ordinary solids. We shall see that the dispersion will be most sensitive to the strength of the fibre–matrix bond when the wavelength of the wave is equal to the interfibre spacing.

5.2 PROPAGATION OF LONG WAVES IN FIBRE COMPOSITES: CONTINUUM MODEL

In this section we shall discuss the propagation of long waves in a composite on the basis of the continuum model. This problem has been discussed in many standard texts on elastic wave propagation (see, for example, Fedorov, 1968). Only a brief review will be given here.

Consider a wave passing through a solid. Each volume element will be subjected to a time-dependent stress and strain caused by the passage of the wave.

Let $\sigma(t)$ and $\varepsilon(t)$ denote, respectively, the stress and the strain in a particular volume element at time t. Let $\mathbf{r}$ denote the radius vector of the volume element and $\mathbf{F}(t)$ and $\mathbf{u}(t)$ denote, respectively, the force per unit volume and the displacement field in that element at time t.

With the help of Eqs. (2.3.3) and (2.3.4) we can write

*Since our main concern in this book is only with elastic waves, we shall henceforth omit the prefix 'elastic'. In what follows, therefore, unless otherwise stated, the word 'wave' will refer to an elastic wave.

$$F_i = -\sum_j \frac{\partial \sigma_{ij}}{\partial r_j}$$

$$= -\sum_{jmn} c_{ijmn} \frac{\partial \varepsilon_{mn}}{\partial r_j} \qquad (5.2.1)$$

where i, j, m, n, etc. denote the Cartesian components.

Substituting for ε from Eq. (2.3.2) in Eq. (5.2.1) we obtain

$$F_i = -\sum_{jmn} c_{ijmn} \frac{\partial^2 u_m}{\partial r_n \, \partial r_j} \qquad (5.2.2)$$

On the other hand, since $\mathbf{u}$ is time-dependent, the force per unit volume exerted on the volume element due to the variation in $\mathbf{u}$ is equal to $-\rho\, d^2u/dt^2$, where ρ is the density of the solid, which has been assumed to be uniform. Equating this value of the force to that given in Eq. (5.2.2), we obtain the following equation of motion:

$$\rho \frac{d^2 u_i}{dt^2} = \sum_{jmn} c_{ijmn} \frac{\partial^2 u_m}{\partial r_n \, \partial r_j} \qquad (5.2.3)$$

We shall seek the solution of Eq. (5.2.3) in the form of plane waves:

$$u_i = e_i \exp i(\mathbf{k} \cdot \mathbf{r} + \omega t) \qquad (5.2.4)$$

where ω denotes the frequency of the wave, $\mathbf{k}$ its wave vector, i.e.

$$\mathbf{k} = \frac{2\pi}{\lambda} \qquad (5.2.5)$$

and e_i, which, in general, will be a function of $\mathbf{k}$, denotes ith component of the polarization of the wave.

If we substitute $\mathbf{u}$ from Eq. (5.2.4) in Eq. (5.2.3) we obtain a set of three simultaneous linear equations in e_i. The condition that a non-trivial solution exists yields the following secular equation:

$$\begin{vmatrix} \Lambda_{11} - \omega^2 & \Lambda_{12} & \Lambda_{13} \\ \Lambda_{21} & \Lambda_{22} - \omega^2 & \Lambda_{23} \\ \Lambda_{31} & \Lambda_{32} & \Lambda_{33} - \omega^2 \end{vmatrix} = 0 \qquad (5.2.6)$$

where $|\ \ |$ denotes the determinant of a matrix and

$$\Lambda_{ij} = \frac{1}{\rho} \sum_{mn} c_{imjn} k_m k_n \qquad (5.2.7)$$

The 3 × 3 matrix Λ whose elements have been defined in Eq. (5.2.7) is the Green–Christoffel matrix, which was introduced in Section 2.3. (It differs from that defined in Eq. 2.3.6 only through the factor $4\pi^2$, which has been absorbed here in $\mathbf{k}$ for notational convenience.)

A knowledge of Λ is sufficient to determine most of the properties of wave propagation in a solid. Its eigenvalues, $\omega^2(\mathbf{k})$, gives the frequencies of the wave and its eigenvectors, $e_i(\mathbf{k})$, define the polarization of the wave characterized by its wave vector $\mathbf{k}$. The phase and the group velocities of the wave, which will be denoted by C_p and C_g, respectively, can be obtained from $\omega(\mathbf{k})$ as follows:

$$C_p = \frac{\omega(\mathbf{k})}{\mathbf{k}} \tag{5.2.8}$$

and

$$C_g = \nabla\omega(\mathbf{k}) \tag{5.2.9}$$

where ∇ denotes the gradient in the $\mathbf{k}$ space, viz.

$$\nabla_i = \frac{d}{dk_i}$$

It may be noted that Λ is a symmetric matrix ($\Lambda_{ij} = \Lambda_{ji}$) and therefore all its eigenvalues are real. The condition that all the eigenvalues are positive, which is the condition of the stability of the solid, leads to the following well-known stability conditions for the elastic constants.

Trigonal crystals:

$$c_{11} > c_{12}$$

$$(c_{11} + c_{12})\, c_{33} > 2c_{13}^2$$

$$(c_{11} - c_{12})\, c_{44} > 2c_{14}^2$$

Tetragonal and hexagonal crystals:

$$c_{11} > c_{12} \qquad (c_{11} + c_{12})\, c_{33} > 2c_{13}^2$$

$$c_{44} > 0, \qquad c_{66} > 0$$

Cubic crystals:

$$c_{11} > c_{12}, \qquad c_{11} + 2c_{12} > 0$$

$$c_{44} > 0$$

We may also note that Λ and therefore $\omega^2(\mathbf{k})$ are quadratic functions of k. We infer, therefore, from Eqs. (5.2.8) and (5.2.9) that C_p and C_g are independent of k, the magnitude of the wave vector. Thus, the wave propagation in the present case is non-dispersive.

The secular equation (5.2.6) is a cubic equation in ω^2. Its three roots, ω_1^2, ω_2^2 and ω_3^2, which are all real provided $\Lambda_{ij} = \Lambda_{ji}$, are given below:

$$\omega_1^2 = -(S_1 + S_2) + \frac{P_2}{3} \tag{5.2.10}$$

$$\omega_2^2 = \tfrac{1}{2}(S_1 + S_2) + \frac{P_2}{3} + i\frac{\sqrt{3}}{2}(S_1 - S_2) \tag{5.2.10a}$$

$$\omega_3^2 = \tfrac{1}{2}(S_1 + S_2) + \frac{P_2}{3} - i\frac{\sqrt{3}}{2}(S_1 - S_2) \tag{5.2.10b}$$

where

$$S_1 = [R + (Q^3 + R^2)^{1/2}]^{1/3}$$

$$S_2 = [R - (Q^3 + R^2)^{1/2}]^{1/3}$$

$$Q = \frac{P_1}{3} - \frac{P_2^2}{9}$$

$$R = \frac{P_1 P_2 - 3P_0}{6} - \frac{P_2^3}{27}$$

$$P_0 = -\Lambda_{11}\Lambda_{22}\Lambda_{33} + \Lambda_{11}\Lambda_{23}^2 + \Lambda_{22}\Lambda_{13}^2 + \Lambda_{33}\Lambda_{12}^2 - 2\Lambda_{12}\Lambda_{23}\Lambda_{31}$$

$$P_1 = \Lambda_{11}\Lambda_{22} + \Lambda_{22}\Lambda_{33} + \Lambda_{33}\Lambda_{11} - \Lambda_{12}^2 - \Lambda_{23}^2 - \Lambda_{31}^2$$

and

$$P_2 = \Lambda_{11} + \Lambda_{22} + \Lambda_{33}$$

The solution of Eq. (5.2.6) is considerably simplified for isotropic crystals and those with high symmetry when **k** is in a symmetry direction. Its solution in the case of cubic isotropy is of no interest in the present case, since fibre composites are highly anisotropic. However, as discussed in Chapter 3, fibre composites can be reasonably well modelled by a solid having hexagonal symmetry which has transverse isotropy. In this case also the solution of Eq. (5.2.6) can be simplified and will be described below.

The number of independent elastic constants for a hexagonal crystal is five—c_{11}, c_{12}, c_{13}, c_{33}, c_{44} and c_{66}—with the relation

$$2c_{66} = c_{11} - c_{12} \tag{5.2.11}$$

The elements of the Green–Christoffel matrix (defined in Eq. 5.2.7) in the present case are given below (Fedorov, 1968):

$$\Lambda_{11} = c_{11}k_1^2 + c_{66}k_2^2 + c_{44}k_3^2$$

$$\Lambda_{22} = c_{66}k_1^2 + c_{11}k_2^2 + c_{44}k_3^2$$

$$\Lambda_{33} = c_{44}k_1^2 + c_{44}k_2^2 + c_{33}k_3^2$$

$$\Lambda_{32} = \Lambda_{23} = (c_{13} + c_{44})\, k_2 k_3$$

$$\Lambda_{13} = \Lambda_{31} = (c_{13} + c_{44})\, k_1 k_3$$

$$\Lambda_{12} = \Lambda_{21} = (c_{11} - c_{66})\, k_1 k_2 \tag{5.2.12}$$

Our object is to determine the eigenvalues, ω^2, of the matrix $\boldsymbol{\Lambda}$. In order to diagonalize $\boldsymbol{\Lambda}$, we look for a transformation $\mathbf{S}$, where $\mathbf{S}$ is an orthogonal matrix, so that the matrix $\boldsymbol{\Lambda}'$ is defined as follows:

$$\boldsymbol{\Lambda}' = \mathbf{S}^{-1}\boldsymbol{\Lambda}\mathbf{S}$$

is diagonal or at least block diagonal.

It can be easily verified that if we choose

$$\mathbf{S} \equiv \begin{pmatrix} \dfrac{k_2}{(k_1^2 + k_2^2)^{1/2}} & \dfrac{k_1}{(k_1^2 + k_2^2)^{1/2}} & 0 \\ -\dfrac{k_1}{(k_1^2 + k_2^2)^{1/2}} & \dfrac{k_2}{(k_1^2 + k_2^2)^{1/2}} & 0 \\ 0 & 0 & 1 \end{pmatrix} \tag{5.2.13}$$

then $\boldsymbol{\Lambda}$ is block diagonalized—the matrix $\boldsymbol{\Lambda}'$ consists of a 1×1 and a 2×2 matrix along its diagonal. The structure of $\boldsymbol{\Lambda}'$ is given below:

$$\boldsymbol{\Lambda}' = \begin{pmatrix} c_{66}(k_1^2 + k_2^2) + c_{44}k_3^2 & 0 & 0 \\ 0 & c_{11}(k_1^2 + k_2^2) + c_{44}k_3^2 & ck_3(k_1^2 + k_2^2)^{1/2} \\ 0 & ck_3(k_1^2 + k_2^2)^{1/2} & c_{44}(k_1^2 + k_2^2) + c_{33}k_3^2 \end{pmatrix} \tag{5.2.14}$$

where $c = c_{13} + c_{44}$. One of the eigenvalues is thus immediately obtained. It is given by

$$\omega_1^2 = c_{66}(k_1^2 + k_2^2) + c_{44}k_3^2 \tag{5.2.15}$$

The eigenvector corresponding to this eigenvalue, as can be seen from Eq. (5.2.13), is given by*

$$\mathbf{e} = \frac{k_2}{(k_1^2 + k_2^2)^{1/2}},\ -\frac{k_1}{(k_1^2 + k_2^2)^{1/2}},\ 0 \tag{5.2.16}$$

Since the direction of the wave propagation is given by the vector $\mathbf{k}$, where

$$\mathbf{k} = k_1, k_2, k_3 \tag{5.2.17}$$

*We shall use the usual notation for a vector and denote any vector $\mathbf{a}$ in terms of its components $\mathbf{a}_1$, $\mathbf{a}_2$ and $\mathbf{a}_3$ as follows:

$$\mathbf{a} = \mathbf{a}_1, \mathbf{a}_2, \mathbf{a}_3.$$

we find that

$$\mathbf{e} \cdot \mathbf{k} = 0$$

Thus, we see that this polarization of the wave is always normal to its direction of propagation, i.e. this is a pure transverse or shear wave. It is an important property of the hexagonal crystal that for any direction of wave propagation at least one of the waves is always transversely polarized. It arises from the transverse isotropy of the crystal. In an anisotropic crystal the waves are neither purely transverse nor purely longitudinal unless **k** lies in a symmetry direction.

The phase velocity of the wave of frequency ω_1 as given by Eq. (5.2.15) can be obtained from Eq. (5.2.8) and is given below:

$$C_{P_1}^2 = c_{66} \sin^2 \theta + c_{44} \cos^2 \theta \tag{5.2.18}$$

where the angular coordinate θ is defined by

$$k_3^2 = k^2 \cos^2 \theta \tag{5.2.19}$$

and

$$k_1^2 + k_2^2 = k^2 \sin^2 \theta$$

since

$$k^2 = k_1^2 + k_2^2 + k_3^2$$

The frequencies and the velocities of the other two waves can be obtained by diagonalizing the 2×2 block matrix in Eq. (5.2.14). The two phase velocities will be denoted by C_{P_2} and C_{P_3} and are given below;

$$C_{P_2}^2 = \tfrac{1}{2}(A + B) + \tfrac{1}{2}[(A - B)^2 + P^2]^{1/2} \tag{5.2.20}$$

and

$$C_{P_3}^2 = \tfrac{1}{2}(A + B) - \tfrac{1}{2}[A - B)^2 + P^2]^{1/2} \tag{5.2.20a}$$

where

$$A = c_{11} \sin^2 \theta + c_{44} \cos^2 \theta$$

$$B = c_{44} \sin^2 \theta + c_{33} \cos^2 \theta$$

and

$$P = (c_{13} + c_{44}) \sin 2\theta$$

It can be verified that the angle between the direction of wave propagation and its polarization vector corresponding to either C_{P_3} or C_{P_2} depends on **k**. These two waves are therefore neither purely transverse nor purely longitudinal unless **k** lies in a symmetry direction.

The wave propagation in anisotropic solids on the basis of the continuum model has been discussed by Musgrave (1954a, b; 1970). For reasons explained in Section 5.1, the formulae obtained on the basis of the continuum model can be applied to wave propagation provided that the wavelength is much larger than the interfibre spacing. The continuum model formulae have been used by Curtis (1969), Reynolds (1969), Markham (1970) and Reynolds and Hancox (1970) for analyzing their experimental results on wave propagation in carbon fibre composites.

The formalism given in this section for wave propagation in the continuum model is essentially a reciprocal space (**k** space) formalism. It does not account for the fibrous nature of the solid, which has been assumed to have been smeared out. A real space calculation of wave propagation in the long-wavelength limit for the purpose of determining the average 'effective' elastic constants has been given by Behrens (1967a) for lamellar composites and by Behrens (1967b) for fibre composites with rectangular symmetry (see Behrens and Kremheller, 1969, for a brief review).

So far in this section we have not discussed the non-linear effects in wave propagation. The non-linear effects can be neglected if the amplitude of the wave is not too large. This is a reasonable approximation in most cases, at least at low temperature. However, in certain cases, as, for example, in the case of shock waves, the amplitude may be very large. Even for ordinary ultrasonic waves, the amplitude may be quite large near a defective fibre in the composite. In such cases the non-linear effects may be quite significant.

In a non-linear medium the waves become attenuated. No attenuation is possible in a linear medium. The attenuation of stress waves can also be easily measured and can be used to derive some information about imperfections in the composite. In this book we shall mainly restrict ourselves to the linear theory. However, a brief description of some work on wave propagation in composites in which the non-linear effects have been included is given below.

A theoretical model for stress wave propagation in plate laminate composites which includes the non-linear effects has been given by Barker (1971). The dispersive effects in the model have been included by establishing an analogy with some viscosity effects. The model has been used to calculate the attenuation of short stress pulses and the steady and transient wave shapes and velocities.

The propagation of stress waves in a laminated plate composite has also been studied—theoretically as well as experimentally—by Lundergan and Drumheller (1971). They have considered dilatational stress waves in a composite containing 10 plates oriented normal to the direction of the wave propagation. A good agreement between the theoretical and the experimental results was obtained. In these calculations the non-linear effects were neglected, which, according to the authors, should be included along with the effect of the debonding of the layers.

The propagation of stress waves in woven fabric composites has been considered by Tauchert (1971). The attenuation of stress pulses in cloth laminate quartz phenolic has been examined by Reed and Munson (1972) in an experimental study of propagation of stress pulses of short duration.

A study of wave surfaces due to impact on anisotropic fibre composite plates has been reported by Moon (1972). The theory of the propagation of 'weak' waves in elastic–plastic solids as given by Janssen et al. (1972) may also be relevant to fibre composites. The term 'weak waves' refers to the wave fronts travelling in continuous media, for which the derivatives of the dependent variables of stress, strain and velocities may be discontinuous.

A theoretical treatment of the response of several composites to the shock waves has been given by Davis and Wu (1972). The steady shock waves in composites have been experimentally studied by Holmes and Tsou (1972). In their experiment on epoxy reinforced by aluminium wires the shock waves were generated by a planar impact. The shock waves and the free surface velocities were measured by an optical technique. The shock front was found to be quite steady in the sense that its velocity and thickness did not change with time. Reasonable agreement with the theoretical results was obtained.

5.3 WAVE PROPAGATION IN FIBRE COMPOSITES: DISCRETE LATTICE MODEL*

In the previous section we discussed wave propagation on the basis of the continuum model of a composite. That treatment, as explained earlier, is valid for long waves. In order to study the propagation of short waves in a fibre composite, we have to take its fibrous structure into account. This can be done with the help of the Born–von Kármán model of a solid, which has been described in Section 2.2.

In the present chapter we shall present a model which is essentially an adaptation of the Born–von Kármán model for a crystal lattice which has been extremely successful in the field of solid state physics (Born and Huang, 1954; Maradudin et al., 1971). The main difference between the present model and the Born–von Kármán model of a crystal lattice is that the present model is only semi-discrete, to allow for the characteristic feature of the composite. On a plane perpendicular to the fibres the present model behaves like a discrete crystal lattice, whereas it behaves like a Debye elastic continuum in a direction parallel to the fibres. This is physically reasonable for a composite, because the only non-continuum effects in a direction parallel to the fibres are those arising from the atomistic structure of the fibres for which a sound wave is too long. The present model for a composite is 'dispersive' for short waves in the same sense as the Born–von Kármán model for electromagnetic waves (neutron beams, X-rays). The limiting case of long waves is obtained by using

*The material presented in Sections 5.3–5.11 and 6.2 has been taken from Tewary and Bullough (1972).

Born's method of long waves (Born and Huang, 1954), when the formulae obtained for the present model reduce to those obtained by using the Green–Christoffel equations for an elastic continuum (Section 5.2).

Two-dimensional translation symmetry has been assumed for the arrangement of the fibres in a plane perpendicular to the fibres, an assumption which does not seem to be too unrealistic for a densely reinforced composite. This assumption has also been made in most of the previously quoted papers. In our model the fibres act as the main 'building blocks' of the composite. The matrix acts as a carrier of interactions between the fibres, i.e. the fibres 'see' one another through the matrix. The interaction between the fibres is assumed to travel through the matrix instantaneously. This assumption is equivalent to the adiabatic approximation in the Born–von Kármán model. Thus, the effect of the matrix is included in the model in terms of Born–von Kármán-type force constants between the fibres and also in the definition of an effective mass for the fibres. It may be mentioned here that the assumption of 'adiabatic approximation' may be of questionable validity for a composite. This would be a major problem if the force constants between the fibres were to be calculated from first principles. However, we shall consider the force constants only as parameters to be determined in a phenomenological fashion. It is therefore not considered worthwhile to bother about the finer details of the effective interaction between the fibres. Finally, the amplitude of the wave is assumed to be sufficiently small for the harmonic approximation equivalent to the linear theory of elasticity to be valid.

In this chapter composites with two different symmetries have been studied—tetragonal and hexagonal, which correspond to an arrangement of fibres on the vertices of a square and a hexagon, respectively, on a plane perpendicular to the fibres. Numerical results have been obtained for a carbon fibre–epoxy resin composite. No experimental measurements are available for any composite using short waves. It is, therefore, not possible at the moment to compare the theoretical results obtained in this paper with any experimental results.

In the present model the force constants serve as a measure of the strength of the fibre–matrix bond. In the framework of this model it is possible to calculate the effect of bonding defects on the velocities of wave propagation. This will be discussed in the following chapter. The present theory predicts an interesting phenomenon—that of reflection of elastic waves of a particular wavelength from the composite, which is analogous to the Bragg reflection of X-rays from crystals. The theory also predicts some very interesting collimating and polarizing properties of the composite for elastic waves which may be important from an application point of view.

For the theoretical treatment given in this paper we shall first assume a perfect composite. A perfect composite has been defined in Chapter 3.

To recapitulate, a perfect composite is defined as one in which all the fibres are of equal mass and length, extend throughout the length of the composite, have equally strong bonding with the matrix, are perfectly aligned and intersect a plane perpendicular to their length in a regular periodic pattern. Note that for the treatment given here it is not essential to assume that the bonding between the fibres and the matrix is perfect; the only assumption is that the bonding strength is equal for all fibres. However, if the bonding strength is equal for all fibres, it is likely that the bonding is perfect or at least the imperfections in the bonding are not random. We shall also assume that such a composite behaves like a solid with well-defined effective bulk properties, such as the elastic constants, density, etc. The fibres are treated as structureless entities which do not absorb elastic energy. It should be emphasized that none of these assumptions are particularly drastic. They are standard assumptions which have been made in most of the theoretical work published so far. Henceforth in this chapter, unless otherwise stated, we shall consider only a perfect composite.

Since all the fibres have been assumed to be well aligned, the composite will behave like a discrete lattice in a plane perpendicular to the fibres and like an elastic continuum in the direction parallel to the fibres. If the z axis is taken along the length of a fibre and the x and y axes in a plane perpendicular to it, then k_z will be continuously distributed from zero up to a certain maximum value, whereas k_x and k_y will be allowed only certain discrete values which will be distributed in the appropriate Brillouin zone. Thus, the final picture of the composite which we have proposed is that it behaves like a Born–von Kármán solid in the xy plane and like a Debye solid in the z direction. In a general direction it will exhibit mixed behaviour. This implies that an elastic wave travelling through the composite will suffer dispersion in most of the directions unless, of course, the wavelength is much larger than the fibre spacing. Only the wave travelling in the z direction will be free from dispersion. The dispersion effects will be most prominent for a wave travelling in the xy plane.

In order to apply the Born–von Kármán method to the present model of a composite, it is essential to define the force constants between the fibres or the change in the potential energy of the composite due to the fibres' displacements. The concept of potential energy itself at equilibrium may be a bit dubious, because, unlike the atoms in a crystal, there is no direct interaction between the fibres. However, the force constants between the fibres and therefore the change in the potential energy of the composite during an elastic disturbance can be defined by noting that a fibre can feel the presence of other fibres through the matrix to which all the fibres are bound. If a fibre were displaced from its original position, it would stretch or compress the matrix in its surrounding region, thus exerting a force on neighbouring fibres. This is precisely the force constant, being the force on a fibre when another fibre is displaced by a unit amount (see Section 2.2). We shall assume that the matrix is stretched or compressed only locally and that the disturbance is transmitted through the composite in a series of local disturbances. Thus, we can restrict the force constants to a few near neighbours. In practice, one may have to include the effect of several neighbours, because the elastic disturbances in the matrix may extend quite far. However, the number of neighbours to be included can only be decided by comparing the results with the experiments.

We have thus been able to define an effective potential energy for the composite during an elastic disturbance. An effective potential between the fibres must exist simply because of the fact that the fibres form a bound assembly in a composite. We are treating the matrix as acting like the Born–von Kármán springs connecting the fibres with an effective mass. The effect of the matrix in the present model of a composite is thus included in terms of the force constants and the effective mass of the fibres. The force constants are obviously a measure of the bonding strength or adhesion between the fibres and the matrix. The effective mass of the fibres will be defined later in terms of the density of the composite.

If the force constants between the fibres were known, we could construct the matrix **D (k)**, as given in Section 2.2, which would completely determine the propagation of an elastic wave through the composite. In the long-wavelenght limit the matrix **D (k)**, reduces to the Green–Christoffel matrix for an elastic continuum. This correspondence enables us to express the effective elastic constants for the composite in terms of the force constants between the fibres. Alternatively, one could obtain the force constants with the help of these relations if the elastic constants were known. In general, however, this is not possible, because there are more force constants than the elastic constants.

Since it is difficult to calculate the force constants from first principles, the usual procedure in the study of the lattice dynamics of crystals is to treat the force constants as parameters and obtain their values in a phenomenological way by comparing the theoretical values with the observed dispersion data in symmetry directions. The same procedure could be used for the composites. Unfortunately, we do not have any experimental data on dispersion from composites for short waves; in most of the published experimental work only long waves have been used, which do not exhibit dispersion, apart from that due to boundary and surface effects. At present, therefore, we have to determine the force constants in terms of the calculated values of the elastic constants. Since the elastic constants are fewer in number, we have to choose arbitrary values for some of the force constants.

As mentioned earlier, two different symmetries for the configuration of fibres in a composite have been assumed—namely, tetragonal and hexagonal. This means that if a plane is cut perpendicular to the fibres, the points of intersection between the fibres and the plane will form a two-dimensional square and a two-dimensional hexagonal lattice, respectively, for the two symmetries (Figures 5.1 and 5.2). The wave motion along this plane will show characteristics of the respective geometry of the lattice. For a wave travelling in the direction of the fibres both the composites will behave like a Debye solid.

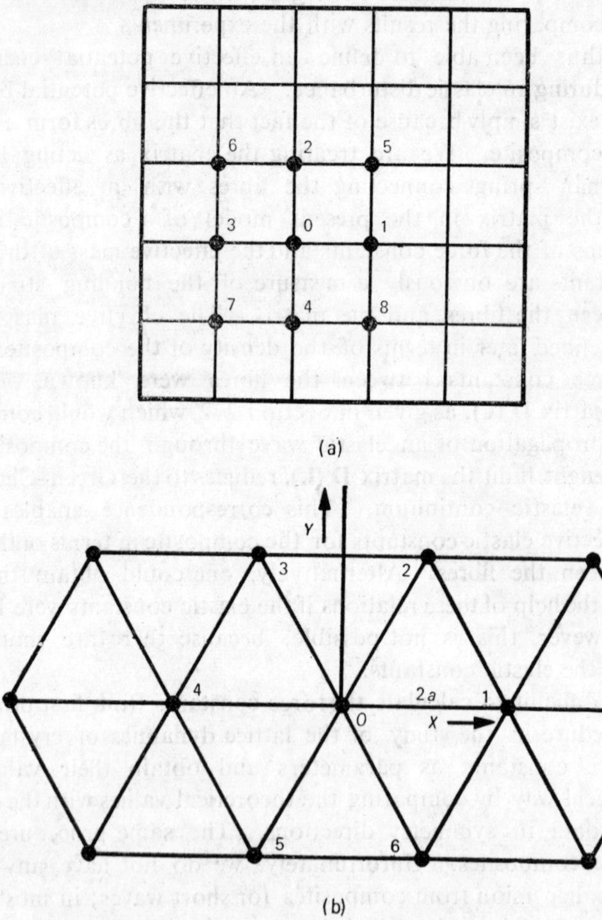

Figure 5.1 Arrangement of the fibres, represented by dots in the *xy* plane forming (a) a square lattice (b) a hexagonal lattice.

5.4 FORCE CONSTANTS FOR A COMPOSITE WITH TETRAGONAL SYMMETRY

Let us take the *z* axis along the fibres and the *x* and *y* axes on a plane perpendicular to the fibres. The composite will behave like a square lattice for a two-dimensional wave travelling parallel to the xy plane and like anelastic continuum for a wave travelling parallel to the z axis. For such a structure it is mathematically convenient to divide the composite into L equally spaced layers parallel to the xy plane separated by a distance c and then consider the limiting case as c becomes very small (of the order of atomic spacing in a fibre). The composite problem is thus reduced to that of an ordinary crystal with tetragonal lattice structure. This structure is shown in Figures 5.1a and 5.2a.

Effectively, we are dividing a fibre into L portions equally distributed in L planes. We shall refer to these portions as objects. Let the mass of each object be m. It should be emphasized that Lm is not supposed to represent the real mass of the fibre. The quantity m is defined in terms of ρ, the average density of the whole composite, viz.

$$m = \rho V \tag{5.4.1}$$

where V is the volume per object.

The unit cell 01520′1′5′2′ is denoted by the shaded region in Figure 5.2a. This is a right prism with a square as its base and having one object at each corner. The number of objects per unit cell is one because the eight objects at each corner of a unit cell are shared by eight unit cells. If a is the length of a side of the square and c is the height of the prism, we have

$$V = a^2c$$

and from Eq. (5.4.1)

$$m = \rho a^2 c \tag{5.4.2}$$

Later, we shall be able to determine c in terms of a Debye temperature. The quantity m includes in its definition the properties of the composite as a whole and is therefore an effective mass.

The symmetry of the lattice and the operations in its point group are given in Appendix I. Let us take the origin at the point occupied by object 0 (Figures 5.1a and 5.2a). For the calculation of D (**k**) we shall include the object 0, its first and second neighbours on the xy plane (1–4 and 5–8, respectively) and the objects which are just above (0′,1′–4′ and 5′–8′, respectively) and just below (0″,1″–4″ and 5″–8″, respectively) the objects, 0, 1–4 and 5–8 respectively. The coordinates of these objects with respect to the chosen system of axes are given in Table 5.1.

By symmetry the force constants between 0 and the objects in the same invariant subspaces will be related as given by Eq. (2.2.7)–(2.2.8a). The general form of the force constant matrices, subject to symmetry restrictions imposed by Eq. (2.2.7)–(2.2.8a) using the group operations given in Appendix I, are given below (the overhead bars denote a minus sign).

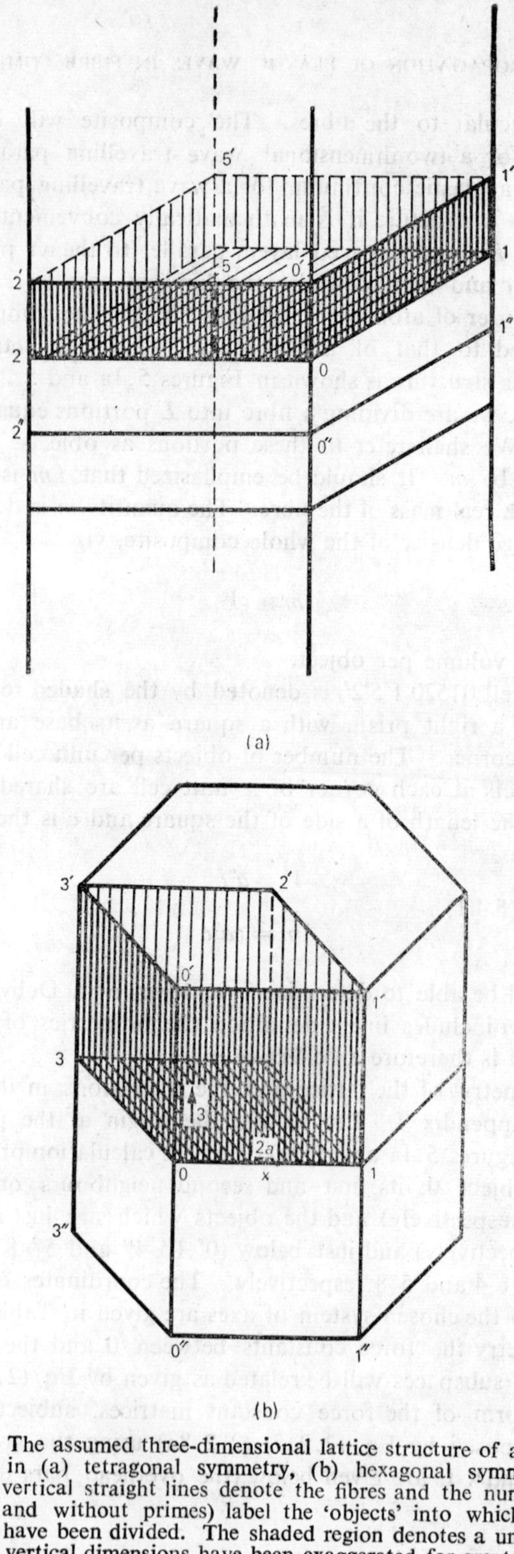

Figure 5.2 The assumed three-dimensional lattice structure of a composite in (a) tetragonal symmetry, (b) hexagonal symmetry. The vertical straight lines denote the fibres and the numbers (with and without primes) label the 'objects' into which the fibres have been divided. The shaded region denotes a unit cell. The vertical dimensions have been exaggerated for greater clarity.

TABLE 5.1

Coordinates of the objects contributing to the dynamical matrix for a composite with tetragonal symmetry

Object	Coordinates x	y	z
0	0	0	0
1	a	0	0
2	0	a	0
3	$-a$	0	0
4	0	$-a$	0
5	a	a	0
6	$-a$	a	0
7	$-a$	$-a$	0
8	a	$-a$	0
0′	0	0	c
1′	a	0	c
2′	0	a	c
3′	$-a$	0	c
4′	0	$-a$	c
5′	a	a	c
6′	$-a$	a	c
7′	$-a$	$-a$	c
8′	a	$-a$	c
0″	0	0	$-c$
1″	a	0	$-c$
2″	0	a	$-c$
3″	$-a$	0	$-c$
4″	0	$-a$	$-c$
5″	a	a	$-c$
6″	$-a$	a	$-c$
7″	$-a$	$-a$	$-c$
8″	a	$-a$	$-c$

$$\phi\,(0,0) = \begin{Bmatrix} \mu_0 & 0 & 0 \\ 0 & \mu_0 & 0 \\ 0 & 0 & \lambda_0 \end{Bmatrix}$$

$$\phi\,(0,1) = \phi\,(0,3) = -\begin{Bmatrix} \mu_1 & 0 & 0 \\ 0 & \nu_1 & 0 \\ 0 & 0 & \lambda_1 \end{Bmatrix}$$

$$\phi\,(0,2) = \phi\,(0,4) = -\begin{Bmatrix} \nu_1 & 0 & 0 \\ 0 & \mu_1 & 0 \\ 0 & 0 & \lambda_1 \end{Bmatrix}$$

$$\phi\,(0,5) = \phi\,(0,7) = -\begin{Bmatrix} \mu_2 & \eta_2 & 0 \\ \eta_2 & \mu_2 & 0 \\ 0 & 0 & \lambda_2 \end{Bmatrix}$$

$$\phi\,(0,6) = \phi\,(0,8) = -\begin{Bmatrix} \mu_2 & \overline{\eta}_2 & 0 \\ \overline{\eta}_2 & \mu_2 & 0 \\ 0 & 0 & \lambda_2 \end{Bmatrix}$$

$$\phi(0,0') = \phi(0,0'') = -\begin{Bmatrix} \mu_0' & 0 & 0 \\ 0 & \mu_0' & 0 \\ 0 & 0 & \lambda_0' \end{Bmatrix}$$

$$\phi(0,1') = \phi(0,3'') = -\begin{Bmatrix} \mu_1' & 0 & \mathcal{E}_1' \\ 0 & \nu_1' & 0 \\ \mathcal{E}_1' & 0 & \lambda_1' \end{Bmatrix}$$

$$\phi(0,1'') = \phi(0,3') = -\begin{Bmatrix} \mu_1' & 0 & \overline{\mathcal{E}}_1' \\ 0 & \nu_1' & 0 \\ \overline{\mathcal{E}}_1' & 0 & \lambda_1' \end{Bmatrix}$$

$$\phi(0,2') = \phi(0,4'') = -\begin{Bmatrix} \nu_1' & 0 & 0 \\ 0 & \mu_1' & \mathcal{E}_1' \\ 0 & \underline{\ }_1' & \lambda_1' \end{Bmatrix}$$

$$\phi(0, 2'') = \phi(0, 4') = -\begin{Bmatrix} \nu_1' & 0 & 0 \\ 0 & \mu_1' & \bar{\mathcal{E}}_1' \\ 0 & \bar{\mathcal{E}}_1' & \lambda_1' \end{Bmatrix}$$

$$\phi(0, 5') = \phi(0, 7'') = -\begin{Bmatrix} \mu_2' & \eta_2' & \mathcal{E}_2' \\ \eta_2' & \mu_2' & \mathcal{E}_2' \\ \mathcal{E}_2' & \mathcal{E}_2' & \mu_2' \end{Bmatrix}$$

$$\phi(0, 5'') = \phi(0, 7') = -\begin{Bmatrix} \mu_2' & \eta_2' & \bar{\mathcal{E}}_2' \\ \eta_2' & \mu_2' & \bar{\mathcal{E}}_2' \\ \bar{\mathcal{E}}_2' & \bar{\mathcal{E}}_2' & \lambda_2' \end{Bmatrix}$$

$$\phi(0, 6') = \phi(0, 8'') = -\begin{Bmatrix} \mu_2' & \bar{\eta}_2' & \bar{\mathcal{E}}_2' \\ \bar{\eta}_2' & \mu_2' & \mathcal{E}_2' \\ \bar{\mathcal{E}}_2' & \mathcal{E}_2' & \lambda_2' \end{Bmatrix}$$

$$\phi(0, 6'') = \phi(0, 8') = -\begin{Bmatrix} \mu_2' & \bar{\eta}_2' & \mathcal{E}_2' \\ \bar{\eta}_2' & \mu_2' & \bar{\mathcal{E}}_2' \\ \mathcal{E}_2' & \bar{\mathcal{E}}_2' & \lambda_2' \end{Bmatrix} \qquad (5.4.3)$$

From Eq. (2.2.2) we obtain

$$\left.\begin{aligned} \mu_0 &= 2(\mu_1 + 2\mu_2 + 2\mu_1' + 4\mu_2' + \mu_0' + \nu_1 + 2\nu_1' \\ \lambda_0 &= 2(2\lambda_1 + 2\lambda_2 + \lambda_0' + 4\lambda_1' + 4\lambda_2') \end{aligned}\right\} \qquad (5.4.4)$$

In terms of these force constants we can now construct the dynamical matrix. We notice from Eqs. (5.4.3) and (5.4.4) that we have 16 independent force constants. However, we shall see in the next section that in the limit of small c not all of them contribute independently to the dynamical matrix. Only 10 independent combinations of the above force constants contribute to the dynamical matrix. We have therefore 10 independent parameters in the present model.

5.5 DYNAMICAL MATRIX FOR A COMPOSITE WITH TETRAGONAL SYMMETRY

The elements of the dynamical matrix as defined in Eqs. (2.2.10) and (2.2.11) are given by ($D_{ij} = D_{ji}$):

$$D_{ij}(\mathbf{k}) = \frac{1}{m} \sum_{l} \phi_{ij}(\mathbf{l}) \exp [2\pi i \mathbf{k} \cdot r(\mathbf{l})] \tag{5.5.1}$$

Using Table 5.1 for the components of $\mathbf{r}(\mathbf{l})$ and Eqs. (5.4.3) and (5.4.4) for the force constant matrices $\phi(\mathbf{l})$, we obtain the following expressions for the elements of the dynamical matrix (the matrix elements are in units of $1/m$; c is in units of a; and the subscripts 1, 2 and 3 refer to x, y and z, respectively):

$$\left.\begin{aligned}
D_{11}(\mathbf{k}) &= 2\mu_1 (1 - \cos k_1) + 4\mu_2 (1 - \cos k_1 \cos k_2) \\
&\quad + 2\nu_1 (1 - \cos k_2) + 2\mu_0' (1 - \cos ck_3) \\
&\quad + 4\mu_1' (1 - \cos k_1 \cos ck_3) + 4\nu_1' (1 - \cos k_2 \cos ck_3) \\
&\quad + 8\mu_2' (1 - \cos k_1 \cos k_2 \cos ck_3) \\[6pt]
D_{22}(\mathbf{k}) &= 2\mu_1 (1 - \cos k_2) + 4\mu_2 (1 - \cos k_1 \cos k_2) \\
&\quad + 2\nu_1 (1 - \cos k_1) + 2\mu_0' (1 - \cos ck_3) \\
&\quad + 4\mu_1' (1 - \cos k_2 \cos ck_3) + 4\nu_1' (1 - \cos k_1 \cos ck_3) \\
&\quad + 8\mu_2' (1 - \cos k_1 \cos k_2 \cos ck_3) \\[6pt]
D_{33}(\mathbf{k}) &= 2\lambda_1 (2 - \cos k_1 - \cos k_2) \\
&\quad + 4\lambda_2 (1 - \cos k_1 \cos k_2) + 2\lambda_0' (1 - \cos ck_3) \\
&\quad + 4\lambda_1' (2 - \cos k_1 \cos ck_3 - \cos k_2 \cos ck_3) \\
&\quad + 8\lambda_2' (1 - \cos k_1 \cos k_2 \cos ck_3) \\[6pt]
D_{12}(\mathbf{k}) &= 4\eta_2 \sin k_1 \sin k_2 + 8\eta_2' \sin k_1 \sin k_2 \cos ck_3 \\
D_{13}(\mathbf{k}) &= 4\xi_1' \sin k_1 \sin ck_3 + 8\xi_2' \sin k_1 \sin ck_3 \cos k_2 \\
D_{23}(\mathbf{k}) &= 4\xi_1' \sin k_2 \sin ck_3 + 8\xi_2' \sin k_2 \sin ck_3 \cos k_1
\end{aligned}\right\} \tag{5.5.2}$$

The Brillouin zone for the the present model is defined by

$$-\frac{\pi}{c} \leqslant k_3 \leqslant \frac{\pi}{c} \tag{5.5.3}$$

$$-\pi \leqslant k_1, k_2 \leqslant \pi \tag{5.5.4}$$

For a discrete lattice k_i only discrete values are allowed. In the present case, to allow for the transition from a discrete lattice to a continuum in the z direction, we allow k_3 to assume continuous values in above range.

For finite, non-zero values of c the composite will behave like a Debye solid rather than an ordinary elastic continuum. In a Debye continuum the components k_i of the wave vector are confined in a finite zone and the maximum value of k_i or the dimensions of the zone are related to the sound velocities in different directions and can be used to define a characteristic temperature known as the Debye temperature. The usual approximation is to replace the zone of $\mathbf{k}$ values by a sphere and thus define a single Debye temperature. For an anisotropic solid it is more realistic to define different Debye temperatures in the x, y and z directions. Such an approach has been used for selenium and tellurium by Kothari and Tewary (1963a), for bismuth by Kothari and Tewary (1963b) and for graphite by Kothari and Singwi (1957). Using the same principle, we shall define θ_3, the Debye temperature in the z direction, in terms of the maximum value of k_3 as given in Eq. (5.5.3). Such a distinction between the z direction and the xy plane is consistent with the basic assumptions of the present model. For the xy plane we do not have to introduce any characteristic temperature, since we are explicitly taking into account the structure of the composite. It is, of course, possible to define a Debye temperature for the xy plane as well, with a parametric dependence on temperature, as is usually done for the interpretation of experimental data on specific heat in the Born–von Kármán theory (see, for example, Kittel, 1976).

We define θ_3 with the help of the following two relations:

$$k_0\theta_3 = \hbar\omega_3^{\max} \tag{5.5.5}$$

$$\omega_3(k_3) = \sqrt{\frac{c_{33}}{\rho}}\,k_3 \tag{5.5.6}$$

where k_0 is the Boltzmann constant, $\hbar$ is the Planck constant and c_{33} is the elastic constant in appropriate units. From Eqs. (5.5.3), (5.5.5) and (5.5.6) we obtain

$$ca = \frac{\hbar\sqrt{c_{33}/\rho}}{2k_0\theta_3} \approx 10^{-10}\ \text{m} \tag{5.5.7}$$

which, as expected, is of the order of the atomic spacing. Substituting this estimate in Eq. (5.4.2), we get an estimate of the effective mass m of 10^{-19} kg.

To make the transition from a Born–von Kármán solid to a Debye solid in the z direction, we shall use Born's method of long waves and expand $\sin ck_3$ and $\cos ck_3$ as follows:

$$\sin ck_3 = ck_3 \tag{5.5.8}$$

$$\cos ck_3 = 1 - \frac{c^2k_3^2}{2} \tag{5.5.9}$$

where cubic and higher powers of k_3 have been neglected. Substituting these expressions in Eqs. (5.5.2) we obtain

$$\begin{aligned} D_{11}(\mathbf{k}) = {} & 2(\mu_1 + 2\mu_1')(1 - \cos k_1) + 4(\mu_2 + 2\mu_2')(1 - \cos k_1 \cos k_2) \\ & + 2(\nu_1 + 2\nu_2')(1 - \cos k_2) \\ & + c^2k_3^2(\mu_0' + 2\mu_1' \cos k_1 + 2\nu_1' \cos k_2 + 4\mu_2' \cos k_1 \cos k_2) \end{aligned} \tag{5.5.10}$$

$$D_{22}(\mathbf{k}) = 2(\mu_1 + 2\mu_1')(1 - \cos k_2) + 4(\mu_2 + 2\mu_2')(1 - \cos k_1 \cos k_2)$$
$$+ 2(\nu_1 + 2\nu_1')(1 - \cos k_1)$$
$$+ c^2 k_3^2 (\mu_0' + 2\mu_1' \cos k_2 + 2\nu_1' \cos k_1 + 4\mu_2' \cos k_1 \cos k_2) \quad (5.5.11)$$

$$D_{33}(\mathbf{k}) = 2(\lambda_1 + 2\lambda_1')(2 - \cos k_1 - \cos k_2) + 4(\lambda_2 + 2\lambda_2')(1 - \cos k_1 \cos k_2)$$
$$+ c^2 k_3^2 (\lambda_0' + 2\lambda_1' \cos k_1 + 2\lambda_1' \cos k_2 + 4\lambda_2' \cos k_1 \cos k_2) \quad (5.5.12)$$

$$D_{12}(\mathbf{k}) = 4(\eta_2 + 2\eta_2' - \eta_2' c^2 k_3^2) \sin k_1 \sin k_2 \quad (5.5.13)$$

$$D_{13}(\mathbf{k}) = 4(\xi_1' + 2\xi_2' \cos k_2)\, c k_3 \sin k_1 \quad (5.5.14)$$

$$D_{23}(\mathbf{k}) = 4(\xi_1' + 2\xi_2' \cos k_1)\, c k_3 \sin k_2 \quad (5.5.15)$$

To simplify the above equations further, let us recall that the distance between two neighbouring objects on the same fibre is of the order of an angstrom whereas on different fibres it is of the order of a micron. Hence, μ_0' and λ_0' will be several orders of magnitude larger than the other force constants. We shall therefore neglect all the terms in the coefficient of k_3^2 except μ_0' in Eqs. (5.5.10) and (5.5.11) and λ_0' in Eq. (5.5.12). Further, we notice in Eq. (5.5.13) that the last term in the parentheses will make a significant contribution only when $k_3 \approx 1/c$, which corresponds to a wavelength of the order of an angstrom. Since in the present paper we are not interested in such short waves, we can neglect the term containing $c^2k_3^2$ in Eq. (5.5.13). With these approximations. we finally obtain the following expressions for the elements of the dynamical matrix:

$$D_{11}(\mathbf{k}) = 2A(1 - \cos k_1) + 2B(1 - \cos k_2)$$
$$+ 4D(1 - \cos k_1 \cos k_2) + Ek_3^2 \quad (5.5.16)$$

$$D_{22}(\mathbf{k}) = 2A(1 - \cos k_2) + 2B(1 - \cos k_1)$$
$$+ 4D(1 - \cos k_1 \cos k_2) + Ek_3^2 \quad (5.5.17)$$

$$D_{33}(\mathbf{k}) = 2F(2 - \cos k_1 - \cos k_2)$$
$$+ 4G(1 - \cos k_1 \cos k_2) + Hk_3^2 \quad (5.5.18)$$

$$D_{12}(\mathbf{k}) = 4K \sin k_1 \sin k_2 \quad (5.5.19)$$

$$D_{13}(\mathbf{k}) = 4L(1 + p \cos k_2)\, k_3 \sin k_1 \quad (5.5.20)$$

$$D_{23}(\mathbf{k}) = 4L(1 + p \cos k_1)\, k_3 \sin k_2 \quad (5.5.21)$$

where

$$A = \mu_1 + 2\mu_1'$$

$$B = \nu_1 + 2\nu_1'$$

$$D = \mu_2 + 2\mu_2'$$

$$E = \mu_0' c^2$$

$$F = \lambda_1 + 2\lambda_1'$$

$$G = \lambda_2 + 2\lambda_2'$$

$$H = \lambda_0' c^2$$

$$K = \eta_2 + 2\eta_2'$$

$$L = \xi_1' c$$

$$p = 2\xi_2'/\xi_1'$$

Thus, we see that **D** (**k**) in its final form contains 10 independent parameters. As we shall see in the next section, seven of these can be determined in terms of the elastic constants.

5.6 DYNAMICAL MATRIX IN THE LONG-WAVELENGTH LIMIT

For a wave much longer than the fibre spacing, the composite will behave like an elastic continuum. In this limit ($|k| \rightarrow 0$) the matrix **D** (**k**) must be identical with the Green–Christoffel matrix. Thus, using the method of long waves, we shall expand the sine and cosine functions in Eqs. (5.5.16)–(5.5.21), keeping only up to second-order terms, and compare the resulting matrix with the Green–Christoffel matrix for a solid with tetragonal symmetry.

Substituting for $\sin k_i = k_i$ and $\cos k_i = 1 - k_i^2/2$, we obtain the following expressions for the elements of **D** (**k**):

$$D_{11}(\mathbf{k}) = (A + 2D)\, k_1^2 + (B + 2D)\, k_2^2 + E k_3^2 \qquad (5.6.1)$$

$$D_{22}(\mathbf{k}) = (B + 2D)\, k_1^2 + (A + 2D)\, k_2^2 + E k_3^2 \qquad (5.6.2)$$

$$D_{33}(\mathbf{k}) = (F + 2G)\,(k_1^2 + k_2^2) + H k_3^2 \qquad (5.6.3)$$

$$D_{12}(\mathbf{k}) = 4K\, k_1 k_2 \qquad (5.6.4)$$

$$D_{13}(\mathbf{k}) = 4L\,(1 + p)\, k_1 k_3 \qquad (5.6.5)$$

$$D_{23}(\mathbf{k}) = 4L\,(1 + p)\, k_2 k_3 \qquad (5.6.6)$$

The elements of the Green–Christoffel matrix $\Lambda(\mathbf{k})$ for a solid with tetragonal symmetry have the form (Fedorov, 1968)

$$\Lambda_{11}(k) = c_{11}k_1^2 + c_{66}k_2^2 + c_{44}k_3^2 \tag{5.6.7}$$

$$\Lambda_{22}(k) = c_{66}k_2^2 + c_{11}k_1^2 + c_{44}k_3^2 \tag{5.6.8}$$

$$\Lambda_{33}(k) = c_{44}(k_1^2 + k_2^2) + c_{33}k_3^2 \tag{5.6.9}$$

$$\Lambda_{12}(k) = (c_{12} + c_{66})\,k_1k_2 \tag{5.6.10}$$

$$\Lambda_{13}(k) = (c_{13} + c_{44})\,k_1k_3 \tag{5.6.11}$$

$$\Lambda_{23}(k) = (c_{13} + c_{44})\,k_2k_3 \tag{5.6.12}$$

Equating the coefficients of k_i^2 in the corresponding elements of the matrices $\mathbf{D}(\mathbf{k})$ and $\Lambda(\mathbf{k})$, we get the following relations between the force constants and the elastic constants:

$$\left.\begin{aligned} A + 2D &= c_{11} \\ B + 2D &= c_{66} \\ E &= c_{44} \\ F + 2G &= c_{44} \\ H &= c_{33} \\ 4K &= c_{12} + c_{66} \\ 4L(1+p) &= c_{13} + c_{44} \end{aligned}\right\} \tag{5.6.13}$$

where the elastic constants are in units of $(4\pi a^2\rho)^{-1}$ and the force constants, as may be recalled, are in units of m^{-1}.

To determine all the 10 force constants, it is necessary to have further information, such as measurements on dispersion in composites, i.e. sound velocities as a function of wavelength in the short-wavelength region. In the absence of such measurements and to provide a basis for the numerical calculations in the present chapter, we shall assume reasonable but arbitrary values for three of the force constants—namely, D, G and p. The remaining force constants will be determined in terms of the elastic constants, using the calculated values of Heaton (1970) (see Chapter 3).

The elastic constants and therefore the force constants, for a composite depend on the elastic constants of the constituents and the loading of the composite, i.e. the relative volume occupied by the fibres, in addition to, of course, several other factors. Heaton's values for the elastic constants of a tetragonal composite are given in Table 5.2 for various percentages of the fibre concentration; the constants are in units of M^m, where M^m is the shear modulus of the matrix. Our final results for the frequencies will therefore be given in units of ω_t, where ω_t is given by

$$\omega_t^2 = \frac{M^m}{4\pi a^2 \rho} \tag{5.6.14}$$

and the velocities are expressed in units of c_t, where

$$c_t = \omega_t \cdot 2\pi a \tag{5.6.15}$$

TABLE 5.2

Elastic constants for a tetragonal composite in units of M^m (Heaton, 1970)*

Elastic constant	30% fibre concentration	50% fibre concentration	70% fibre concentration
c_{11}	6.118	8.164	11.576
c_{12}	2.852	3.204	3.892
c_{13}	3.599	4.917	7.170
c_{33}	86.359	141.560	197.800
c_{44}	1.712	2.593	4.738
c_{66}	1.423	1.856	2.817

*Units, 10 GN/m^2.

5.7 WAVE PROPAGATION IN A GENERAL DIRECTION

A knowledge of the dynamical matrix is sufficient to define most of the dynamical properties of the composite. The three eigenvalues of the dynamical matrix for a value of **k** give the frequencies of the three types of waves and the corresponding eigenvectors describe the polarization of those three waves. If e_i denote the three Cartesian components ($i = 1, 2$ and 3) of an eigenvector corresponding to an eigenvalue ω^2, we may write

$$D_{11}\, e_1 + D_{12}\, e_2 + D_{13}\, e_3 = \omega^2\, e_1 \tag{5.7.1}$$

$$D_{21}\, e_1 + D_{22}\, e_2 + D_{23}\, e_3 = \omega^2\, e_2 \tag{5.7.1a}$$

$$D_{31}\, e_1 + D_{32}\, e_2 + D_{33}\, e_3 = \omega^2\, e_3 \tag{5.7.1b}$$

with the condition

$$\begin{vmatrix} D_{11} - \omega^2 & D_{12} & D_{13} \\ D_{21} & D_{22} - \omega^2 & D_{23} \\ D_{31} & D_{32} & D_{33} - \omega^2 \end{vmatrix} = 0 \tag{5.7.2}$$

for a non-trivial solution. The frequencies of the three waves are given as the solution of the secular equation (5.7.2). The solution of the secular equation is considerably simplified for a symmetry direction in **k** space and, fortunately, most of the information of interest about the composite can be obtained from the waves travelling in a symmetry direction or a symmetry plane. We shall discuss these special directions in later sections. In a general direction the formal solution of the secular equation which is valid for any value of **k** can be obtained by using the formulae given in Section 5.2 (Eqs. 5.2.10–5.2.10b) for the solution of a 3×3 secular equation.

5.8 WAVE PROPAGATION IN THE *xy* PLANE: BONDING STRENGTH OF THE FIBRES

We shall see in this section that the propagation of an elastic wave in the *xy* plane can yield complete information about the bonding strength of the fibres. It is obvious that only a wave travelling across the fibres will feel the discrete nature of the composite and will therefore be most sensitive to the bonding strength of the fibres. For a wave travelling in the *z* direction, the discrete nature of the composite will manifest itself only through the distortions in the wave front. This distortion is taken into account in the present model only in an average way by defining an effective value of the elastic constants for the whole composite. In any case, such a distortion will depend more on the fibre misalignments than on the bonding strength.

For a wave travelling in the *xy* plane $k_3 = 0$, the matrix $\mathbf{D}(\mathbf{k})$ assumes a block diagonal form with the following structure

$$\mathbf{D}(\mathbf{k}) = \begin{pmatrix} \mathbf{D}^p & 0 \\ 0 & \mathbf{D}^z \end{pmatrix} \tag{5.8.1}$$

where the block $\mathbf{D}^p$ is a 2×2 matrix and $\mathbf{D}^z$ is a 1×1 matrix, i.e. a pure number. The elements of these matrices, obtained from Eqs. (5.5.16)–(5.5.21), are given below:

$$\left.\begin{aligned} D^p_{11}(\mathbf{k}) &= 2A(1-\cos k_1) + 2B(1-\cos k_2) + 4D(1-\cos k_1 \cos k_2) \\ D^p_{12}(\mathbf{k}) &= D^p_{21}(k) = 4K \sin k_1 \sin k_2 \\ D^p_{22}(\mathbf{k}) &= 2A(1-\cos k_2) + 2B(1-\cos k_1) + 4D(1-\cos k_1 \cos k_2) \end{aligned}\right\} \tag{5.8.2}$$

and

$$D^z(\mathbf{k}) = 2F(2-\cos k_1 - \cos k_2) + 4G(1-\cos k_1 \cos k_2) \tag{5.8.3}$$

The eigenvalue corresponding to the element $\mathbf{D}^z(\mathbf{k})$ is immediately known and is given by

$$\omega_z^2(k) = 2F(2 - \cos k_1 - \cos k_2) + 4G(1 - \cos k_1 \cos k_2) \tag{5.8.4}$$

The eigenvector corresponding to this eigenvalue is [0, 0, 1]. This wave is always polarized perpendicular to the xy plane and is therefore a true shear wave for all values of k_1 and k_2.

The frequencies associated with the other two waves, polarized in the xy plane, can be obtained by diagonalizing the matrix $\mathbf{D}^p(\mathbf{k})$. If ω_{p_1} and ω_{p_2} represent the frequencies of the two waves polarized at angles θ_p and $\theta_p + \pi/2$, respectively, from the x axis in the xy plane, then we have

$$\begin{aligned}\omega_{p_1}^2 &= (A+B)(2 - \cos k_1 - \cos k_2) + 4D(1 - \cos k_1 \cos k_2)\\ &\quad + [(A-B)^2(\cos k_2 - \cos k_1)^2 + 16K^2 \sin^2 k_1 \sin^2 k_2]^{1/2}\end{aligned} \tag{5.8.5}$$

$$\left.\begin{aligned}e_{p_1} &= [\cos\theta_p, \sin\theta_p, 0]\\ \omega_{p_2}^2 &= (A+B)(2 - \cos k_1 - \cos k_2) + 4D(1 - \cos k_1 \cos k_2)\\ &\quad - [(A-B)^2(\cos k_2 - \cos k_1)^2 + 16K^2 \sin^2 k_1 \sin^2 k_2]^{1/2}\end{aligned}\right\} \tag{5.8.6}$$

and

$$e_{p_2} = [-\sin\theta_p, \cos\theta_p, 0]$$

where $\mathbf{e}_{pi}$ denote the polarization vectors and

$$\tan 2\theta_p = \frac{4K}{A-B}\,\frac{\sin k_1 \sin k_2}{\cos k_2 - \cos \mathbf{k}_1} \tag{5.8.7}$$

we notice from Eq. (5.8.7) that for a general value of $\mathbf{k}$ the waves p_1 and p_2 are not truly longitudinal or transverse waves, since the displacements given by the polarization vectors are in a direction different from that of the incident wave, i.e.

$$\theta_p \neq \theta_k$$

where

$$\tan\theta_k = \frac{k_2}{k_1}$$

However, it is customary to refer to the waves p_1 and p_2 as the quasi-longitudinal and quasi-transverse waves, respectively, because the angle between $\mathbf{e}_{p1}$ and $\mathbf{k}$ is usually small. This is true in the present case but not, as we shall see later, for the wave travelling in a plane perpendicular to the xy plane.

The phase velocities for the three waves can be obtained from the above equations very simply by using Eqs. (5.2.8). The group velocities, using Eqs. (5.2.9) are given below.

Quasi-longitudinal wave (p_1–wave):

$$\left.\begin{aligned} C_{gx} &= \frac{\sin k_1}{2\omega_{p1}}\Big[A+B+4D\cos k_2 \\ &\quad + \frac{1}{X}\Big\{(A-B)^2(\cos k_2-\cos k_1)+16K^2\sin^2 k_2\cos k_1\Big\}\Big] \\ C_{gy} &= \frac{\sin k_2}{2\,\omega_{p1}}\Big[A+B+4D\cos k_1 \\ &\quad + \frac{1}{X}\Big\{(A-B)^2(\cos k_1-\cos k_2)+16K^2\sin^2 k_1\cos k_2\Big\}\Big] \\ C_{gz} &= 0 \end{aligned}\right\} \tag{5.8.8}$$

Quasi-transverse wave (p_2-wave):

$$\left.\begin{aligned} C_{gx} &= \frac{\sin k_1}{2\,\omega_{p2}}\Big[A+B+4D\cos k_2 \\ &\quad - \frac{1}{X}\Big\{(A-B)^2(\cos k_2-\cos k_1)+16K^2\sin^2 k_2\cos k_1\Big\}\Big] \\ C_{gy} &= \frac{\sin k_2}{2\,\omega_{p2}}\Big[A+B+4D\cos k_1 \\ &\quad - \frac{1}{X}\Big\{(A-B)^2(\cos k_1-\cos k_2)+16K^2\sin^2 k_1\cos k_2\Big\}\Big] \\ C_{gz} &= 0 \end{aligned}\right\} \tag{5.8.9}$$

where

$$X^2=(A-B)^2(\cos k_2-\cos k_1)^2+16K^2\sin^2 k_1\sin^2 k_2$$

Transverse wave (z-wave):

$$\left.\begin{aligned} C_{gx} &= \frac{\sin k_1}{\omega_z}[F+2G\cos k_2] \\ C_{gy} &= \frac{\sin k_2}{\omega_z}[F+2G\cos k_1] \\ C_{gz} &= 0 \end{aligned}\right\} \tag{5.8.10}$$

In the above expressions, the suffixes x and y represent the x and y components of the group velocity. The z component in each case is zero. The magnitude C_g and the angle θ_g from the x axis in each case are given by

$$C_g^2 = C_{gx}^2 + C_{gy}^2 \tag{5.8.11}$$

and

$$\tan \theta_g = \frac{C_{gy}}{C_{gx}} \tag{5.8.12}$$

The group velocity C_g is the velocity with which the energy flux actually travels through the composite and the angle θ_g defines the direction of the flux. We notice that, in general, θ_g differs from θ_k, the direction of the incident wave.

Experimentally, it may often be more convenient to measure the velocities as a function of θ_k for a fixed wavelength rather than as a function of wavelength itself. Such measurements will also yield the values of the force constants. For this purpose, we have plotted in Figures 5.3 and 5.4 ω^2 (**k**) and C_g as a function of θ_k for $|\mathbf{k}|/\pi = 0.25, 0.5$ and 1.0. The angles θ_p and θ_g are also plotted in Figure 5.5 as a function of θ_k for $|\mathbf{k}|/\pi = 0.25$. The force constants have been obtained with the help of Eqs. (5.6.13) and Table 5.2. The constant p does not contribute independently to the results given here. For D and G we have used the following arbitrary values: $D = A/4$, $G = F/4$.

The above equations have particularly simple forms in the symmetry directions and provide convenient directions for the dispersion measurements. We shall give here explicit expressions for the frequency and the group velocity as a function of **k** in the ⟨100⟩ and ⟨110⟩ directions. In these directions p_1 and p_2 waves become true longitudinal and transverse waves respectively. The p_2 wave will be referred to as the p-transverse wave and the wave polarized perpendicular to the xy plane will be referred to as the z-transverse wave.

⟨100⟩ *direction.* In this direction $k_1 = k, k_2 = 0, k_3 = 0$ and $\theta_k = \theta_p = \theta_g = 0$. The direction ⟨010⟩ is equivalent by symmetry. The frequencies and the group velocities of different waves are given below.

Longitudinal wave:

$$\left.\begin{aligned} \omega^2 &= 2\,(A + 2D)\,(1 - \cos k) \\ C_g &= \sqrt{A + 2\,D}\,\cos \frac{k}{2} \end{aligned}\right\} \tag{5.8.13}$$

p-transverse wave:

$$\left.\begin{aligned} \omega^2 &= 2\,(B + 2D)\,(1 - \cos k) \\ C_g &= \sqrt{B + 2\,D}\,\cos \frac{k}{2} \end{aligned}\right\} \tag{5.8.14}$$

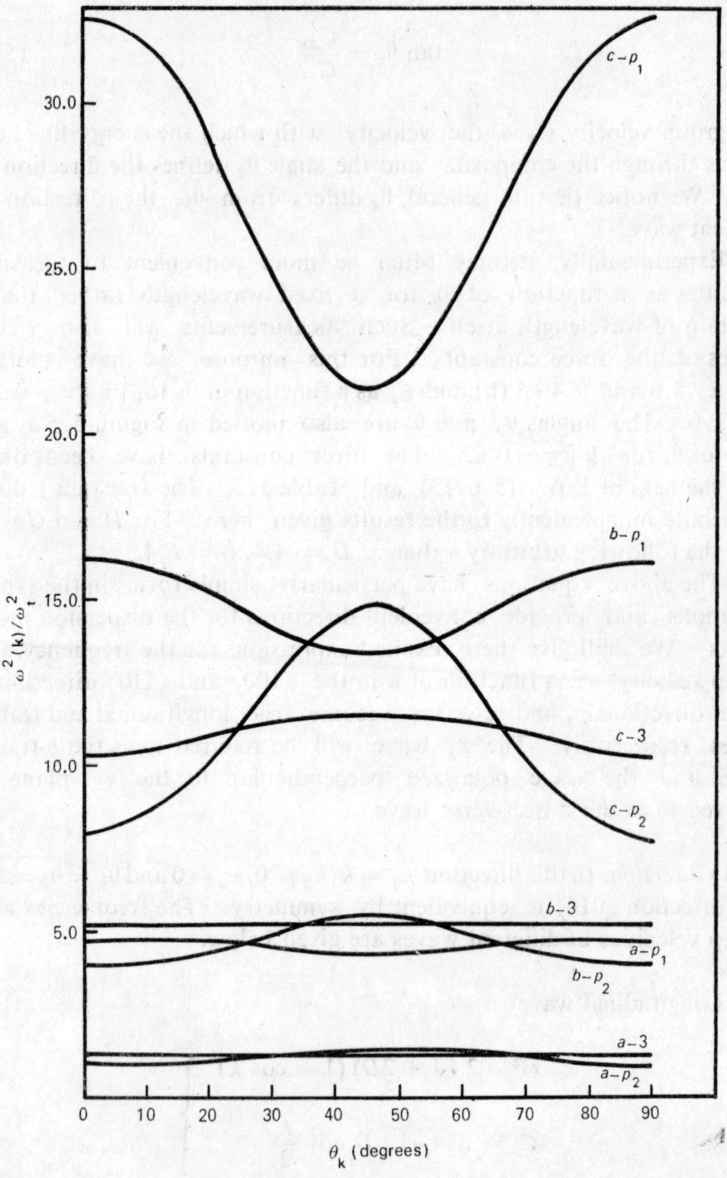

Figure 5.3 Angular dependence of the dispersion relations for a tetragonal composite in the xy plane. The values of $|\mathbf{k}|/\pi$ are (a) 0.25, (b) 0.5 and (c) 1.0. The angles are measured from the x axis.

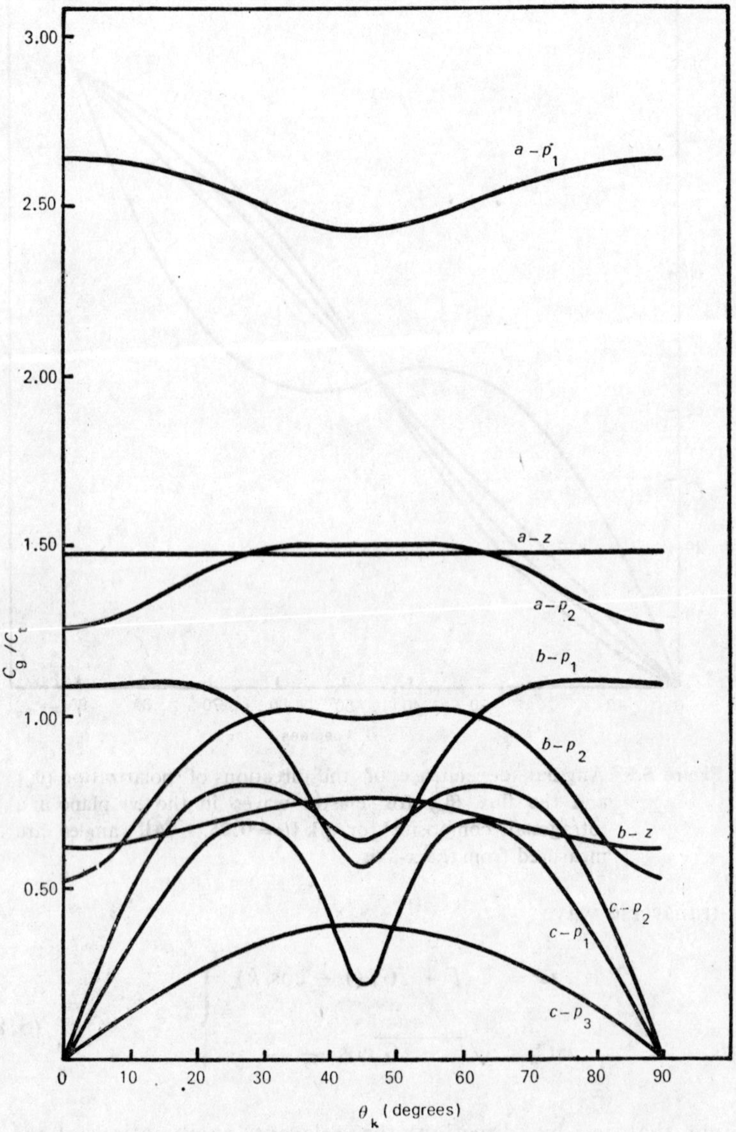

Figure 5.4 Angular dependence of the group velocities of the elastic waves in the xy plane in a tetragonal composite. The values of $|\mathbf{k}|/\pi$ are for (a) 0.25, (b) 0.5 and (c) 1.0. The angles are measured from the x-axis.

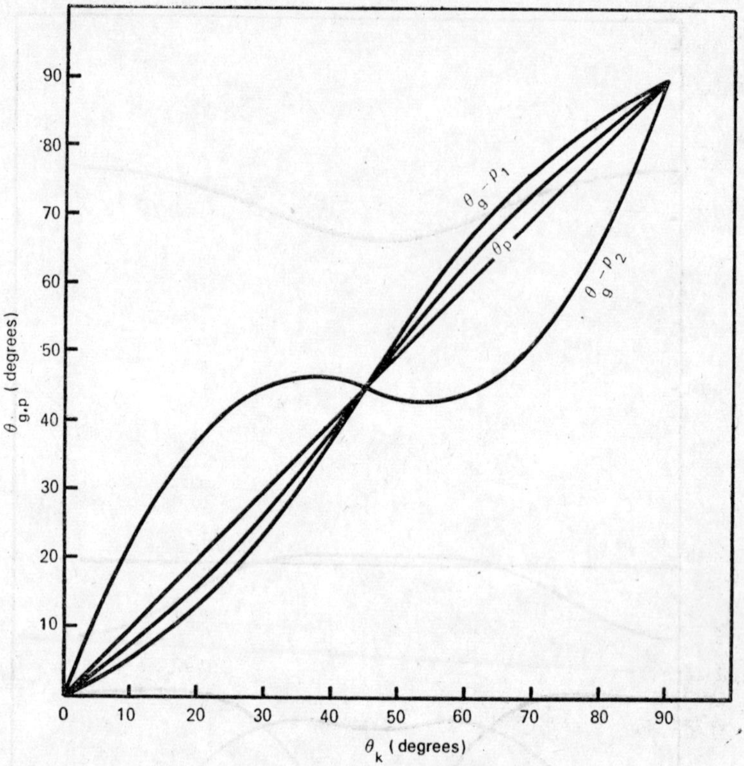

Figure 5.5 Angular dependence of the directions of polarization (θ_p) and the flux (θ_g) for elastic waves in the xy plane in a tetragonal composite for $|\mathbf{k}|/\pi = 0.25$. All angles are measured from the x-axis.

z-transverse wave:

$$\left.\begin{aligned} \omega^2 &= 2\,(F + 2G)\,(1 - \cos k) \\ C_g &= \sqrt{F + 2G}\,\cos\frac{k}{2} \end{aligned}\right\} \qquad (5.8.15)$$

We note that in this direction the relevant combinations of the force constants are completely determined by the elastic constants. This is a consequence of taking only up to second-neighbour interactions. The calculated values of ω^2 (**k**) and C_g are plotted as a function of k in Figures 5.6 and 5.7, respectively.

⟨110⟩ *direction.* In this direction $k_1 = k_2 = k$, $k_3 = 0$ and $\theta_k = \theta_p = \pi/4$ and the following expressions are obtained for the frequencies and group velocities of the three waves.

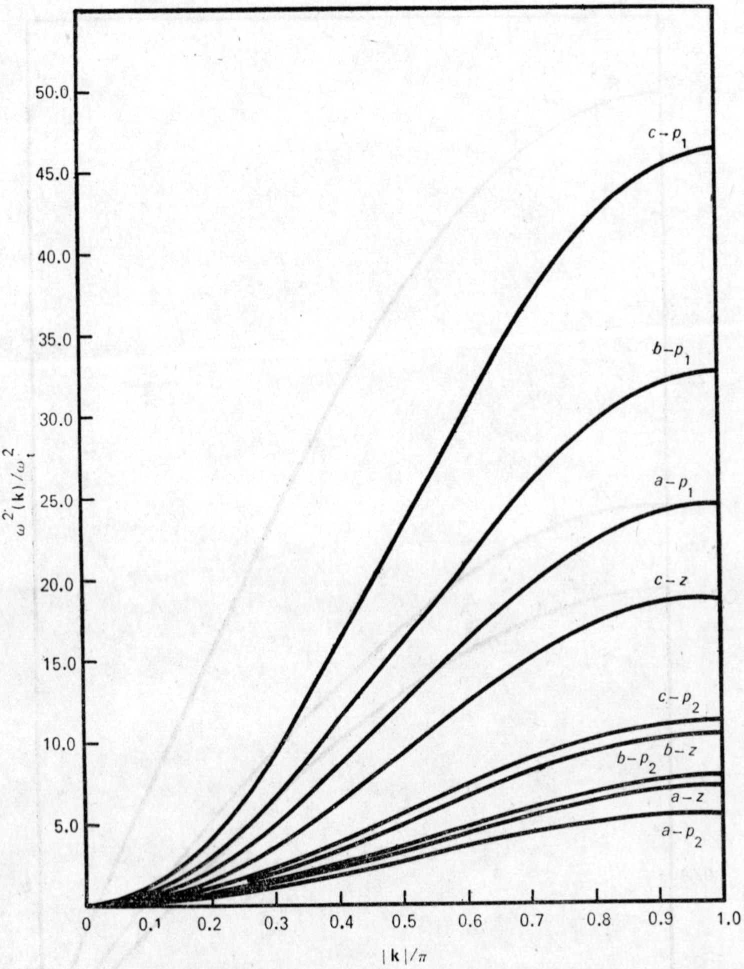

Figure 5.6 Dispersion relations in <100> direction for a tetragonal composite with (a) 30% fibre loading; (b) 50% fibre loading; and (c) 70% fibre loading.

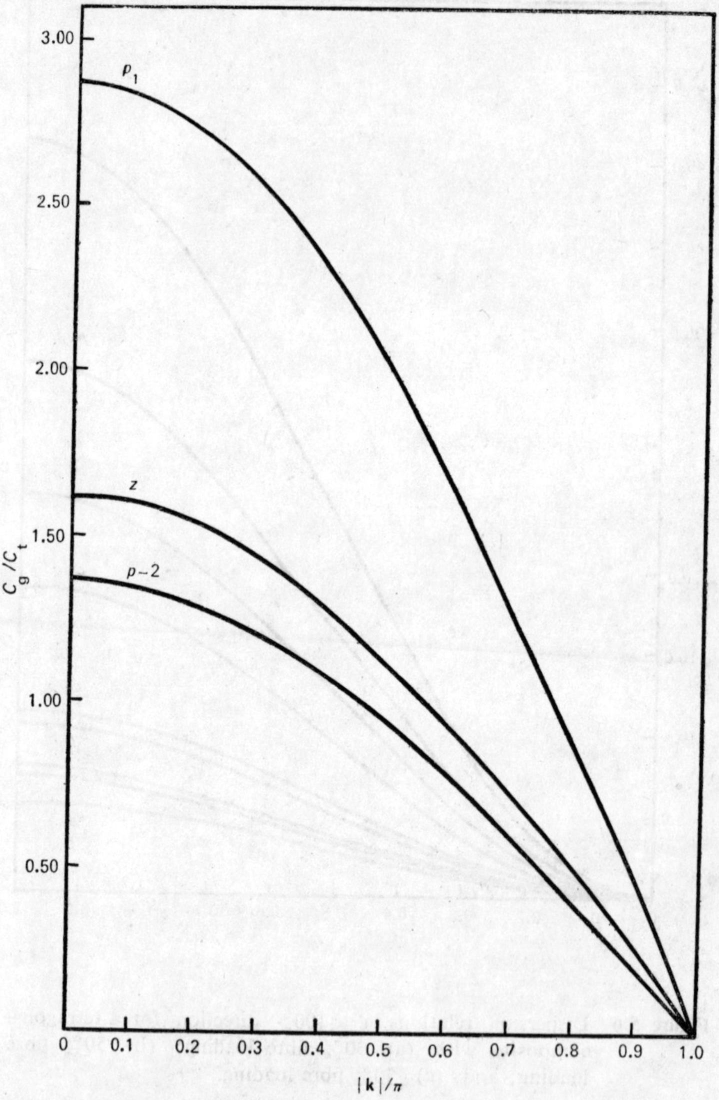

Figure 5.7 Group velocities of the three elastic waves travelling in <100> direction in a tetragonal composite.

Longitudinal wave

$$\left.\begin{aligned} \omega^2 &= 2\,(A+B)\,(1-\cos k) + 4\,(D+K)\sin^2 k \\ C_g &= V_1 \cos\frac{k}{2} \end{aligned}\right\} \quad (5.8.16)$$

where

$$V_1 = \frac{A+B+4\,(D+K)\cos k}{[2\,(A+B)+4\,(D+K)\,(1+\cos k)]^{1/2}}$$

p-transverse wave

$$\left.\begin{aligned} \omega^2 &= 2\,(A+B)\,(1-\cos k) + 4\,(D-K)\sin^2 k \\ C_g &= V_2 \cos\frac{k}{2} \end{aligned}\right\} \quad (5.8.17)$$

where

$$V_2 = \frac{A+B+4\,(D-K)\cos k}{[2\,(A+B)+4\,(D-K)\,(1+\cos k)]^{1/2}}$$

z-transverse wave

$$\left.\begin{aligned} \omega^2 &= 4\,F\,(1-\cos k) + 4\,G\sin^2 k \\ C_g &= V_z \cos\frac{k}{2} \end{aligned}\right\} \quad (5.8.18)$$

where

$$V_z = \frac{F+2\,G\cos k}{[F+G\,(1+\cos k)]^{1/2}}$$

ω^2 (**k**) and C_g are plotted as a function of k for this direction in Figures 5.8 and 5.9, respectively.

It is interesting to note in the above expressions that the group velocities C_g all vanish when $k=\pi$. This means that a wave of this wave number will not be able to travel through the composite and will be reflected. This phenomenon is analogous to the Bragg reflection of X-rays from crystals. This reflection is of course a consequence of the assumed translational symmetry and may not occur in a real composite. However, for a densely reinforced composite the assumption of translational symmetry is reasonable and therefore, the group velocity should become quite small if not zero for $k=\pi$.

Bonding strength of the fibres

As has been remarked earlier, we can get useful information about the

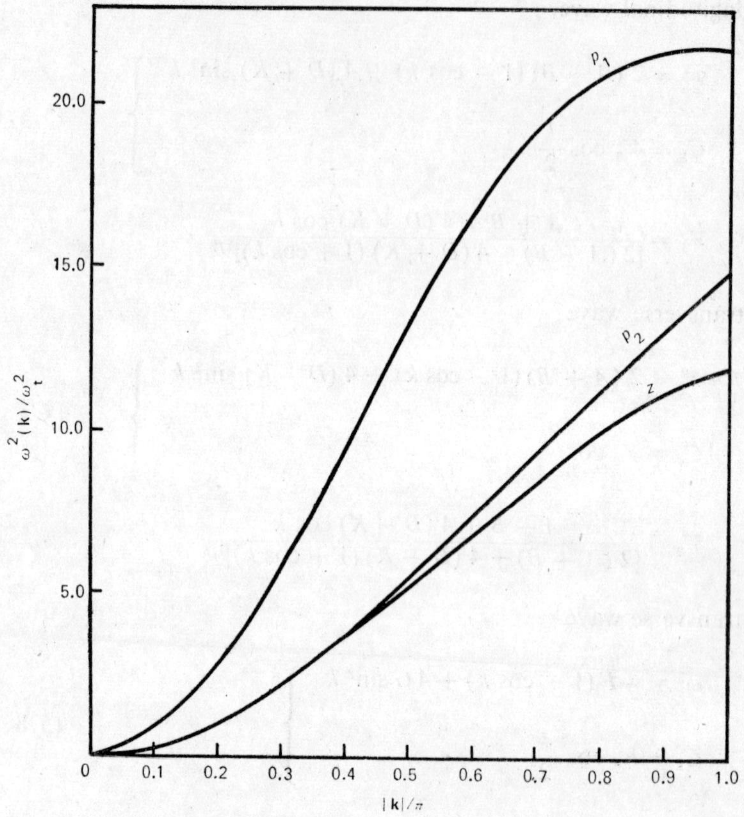

Figure 5.8 Dispersion relations in $\langle 110 \rangle$ direction for a tetragonal composite. $|\mathbf{k}|=\pi$ is not the maximum allowed value of $|\mathbf{k}|$ in this direction.

bonding strength of the fibres from an analysis of the dispersion data for waves travelling in the *xy* plane. In the present model the bonding strength will be defined in terms of the force constants between the fibres. The force constants can be obtained from a measurement of the dispersion in the symmetry directions (or the directional dependence of the velocity in the *xy* plane), as is usually done in the study of the lattice dynamics of crystals by using neutron scattering techniques.

First, let us note that the matrix $\mathbf{D}^p$ is exactly similar to the dynamical matrix of a two-dimensional square lattice of lattice constant equal to *a*, with nearest and next-nearest neighbour interactions which are given by the following force constant matrices:

$$\phi(a, 0) = \phi(\bar{a}, 0) = -\begin{pmatrix} A & 0 \\ 0 & B \end{pmatrix}$$

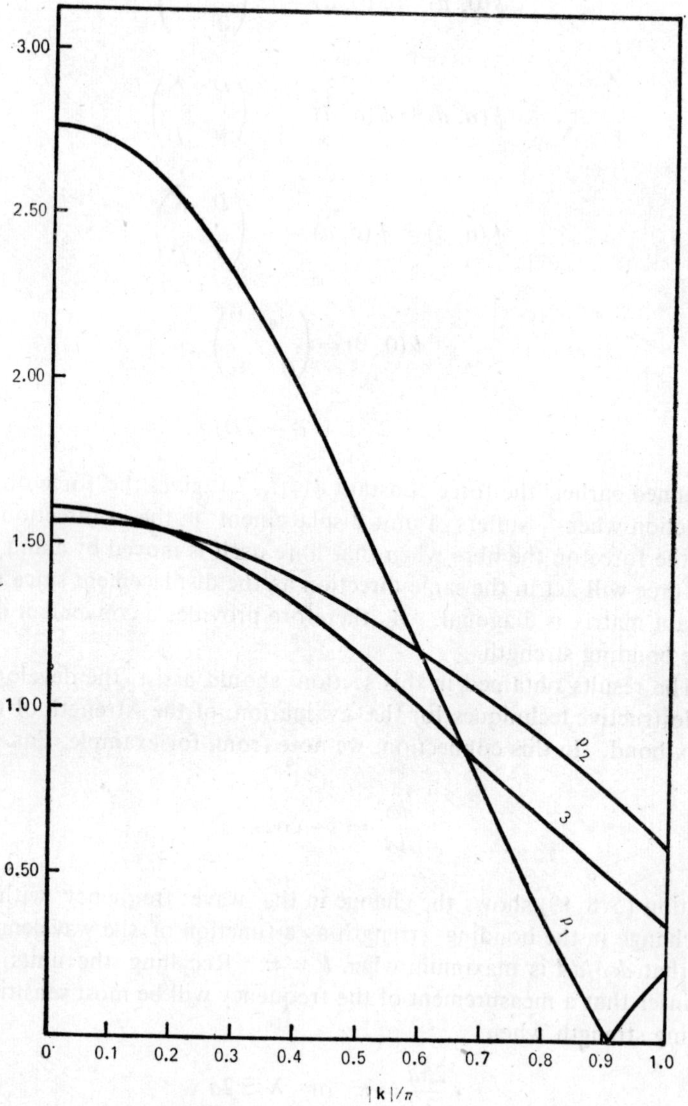

Figure 5.9 Group velocities of the three elastic waves travelling in ⟨100⟩ direction in a tetragonal composite.

$$\phi(0, a) = \phi(0, \bar{a}) = -\begin{pmatrix} B & 0 \\ 0 & A \end{pmatrix}$$

$$\phi(a, a) = \phi(\bar{a}, \bar{a}) = -\begin{pmatrix} D & K \\ K & D \end{pmatrix}$$

$$\phi(a, \bar{a}) = \phi(\bar{a}, a) = -\begin{pmatrix} D & \overline{K} \\ \overline{K} & D \end{pmatrix}$$

and

$$\phi(0, 0) = \begin{pmatrix} A_0 & 0 \\ 0 & A_0 \end{pmatrix}$$

where

$$A_0 = 2(A + B + 2D)$$

As defined earlier, the force constant $\phi_{xy}(l_1, l_2)$ gives the force on l_2 in the y direction when l_1 suffers a unit displacement in the x direction. Thus, A_0 is the force on the fibre when that fibre itself is moved by a unit amount. This force will act in the same direction as the displacement since the force constant matrix is diagonal. A_0 therefore provides a convenient definition of the bonding strength.

The results obtained in this section should assist the development of non-destructive techniques for the evaluation of the strength of the fibre-matrix bond. In this connection, we note from, for example, Eqs. (5.8.13) that

$$\frac{d\omega^2}{dA} = 1 - \cos k \qquad (5.8.19)$$

Equation (5.8.19) shows the change in the wave frequency with respect to a change in the bonding strength as a function of the wavelength. We note that $d\omega^2/dA$ is maximum when $k = \pi$. Recalling the units of k, we may infer that a measurement of the frequency will be most sensitive to the bonding strength when

$$\frac{2\pi a}{\lambda} = \pi \quad \text{or} \quad \lambda \approx 2a \qquad (5.8.20)$$

i.e. when the wavelength of the incident wave is of the order of the inter-fibre spacing. This result, of course, could have been expected intuitively.

5.9 WAVE PROPAGATION IN THE *yz* PLANE

In this section we shall study the propagation of a wave in the yz plane. By symmetry the xz plane is equivalent to the yz plane.

The elements of the dynamical matrix for the present case are obtained from Eqs. (5.5.16)–(5.5.21) by putting $k_1 = 0$. We obtain the block diagonal form for $\mathbf{D}(\mathbf{k})$:

$$\mathbf{D}(\mathbf{k}) = \begin{pmatrix} D^x & 0 \\ 0 & \mathbf{D}^p \end{pmatrix}$$

where

$$D^x = 2c_{66}(1 - \cos k_2) + c_{44} k_3^2 \tag{5.9.1}$$

is a number and the elements of the 2 × 2 matrix $\mathbf{D}^p$ are

$$D_{11}^p = 2c_{11}(1 - \cos k_2) + c_{44} k_3^2 \tag{5.9.2}$$

$$D_{12}^p = D_{21}^p = (c_{13} + c_{44}) k_3 \sin k_2 \tag{5.9.3}$$

$$D_{22}^p = 2c_{44}(1 - \cos k_2) + c_{33} k_3^2 \tag{5.9.4}$$

As in Section 5.8, one of the eigenvalues, the one corresponding to D^x, is immediately known. It is given by

$$\omega_x^2 = 2c_{66}(1 - \cos k_2) + c_{44} k_3^2 \tag{5.9.5}$$

The eigenvector corresponding to this eigenvalue is [1, 0, 0]. This wave is always polarized perpendicular to the *yz* plane for all values of *k* and is a true transverse wave.

If ω_{p_1} and ω_{p_2} represent the frequencies of the two waves polarized at angles θ_p and $\theta_p + \pi/2$, respectively, from the *z* axis in the *yz* plane, then we have

$$\omega_{p_1}^2 = (c_{11} + c_{44})(1 - \cos k_2) + (c_{44} + c_{33}) \frac{k_3^2}{2} + \left[\left\{ (c_{11} - c_{44})(1 - \cos k_2) + (c_{44} - c_{33}) \frac{k_3^2}{2} \right\}^2 + (c_{13} + c_{44})^2 k_3^2 \sin^2 k_2 \right]^{1/2} \tag{5.9.6}$$

$$e_{p_1} = [\sin \theta_p, \cos \theta_p, 0]$$

$$\omega_{p_2}^2 = (c_{11} + c_{44})(1 - \cos k_2) + (c_{44} + c_{33}) \frac{k_3^2}{2} - \left[\left\{ (c_{11} - c_{44})(1 - \cos k_2) + (c_{44} - c_{33}) \frac{k_3^2}{2} \right\}^2 + (c_{13} + c_{44})^2 k_3^2 \sin^2 k_2 \right]^{1/2} \tag{5.9.7}$$

and

$$e_{p_2} = [-\cos \theta_p, \sin \theta_p, 0]$$

where

$$\theta_p = \frac{\pi}{2} - \tfrac{1}{2}\tan^{-1}\frac{(c_{13}+c_{44})\,k_3 \sin k_2}{(c_{11}-c_{44})(1-\cos k_2)+(c_{44}-c_{33})\,k_3^2/2} \tag{5.9.8}$$

We notice that, owing to a relatively large value of c_{33}, the propagation of waves in the yz plane in the present model is almost completely non-dispersive. A measurement of dispersion for waves in this plane is, therefore, not likely to yield any information about the bonding strength of the fibres. However, such a measurement can yield some useful information about the alignment of the fibres, which we shall discuss in the next chapter.

Owing to a relatively large value of c_{33}, θ_p is nearly zero unless $k_3 \approx 0$. This means that for most of the angles of incidence the waves through the composite will be polarised in a direction close to the z axis. Thus, in the present case we cannot label the waves as even quasi-longitudinal and quasi-transverse. We shall therefore label them simply as p_1 and p_2. The wave given by Eq. (5.9.6) will be labelled as the x transverse wave and is always a true transverse wave.

The group velocities for these waves are given below.

p_1 wave:

$$\left.\begin{aligned} C_{g_y} &= \frac{\sin k_2}{2\omega_{p_1}}\left[c_{11}+c_{44}+\frac{P_y}{X}\right] \\ C_{g_z} &= \frac{k_3}{2\omega_{p_1}}\left[c_{44}+c_{33}+\frac{P_z}{X}\right] \\ C_{g_x} &= 0 \end{aligned}\right\} \tag{5.9.9}$$

p_2 wave:

$$\left.\begin{aligned} C_{g_y} &= \frac{\sin k_2}{2\omega_{p_1}}\left[c_{11}+c_{44}-\frac{P_y}{X}\right] \\ C_{g_z} &= \frac{k_3}{2\omega_{p_2}}\left[c_{44}+c_{33}-\frac{P_z}{X}\right] \\ C_{g_x} &= 0 \end{aligned}\right\} \tag{5.9.10}$$

where

$$X^2 = \left[(c_{11}-c_{44})(1-\cos k_2)+(c_{44}-c_{33})\frac{k_3^2}{2}\right]^2 + (c_{13}+c_{44})^2\, k_3^2 \sin^2 k_2$$

$$P_y = \left[(c_{11}-c_{44})(1-\cos k_2)+(c_{44}-c_{33})\frac{k_3^2}{2}\right](c_{11}-c_{44}) + (c_{13}+c_{44})^2\, k_3^2 \cos k_2$$

and

$$P_z = \left[(c_{11} - c_{44})(1 - \cos k_2) + (c_{44} - c_{33})\frac{k_3^2}{2}\right](c_{44} - c_{33})$$
$$+ (c_{13} + c_{44})^2 \sin^2 k_2$$

x transverse wave:

$$\left.\begin{aligned} C_{g_y} &= \frac{c_{66} \sin k_2}{\omega_x} \\ C_{g_z} &= \frac{c_{44}k_3}{\omega_x} \\ C_{g_x} &= 0 \end{aligned}\right\} \qquad (5.9.11)$$

As before, the magnitude and direction relative to the z axis of the group velocity are given by

$$C_g^2 = C_{g_y}^2 + C_{g_z}^2$$

and

$$\tan \theta_g = \frac{C_{g_y}}{C_{g_z}} \qquad (5.9.12)$$

Writing $k_3 = k \cos \theta_k$ and $k_2 = k \sin \theta_k$, we have plotted C_g, θ_p and θ_g as a function of k for the above three waves in Figures 5.10 and 5.11 for $\theta_k = 45°$. The dependence of the above quantities on θ_k is shown in Figures 5.12 and 5.13 for $|\mathbf{k}|/\pi = 0.25$. Figure 5.13 showing variations of θ_p and θ_g with θ_k, is very interesting from an application point of view. First, we notice from the curve for θ_p that for most of the directions of the incident wave one of the outgoing waves is polarized in a direction close to the z axis. Thus, a composite can be used as a polarizer for the stress waves. This effect arises because of a relatively large value of c_{33}, as discussed earlier.

Similarly, we notice from the curves for θ_g that for most of the directions of the incident wave the outgoing wave travels close to the z direction. If we impose a disturbance at one end of a composite, characterized by a set of waves incident at all angles with respect to the z axis, the outgoing disturbance will be largely confined in a narrow cone around the z axis. Thus, the intensity of the disturbance or the energy will increase by a large amount in this narrow region and will be practically zero across the fibres. Thus, let us assume that we have a set of waves incident at all angles ranging from 0 to $\pi/2$ with respect to the z axis at one end of the composite and with a uniform angular density. For the outgoing waves the angular distribution function, i.e. the number of waves emerging at an angle θ from the z axis, is given by

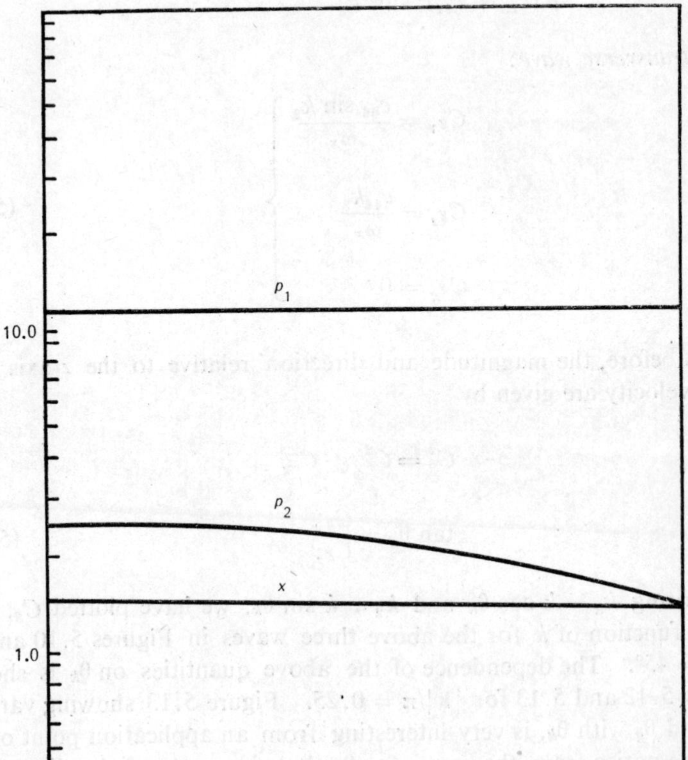

Figure 5.10 Group velocities of the three elastic waves travelling in the yz plane in a tetragonal composite for $\theta_k = 45°$.

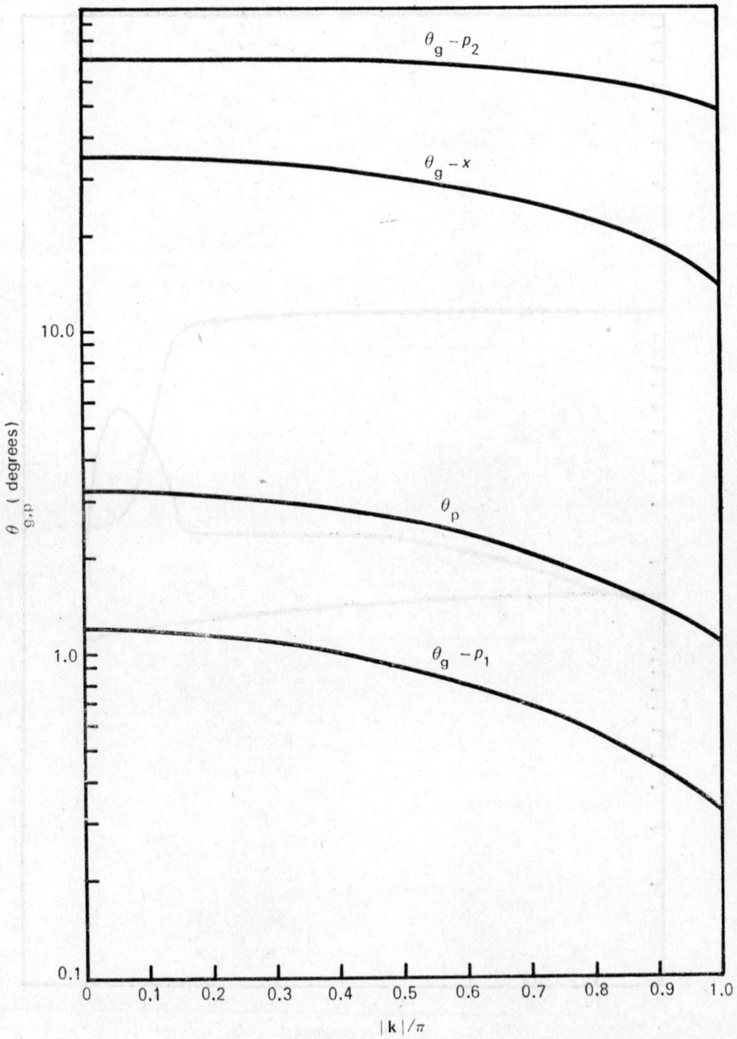

Figure 5.11 k-dependence of the directions of polarization (θ_p) and flux travel (θ_g) for the elastic waves travelling in the yz plane in a tetragonal composite for $\theta_k = 45°$. The angles are measured from the z axis.

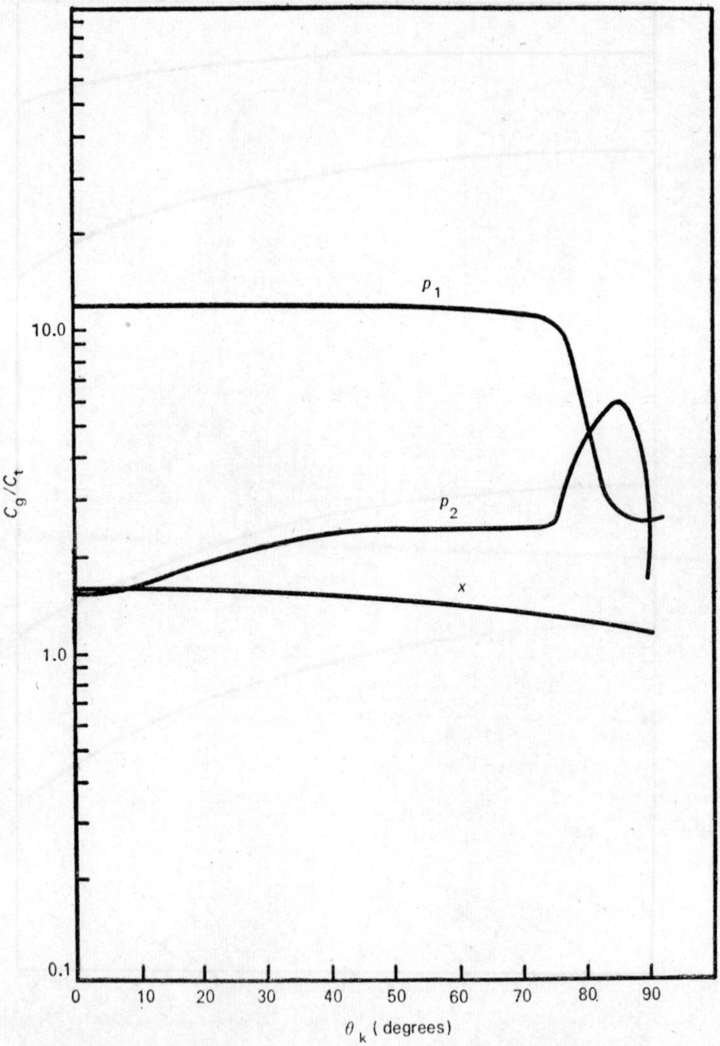

Figure 5.12 Angular dependence of the group velocities of the elastic waves travelling in the *yz* plane in a tetragonal composite for the $|k|/\pi = 0.25$. The angles are measured from the *z* axis.

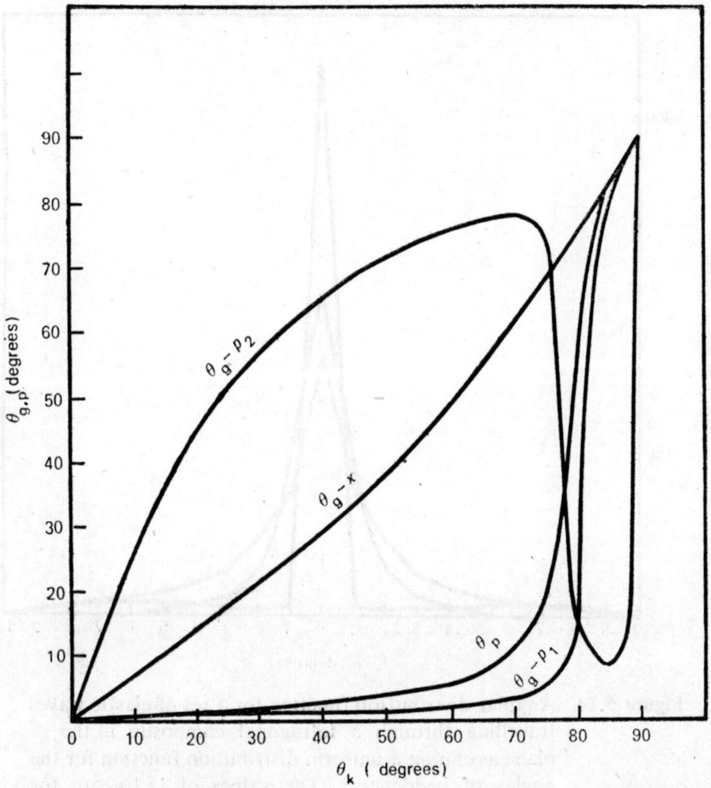

Figure 5.13 **Angular dependence of the directions of polarization (θ_p) and flux (θ_g) for elastic waves in the *yz* plane in tetragonal composite for $|k|/\pi = 0.25$. All the angles are measured from the *z* axis.**

$$g(\theta) = \sum_k \delta(\theta - \theta_g(\theta_k)) \tag{5.9.13}$$

where $\theta_g(\theta_k)$ shows the functional dependence of θ_g on θ_k. In this expression we shall consider only the p_1 wave, which, being mostly polarized along the *z* axis, is of maximum interest.

The function $g(\theta)$, obtained numerically from Eqs. (5.9.9), (5.9.12) and (5.9.13), is plotted against θ in Figure 5.14 for $|k|/\pi = 0.001$, 0.5 and 1.0. To see qualitatively the behaviour of $g(\theta)$, let us consider a case when c_{33} is very large and when the off-diagonal terms in $\mathbf{D}^p$ are negligible. In this case the p_1 wave is always polarized along the *z* axis, Its frequency is given by

$$\omega^2 = 2c_{11}(1 - \cos k_2) + c_{33}k_3^2$$

and the group velocity by

$$C_{g_y} = c_{11}\frac{\sin k_2}{\omega} \tag{5.9.14}$$

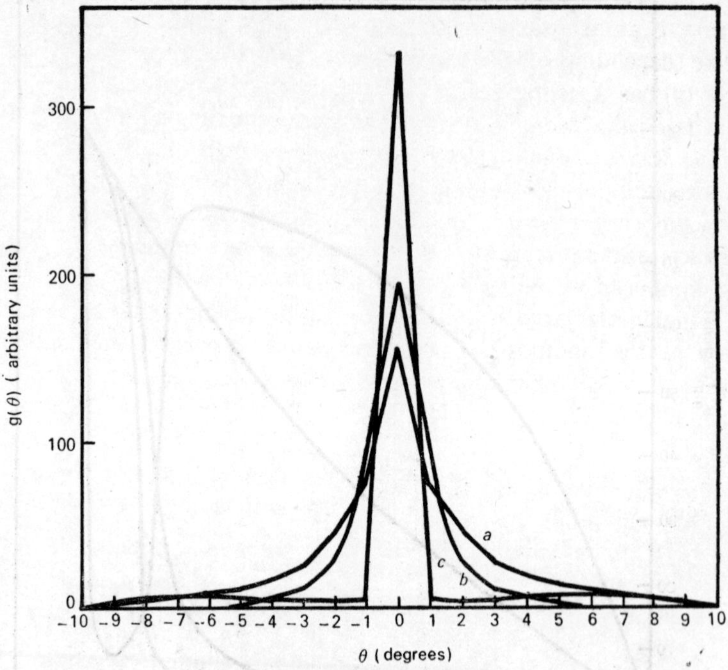

Figure 5.14 Angular distribution function for a set of elastic waves travelling through a tetragonal composite in the yz plane assuming a uniform distribution function for the angles of incidence. The values of $|k|/\pi$ are for (a) 0.0001, (b) 0.5 and (c) 1.0. For $\theta > 10°$, $g(\theta) \approx 0$. The angle θ is measured from the z axis. Only the p_1 waves have been included.

$$C_{g_z} = \frac{c_{33}k_3}{\omega} \tag{5.9.15}$$

and

$$\tan\theta_g(\theta_k) = \frac{c_{11}}{c_{33}} \frac{\sin(\pi \sin\theta_k)}{\pi\cos\theta_k} \tag{5.9.16}$$

where

$$|k| = \pi$$

From Eq. (5.9.16) we see that, since $c_{33} \gg c_{11}$, $\tan\theta_g \approx \theta_g$ is nearly zero unless $\theta_k = \pi/2$ when $\theta_g = \pi/2$. Replacing the sum in Eq. (5.9.13) by an integral, we have

$$g(\theta) \approx \int_0^{\pi/2-\eta} \delta(\theta - \theta_g(\theta_k))\, d\theta_k + \int_{\pi/2-\eta}^{\pi} \delta(\theta - \theta_g(\theta_k))\, d\theta_k$$

$$\approx \left(\frac{\pi}{2} - \eta\right)\delta(\theta) + \eta\,\delta\left(\theta - \frac{\pi}{2}\right) \tag{5.9.17}$$

where η is an arbitrary small number. Since η can be made as small as we like (depending on the magnitude of c_{33}/c_{11}), it follows from Eq. (5.9.17) that $g(\theta)$ has a strong peak at $\theta=0$.

These calculations show that the composite can be used as a collimator of stress waves to obtain stress waves of very high intensity and polarized in a direction parallel to the fibres. It can also be used as a filter of stress waves, since the amount of energy passing across the fibres is small. A possible practical application of the 'filtering' properties of the carbon fibre composite would be in a noise reduction device (Bedwell, 1973).

Finally, the large accumulation of strain energy within a small angular region in the composite may have some important implications. For example, stress waves, after entering a closed continuous structure of carbon fibres (such as structures of cylindrical, spherical or conical shape) cannot come out of it (Tewary and Bullough, 1971). Thus, a large amount of strain energy may build up inside the structure, causing an early failure. It is important, therefore, to study this property of a composite in detail before using it for such structural applications.

5.10 FIBRE COMPOSITE WITH HEXAGONAL SYMMETRY

In this section we shall study the wave propagation in a composite with hexagonal symmetry. In this symmetry, the fibres are assumed to be arranged on the vertices of a regular hexagonal in a plane perpendicular to the fibres. The basic assumptions and the technique for obtaining the the dynamical matrix in this section will be the same as in the preceding sections on tetragonal symmetry. Most of the discussion given in the preceding sections also applies to this section and in this section, therefore, we shall usually give the formulae without any discussion.

The operations of the point group for hexagonal symmetry are given in Appendix 2. As before, the z axis is taken in a direction parallel to the fibres and the x and y axes in a plane perpendicular to the fibres. Figure 5.1b shows the arrangements of the fibres in the xy plane and Figure 5.2b shows the three dimensional lattice structure. The unit cell for the present model is a rhombohedral prism, shown as the shaded region in Figure 5.2b and each unit cell contains one object. If the length of each side of the rhombus base is $2a$ and c is the height of the prism (distance between two adjacent layers) then the volume of a unit cell is given by

$$V = 2\sqrt{3}\, a^2 c \tag{5.10.1}$$

The effective mass of each object is defined in terms of V and ρ, the average density of the composite and c is defined in terms of the Debye temperature for the composite in the z direction.

For the calculation of the dynamical matrix we have assumed only nearest neighbour interactions between the fibres. The coordinates of the objects contributing to the dynamical matrix are given in Table 5.3. In the present case the dynamical matrix depends on only six independent parameters which can be determined in terms of the elastic constants. For a more realistic calculation, one should include the effect of further neighbours. However, in the absence of any short wavelength dispersion measurements it does not seem worthwhile to introduce further unknown parameters. In this section numerical results have been obtained for a carbon fibres–epoxy resin composite using the elastic constants as obtained by Heaton (1970).

TABLE 5.3

Coordinates of the 'objects' contributing to the dynamical matrix for a composite with hexagonal symmetry

Number	Object	Coordinates		
		x	y	z
1	0	0	0	0
2	1	$2a$	0	0
3	2	a	$a\sqrt{3}$	0
4	3	$-a$	$a\sqrt{3}$	0
5	4	$-2a$	0	0
6	5	$-a$	$-a\sqrt{3}$	0
7	6	a	$-a\sqrt{3}$	0
8	0′	0	0	c
9	1′	$2a$	0	c
10	2′	a	$a\sqrt{3}$	c
11	3′	$-a$	$a\sqrt{3}$	c
12	4′	$-2a$	0	c
13	5′	$-a$	$-a\sqrt{3}$	c
14	6′	a	$-a\sqrt{3}$	c
15	0″	0	0	$-c$
16	1″	$2a$	0	$-c$
17	2″	a	$a\sqrt{3}$	$-c$
18	3″	$-a$	$a\sqrt{3}$	$-c$
19	4″	$-2a$	0	$-c$
20	5″	$-a$	$-a\sqrt{3}$	$-c$
21	6″	a	$-a\sqrt{3}$	$-c$

Force Constants

$$\phi\,(0,1)=\phi\,(0,4)=-\begin{pmatrix}\mu & 0 & 0\\ 0 & \lambda & 0\\ 0 & 0 & \delta\end{pmatrix}$$

$$\phi\,(0,2)=\phi\,(0,5)=-\tfrac{1}{4}\begin{pmatrix}\mu+3\lambda & \sqrt{3}\,(\mu-\lambda) & 0\\ \sqrt{3}\,(\mu-\lambda) & 3\mu+\lambda & 0\\ 0 & 0 & 4\delta\end{pmatrix}$$

$$\phi\,(0,3)=\phi\,(0,6)=-\tfrac{1}{4}\begin{pmatrix}\mu+3\lambda & -\sqrt{3}\,(\mu-\lambda) & 0\\ -\sqrt{3}\,(\mu-\lambda) & 3\mu+\lambda & 0\\ 0 & 0 & 4\delta\end{pmatrix}$$

$$\phi\,(0,0')=\phi\,(0,0'')=-\begin{pmatrix}\mu_0' & 0 & 0\\ 0 & \mu_0' & 0\\ 0 & 0 & \delta_0'\end{pmatrix}$$

$$\phi(0,1')=\phi(0,4'')=-\begin{pmatrix}\mu' & 0 & \nu'\\ 0 & \lambda' & 0\\ \nu' & 0 & \delta'\end{pmatrix}$$

$$\phi(0,1'')=\phi(0,4')=-\begin{pmatrix}\mu' & 0 & -\nu'\\ 0 & \lambda' & 0\\ -\nu' & 0 & \delta'\end{pmatrix}$$

$$\phi(0,2')=\phi(0,5'')=-\tfrac{1}{4}\begin{pmatrix}\mu'+3\lambda' & \sqrt{3}(\mu'-\lambda') & 2\nu'\\ \sqrt{3}(\mu'-\lambda') & 3\mu'+\lambda' & 2\nu'\sqrt{3}\\ 2\nu' & 2\nu'\sqrt{3} & 4\delta'\end{pmatrix}$$

$$\phi(0,2'')=\phi\,(0,5')=-\tfrac{1}{4}\begin{pmatrix}\mu'+3\lambda' & \sqrt{3}\,(\mu'-\lambda') & -2\nu'\\ \sqrt{3}(\mu'-\lambda') & 3\mu'+\lambda' & -2\nu'\sqrt{3}\\ -2\nu' & -2\nu'\sqrt{3} & 4\delta'\end{pmatrix}$$

$$\phi(0,3') = \phi(0,6'') = -\tfrac{1}{4}\begin{pmatrix} \mu' + 3\lambda' & -\sqrt{3}(\mu' - \lambda') & -2\nu' \\ -\sqrt{3}(\mu' - \lambda') & 3\mu' + \lambda' & 2\nu'\sqrt{3} \\ -2\nu' & 2\nu'\sqrt{3} & 4\delta' \end{pmatrix}$$

$$\phi(0,3'') = \phi(0,6') = -\tfrac{1}{4}\begin{pmatrix} \mu' + 3\lambda' & -\sqrt{3}(\mu' - \lambda') & 2\nu' \\ -\sqrt{3}(\mu' - \lambda') & 3\mu' + \lambda' & -2\nu'\sqrt{3} \\ 2\nu' & -2\nu'\sqrt{3} & 4\delta' \end{pmatrix}$$

and

$$\phi(0,0) = \begin{pmatrix} \mu_0 & 0 & 0 \\ 0 & \mu_0 & 0 \\ 0 & 0 & \delta_0 \end{pmatrix}$$

where

$$\mu_0 = 2\mu_0' + 3\mu + 3\lambda + 6\mu' + 6\lambda'$$

and

$$\delta_0 = 2\delta_0' + 6\delta + 12\delta'$$

Dynamical matrix. In what follows, c is in units of a, the force constants are in units of $1/m$, k_1 and k_3 replace $2\pi a k_1$ and $2\pi a k_3$ respectively and k_2 replaces $2\pi a\sqrt{3}\, k_2$:

$$\left.\begin{aligned} D_{11}(\mathbf{k}) &= 2A(1 - \cos 2k_1) + (A + 3B)(1 - \cos k_1 \cos k_2) + Ek_3^2 \\ D_{22}(\mathbf{k}) &= 2B(1 - \cos 2k_1) + (3A + B)(1 - \cos k_1 \cos k_2) + Ek_3^2 \\ D_{33}(\mathbf{k}) &= 2D(1 - \cos 2k_1) + 4D(1 - \cos k_1 \cos k_2) + Hk_3^2 \\ D_{12}(\mathbf{k}) &= D_{21}(\mathbf{k}) = \sqrt{3}(A - B)\sin k_1 \sin k_2 \\ D_{13}(\mathbf{k}) &= D_{31}(\mathbf{k}) = Fk_3 \sin k_1 (2\cos k_1 + \cos k_2) \\ D_{23}(\mathbf{k}) &= D_{32}(\mathbf{k}) = F\sqrt{3}\, k_3 \sin k_2 \cos k_1 \end{aligned}\right\} \quad (5.10.2)$$

where

$$\left.\begin{aligned} A &= \mu + 2\mu' \\ B &= \lambda + 2\lambda' \\ E &= \mu_0' c^2 \\ H &= \delta_0' c^2 \\ D &= \delta + 2\delta' \\ F &= 4\nu' \end{aligned}\right\} \quad (5.10.3)$$

Dynamical matrix in the long wavelength limit

$$\left.\begin{aligned}
D^0_{11}(\mathbf{k}) &= \frac{3}{2}(3A+B)\,k_1^2 + \frac{1}{2}(A+3B)\,k_2^2 + Ek_3^2 \\
D^0_{22}(\mathbf{k}) &= \frac{3}{2}(A+3B)\,k_1^2 + \frac{1}{2}(3A+B)k_2^2 + Ek_3^2 \\
D^0_{33}(\mathbf{k}) &= 6\,Dk_1^2 + 2D\,k_2^2 + Hk_3^2 \\
D^0_{12}(\mathbf{k}) &= D^0_{21}(\mathbf{k}) = \sqrt{3}\,(A-B)\,k_1k_2 \\
D^0_{13}(\mathbf{k}) &= D^0_{31}(\mathbf{k}) = 3F\,k_1k_3 \\
D^0_{23}(\mathbf{k}) &= D^0_{32}(\mathbf{k}) = \sqrt{3}F\;k_2k_3
\end{aligned}\right\} \quad (5.10.4)$$

Relations between the force constants and the elastic constants

$$\left.\begin{aligned}
A &= \frac{3c_{11}-c_{66}}{12} \\
B &= \frac{3c_{66}-c_{11}}{12} \\
E &= c_{44} \\
D &= \frac{c_{44}}{6} \\
H &= c_{33} \\
F &= \frac{c_{13}+c_{44}}{3}
\end{aligned}\right\} \quad (5.10.5)$$

Wave propagation in the xy plane ($k_3 = 0$)

$$D(\mathbf{k}) = \begin{pmatrix} D^p & 0 \\ 0 & D^z \end{pmatrix} \quad (5.10.6)$$

where

$$D^p_{11} = 2A(1-\cos 2k_1) + (A+3B)(1-\cos k_1 \cos k_2)$$

$$D^p_{12} = D^p_{21} = \sqrt{3}(A-B)\sin k_1 \sin k_2$$

$$D^p_{22} = 2B(1-\cos 2k_1) + (B+3A)(1-\cos k_1 \cos k_2)$$

and

$$D^z = 2D(1-\cos 2k_1) + 4D(1-\cos k_1 \cos k_2)$$

Quasi-longitudinal wave

$$\omega_{p_1}^2 = (A + B)(3 - 2\cos k_1 \cos k_2 - \cos 2k_1) + (A - B)X \quad (5.10.7)$$

$$e_{p_1} = [\cos\theta_p, \sin\theta_p, 0] \quad (5.10.8)$$

$$Cg_x = \frac{\sin k_1}{2\omega_{p_1}}\left[2(A + B)(1 + \cos k_2) + \frac{(A - B)}{X}P_x\right] \quad (5.10.9)$$

$$Cg_y = \frac{\sin k_2}{2\omega_{p_1}}\left[2(A + B)\cos k_1 + \frac{A - B}{X}P_y\right] \quad (5.10.10)$$

$$Cg_z = 0$$

Quasi-transverse wave

$$\omega_{p_2}^2 = (A + B)(3 - 2\cos k_1 \cos k_2 - \cos 2k_1) - (A - B)X \quad (5.10.11)$$

$$e_{p_2} = [-\sin\theta_p, \cos\theta_p, 0] \quad (5.10.12)$$

$$Cg_x = \frac{\sin k_1}{2\omega_{p_2}}\left[2(A + B)(2\cos k_1 + \cos k_2) - \frac{(A - B)}{X}P_x\right] \quad (5.10.13)$$

$$Cg_y = \frac{\sin k_2}{2\omega_{p_2}}\left[2(A + B)\cos k_1 - \frac{(A - B)}{X}P_y\right] \quad (5.10.14)$$

$$Cg_z = 0$$

where

$$X^2 = (\cos k_1 \cos k_2 - \cos 2k_1)^2 + 3\sin^2 k_1 \sin^2 k_2$$

$$P_x = (4\cos k_1 - \cos k_2)(\cos k_1 \cos k_2 - \cos 2k_1) + 3\sin^2 k_2 \cos k_1$$

$$P_y = -\cos k_1 (\cos k_1 \cos k_2 - \cos 2k_1) + 3\sin^2 k_1 \cos k_2$$

and θ_p, the angle of polarization from the x axis is given by

$$\tan 2\theta_p = \frac{\sqrt{3}\sin k_1 \sin k_2}{\cos k_1 \cos k_2 - \cos 2k_1} \quad (5.10.15)$$

z transverse wave :

$$\omega_z^2 = 2D(1 - \cos 2k_1) + 4D(1 - \cos k_1 \cos k_2) \quad (5.10.16)$$

$$e_z = [0,0,1] \quad (5.10.17)$$

$$Cg_x = \frac{2D\sin k_1}{\omega_z}\left[2\cos k_1 - \cos k_2\right] \quad (5.10.18)$$

$$C_{g_y} = \frac{2D \sin k_2 \cos k_1}{\omega_z} \tag{5.10.19}$$

$$C_{g_z} = 0$$

We notice from Eq. (5.10.15) that the angle of polarization is independent of the force constants. In the continuum limit, using the method of long waves, we find that $\theta_p = \theta_k$. In this limit the waves p_1 and p_2 are true longitudinal and transverse waves, respectively, for all values of **k**. This fact is a consequence of the transverse elastic isotropy of a hexagonal crystal.

⟨100⟩ *direction* ($k_1 = k,\ k_2 = k_3 = 0$)

Longitudinal wave :

$$\omega^2 = (1 - \cos k)\,[A\,(4 \cos k + 5) + 3B]$$

$$C_g = \frac{A\,(1 + 8 \cos k) + 3B}{[2A\,(4 \cos k + 5) + 6B]^{1/2}} \cos \frac{k}{2}$$

p transverse wave :

$$\omega^2 = (1 - \cos k)\,[B\,(4 \cos k + 5) + 3A]$$

$$C_g = \frac{B\,(1 + 8 \cos k) + 3A}{[2B\,(4 \cos k + 5) + 6A]^{1/2}} \cos \frac{k}{2}$$

z transverse wave :

$$\omega^2 = 4D\,(1 - \cos k)\,(\cos k + 2)$$

$$C_g = \frac{2D\,(2 \cos k + 1)}{[2D\,(\cos k + 2)]^{1/2}} \cos \frac{k}{2}$$

⟨010⟩ *direction* ($k_1 = 0,\ k_2 = k,\ k_3 = 0$)

Longitudinal wave :

$$\omega^2 = (A + 3B)\,(1 - \cos k)$$

$$C_g = \sqrt{\frac{A + 3B}{2}} \cos \frac{k}{2}$$

p transverse wave :

$$\omega^2 = (B + 3A)\,(1 - \cos k)$$

$$C_g = \sqrt{\frac{B + 3A}{2}} \cos \frac{k}{2}$$

z transverse wave :

$$\omega^2 = 4D\,(1 - \cos k)$$

$$C_g = \sqrt{2D} \cos \frac{k}{2}$$

Wave propagation in yz plane

$$D\,(k) = \begin{pmatrix} D^x & 0 \\ 0 & D^p \end{pmatrix}$$

where

$$D^x = (A + 3B)\,(1 - \cos k_2) + Ek_3^2$$

$$D^p_{11} = (3A + B)\,(1 - \cos k_2) + Ek_3^2$$

$$D^p_{12} = D^p_{21} = F\sqrt{3}\, k_3 \sin k_2$$

$$D^p_{22} = 4D\,(1 - \cos k_2) + Hk_3^2$$

x transverse wave :

$$\omega^2_x = (A + 3B)\,(1 - \cos k_2) + Ek_3^2$$

$$e_x = [1,0,0]$$

$$C_{g_y} = \frac{(A + 3B) \sin k_2}{2\,\omega_x}$$

$$C_{g_z} = \frac{Ek_3}{\omega_x}$$

$$C_{g_x} = 0$$

p_1 *wave* :

$$\omega^2_{p_1} = R_1\,(1 - \cos k_2) + T_1 k_3^2 + X$$

$$e_{p_1} = [0, \sin \theta_p, \cos \theta_p]$$

$$C_{g_y} = \frac{\sin k_2}{2\omega_{p_1}}\left[R_1 + \frac{P_y}{X} \right]$$

$$C_{g_z} = \frac{k_3}{2\omega_{p_1}}\left[2T_1 + \frac{P_z}{X} \right]$$

$$C_{g_x} = 0$$

p_2 *wave:*

$$\omega_{p_2}^2 = R_1 (1 - \cos k_2) + T_1 k_3^2 - X$$

$$c_{p_2} = [0, -\cos \theta_p, \sin \theta_p]$$

$$C_{g_y} = \frac{\sin k_2}{2\omega_{p_2}} \left[R_1 - \frac{P_y}{X} \right]$$

$$C_{g_z} = \frac{k_3}{2\omega_{p_2}} \left[2T_1 - \frac{P_z}{X} \right]$$

$$C_{g_x} = 0$$

where

$$R_1 = \tfrac{1}{2} (3A + B + 4D)$$

$$T_1 = \tfrac{1}{2} (E + H)$$

$$R_2 = \tfrac{1}{2} (3A + B - 4D)$$

$$T_2 = \tfrac{1}{2} (E - H)$$

$$X^2 = [R_2 (1 - \cos k_2) + T_2 k_3^2]^2 + 3F^2 k_3^2 \sin^2 k_2$$

$$P_y = R_2 [R_2 (1 - \cos k_2) + T_2 k_3^2] + 3F^2 k_3^2 \cos k_2$$

$$P_z = 2T_2 [R_2 (1 - \cos k_2) + T_2 k_3^2] + 3F^2 \sin^2 k_2$$

and θ_p, the angle of polarisation with respect to the z axis in the yz plane, is given by

$$\theta_p = \frac{\pi}{2} - \frac{1}{2} \tan^{-1} \frac{\sqrt{3}\, F k_3 \sin k_2}{R_2 (1 - \cos k_2) + T_2 k_3^2}$$

5.11 DISCUSSION

In the theory of elastic wave propagation in composite materials, as described in this chapter, we have neglected three effects which may be important—namely, the effects due to (1) anharmonic forces, (2) long-range elastic interactions between the fibres and (3) the finite thickness of the fibres. Of these three, the contribution of the anharmonic forces may be most significant (see Curtis et al. 1968 for the non-Hookean behaviour of carbon fibres). The anharmonic effects in the present model will lead to the attenuation of the elastic waves which we have completely neglected. The effect of finite thickness of the fibres has been partially included in the present theory by our defining an effective mass of the fibres and using the averaged elastic constants and the density of the whole composite.

However, the finite thickness of the fibres will also be manifest through the effects of phonon–phonon interaction and the absorption of elastic waves due to various defects in the atomic structure of the fibres. The contribution due to these effects will increase with an increase in the fibre concentration of the composite and a decrease in the wavelength of the elastic wave. A proper determination of the magnitude of these and other corrections has to await a rigorous experimental study of shortwave propagation in composites.

Nevertheless, some interesting qualitative inferences can be derived from the numerical results reported in this paper. We notice from Figures 5.3–5.9 that, as previously asserted, wave propagation in the xy plane is strongly dispersive and quite non-dispersive in the yz plane, as seen from Figures 5.10–5.13. The effect of the fibre concentration, and therefore a change in the force constants, on dispersion for the wave travelling in the xy plane can be estimated from Figure 5.6. We notice that the change in $\omega^2(k)$ is a maximum when $|k| = \pi$ or the wavelength is equal to twice the spacing between the fibres in a tetragonal composite. The vanishing of the group velocity at $|k| = \pi$, predicting a Bragg-type reflection of the elastic waves from a composite, is shown in Figure 5.7. The group velocity may also vanish at other points in k space, such as, for example, those in Figure 5.9. This, however, is accidental. It depends upon the choice of the force constants and is not a basic property of the composite.

It can be verified that the nature of the elastic wave propagation in the yz plane is quite similar for a tetragonal and for a hexagonal composite (Tewary and Bullough, 1972). This is obvious on physical grounds, since the wave travelling along the fibres will not notice the arrangement of the fibres in the xy plane. Similarly, a long elastic wave, even if travelling in the xy plane, will also not be very sensitive to the arrangement of the fibres, since it sees only a quasi-homogeneous structure of the composite. The situation is different for a short elastic wave travelling in the xy plane, which will be extremely sensitive to the geometry of the arrangement of the fibres in the composite. Thus, information about the arrangement of the fibres can be obtained from a study of the propagation of short elastic waves in the xy plane.

As can be seen from Figures 5.12 and 5.13, the dependence of C_g, θ_p and θ_g on θ_k for a wave travelling in the yz plane shows a strong transitional behaviour near $\theta_k = 90°$. This occurs because c_{33} is very much larger than c_{11} or c_{44}. For most values of θ_k, the wave propagation is determined by c_{33}, which has a non-dispersive contribution. For $\theta_k \approx 90°$, the contribution of c_{11} or c_{44} becomes comparable to that of c_{33}. For short elastic waves it marks the onset of the dispersive effects. This can be easily seen from, for example, Eq. (5.9.8), where the contribution of the dispersive term containing $(1 - \cos k_2)$ is appreciable only when k_3, i.e. $\cos \theta_k$ is of the order of $(c_{11}/c_{33})^{1/2}$. The denominator in Eq. (5.9.8) will become zero when θ_k satisfies the equation

$$\frac{1-\cos(k \sin \theta_k)}{k^2 \cos^2 \theta_k} = \frac{c_{33}-c_{44}}{2(c_{11}-c_{44})}$$

On approaching this value, the curve for θ_p shows a steep rise, as in Figure 5.13; the transitional behaviour of other curves near $\theta_k = 90°$ can also be interpreted similarly.

The collimation properties of a tetragonal composite for elastic waves are shown in Figure 5.14. The corresponding curve for a hexagonal composite will be quite similar, in view of the earlier discussion, since we are dealing here with wave propagation in the yz plane. Figure 5.14 shows a plot of the angular distribution function $g(\theta)$ vs. θ for the p_1 wave; $g(\theta)\, d\theta$ defines the number of waves emerging between the angles θ and $\theta + d\theta$ from the z axis, when a set of elastic waves is incident at one end of the composite at angles of incidence ranging from $-90°$ to $90°$ from the z axis. For the calculation of $g(\theta)$, we have assumed that $|\mathbf{k}|$ is the same for all the incident waves and that the angles of incidence are uniformly distributed, i.e. the number of waves incident between the angles θ and $\theta + d\theta$ is the same for all values of θ, where the range of θ is from $-90°$ to $90°$.

We notice from Figure 5.14 that $g(\theta)$ is very small for $|\theta| > 5°$ compared with the height of the peak at $\theta = 0°$. This means that most of the elastic energy will travel within the angles $\pm 5°$ from the fibres and very little across the fibres. The height and the inverse width of the peak at $\theta = 0°$ increase with a decrease in the wavelength of the incident waves. For $|\mathbf{k}| = \pi$, nearly 93% of the incident energy travels within the angles $\pm 1°$ from the fibres. The distribution function for the other two waves is more or less constant with respect to θ; the p_2 waves do show a little accumulation near $\theta = 0$.

For most angles of incidence the p_1 wave is polarized in a direction close to the fibre axes (Figure 5.13); thus, on travelling through a composite, the original set of elastic waves will emerge as a highly collimated beam of more or less longitudinally polarized elastic waves. Such a large accumulation of energy seems to be very interesting and significant from an application point of view.

Alternatively, the composite can be used as a filter for elastic waves which may be useful for noise reduction devices. We have seen that the amount of elastic energy travelling across the fibres is relatively small. Some structural applications of this principle are also possible. For example, if two structures are joined to each other through a composite in such a way that the line joining the two structures is perpendicular to the fibres, each structure will be more or less unaffected by the stress waves produced in the other one.

An important inference which can be derived from these calculations is that from a theoretical as well as an application point of view a study of the propagation of short elastic waves in a composite would be worthwhile. To estimate the frequency of the ultrasonic generator required for this purpose we can use Eqs. (5.8.13)–(5.8.15). The maximum frequency for the longitudinal wave is given by

$$\omega^2 = 4\,(A + 2D)$$

or

$$\nu^2 = \frac{4 \times c_{11}}{2\pi^2 \times \rho \times 4\pi^2 \times a^2}$$

Taking c_{11}=10 GN/m², $a = 10^{-6}$ m and $\rho = 2 \times 10^3$ kg/m³, we get

$$\nu = 11.3 \text{ MHz}.$$

The corresponding frequency will be even less for the two transverse waves. It should not be too difficult to generate ultrasonic waves of this frequency, and, using such short waves, one can measure the dispersion relations in the composite. With the help of such observed dispersion relations the force constants and therefore the bonding strength of the fibres can be estimated. Some estimate of the force constants can also be obtained by a measurement of the thermodynamic quantities. The calculation of thermodynamic quantities is outside the scope of this book and the effect of bonding defects on the dispersion relations using the present model will be given in the next chapter.

It may be mentioned here that, as shown by Leigh et al. (1971), it may not be possible to determine the force constants uniquely by a measurement of the frequencies alone, and one has to measure the polarization vectors as well. This conclusion has been derived for dispersion curves obtained by neutron scattering from a crystal lattice and will also apply to some extent to the present case of the scattering of elastic waves from composites. However, it may be that the uncertainty in the force constants between fibres in a composite due to an uncertainty in the eigenvectors is much less than the corresponding uncertainty in a crystal lattice. This is because the elastic displacements of fibres can be determined from the theory of elasticity, unlike the atomic displacements in a crystal lattice.

Finally, we shall summarise the main results of the present discussion.

1. The bonding strength of the fibres in a composite can be measured by using short elastic waves. The measurement will be most sensitive to the bonding strength if the wave is travelling in a plane perpendicular to the fibres and its wavelength is of the order of the spacing between the

fibres. Such a measurement will also give some information about the arrangement of the fibres in the composite.

2. A wave of wavelength equal to twice the fibre spacing, travelling in a plane perpendicular to the fibres, will be reflected from the composite, a phenomenon which is analogous to the Bragg reflection of X-rays from crystals. This, however, is a consequence of the assumption of the periodic arrangement of the fibres, and may not be exactly obeyed by a real composite. Certainly, for a densely reinforced composite one should be able to observe an appreciable reduction in the group velocity for a wave of wavelength equal to *twice the average* fibre spacing.

3. If a set of elastic waves is incident at one end of a composite, a large percentage of the incident elastic energy will travel along the fibres and very little across the fibres. Thus, a composite can be used as a collimator and filter for elastic waves.

The qualitative results as summarized above are quite independent of the details of the model. They can be attributed to the strongly anisotropic and fibrous nature of the composite. Most of the experimental work reported in the literature so far is confined to long waves and therefore it is not possible at present to verify the quantitative predictions of the present model for the propagation of short waves. Recently, however, an experimental study of the geometrical dispersion of acoustic waves in a tungsten wire/aluminium matrix composite has been carried out by Sutherland and Lingle (1972). Their results are in good qualitative agreement with those predicted by the theory given in this chapter.

We shall now give a brief description of some more recent work on wave propagation in composite materials.

Rose et al. (1974) have presented analysis of wave profiles in a unidirectionally reinforced graphite-epoxy plate. Wave velocities in graphite epoxy shells have been determined by Rose and Mortimer (1974). Nemat-Nasser and Fu (1974) have calculated bounds on frequencies for harmonic waves in layered composites. Bose and Mal (1974) have studied propagation of elastic waves in a fibre composite on the continuum model. The model is based on long parallel randomly distributed fibres embedded in homogeneous isotropic medium.

Propagation of low frequency elastic disturbances in a composite material has been discussed by Kohn (1974). Similar study for a three dimensional composite material with a detailed mathematical analysis has been reported by Kohn (1975). Measurement of elastic moduli of continuous filament and eutectic composite materials using ultrasonic wave propagation has been described by Sachse (1974).

Nelson and Navi (1975) have discussed the propagation of harmonic waves in composite materials. The harmonic waves in cases of one, two and three dimensional composites have been studied by Nemat-Nasser et al. (1975). In this paper bounds for eigen frequencies have been obtained.

Drumheller and Norwood (1975) have discussed the behaviour of stress waves in composite materials and derived a 'universal' set of boundary conditions. A theoretical and experimental study on the effect of constituents debonding on the stress waves in composite materials has been reported by Drumheller and Lundergan (1975).

Turhan (1975) has used a technique based on the dispersive wave propagation as given earlier in this section to study the mechanical behaviour of fibre reinforced composites. A crystal lattice type model has been used. The parameters of this model are related to the overall effective moduli of the composite in the limit of long waves.

Mukherjee and Lee (1975) have calculated dispersion relations and mode shapes for waves using Fluquet theory in laminated viscoelastic composites by finite difference methods.

The propagation of elastic waves along the fibres in oriented glass reinforced plastic has been studied by Davydov et al. (1975). In this paper the propagation of axisymmetric longitudinal and torsional waves have been considered. The dispersion equations are obtained for a three dimensional network of fibres. Two main results are obtained—(i) the longitudinal wave velocity is not sensitive to the fibre–fibre spacing in the low frequency limit and (ii) the torsional wave velocity is not sensitive to the fibre–fibre interaction.

A review of elastic wave propagation in fibre composites has been given by Ross et al. (1975). The authors indicate that the elastic wave velocities in one of the principal directions can be determined by using either elastic impact or ultrasonic methods. However, in directions other than the principal, this measurement is quite different.

The attenuation and dispersion characteristics of various reinforced plastics in the frequency range 1–30 MHz. have been described by Felix (1974). Pompe and Schultrich (1974) have studied interface damping of composite materials with discontinuous fibres. No experimental verifications are reported by these authors on interface damping.

5.12 EXPERIMENTAL STUDY OF WAVE PROPAGATION IN FIBRE COMPOSITES: NON-DESTRUCTIVE TESTING

In the previous sections of this chapter we have showed that wave propagation can yield useful information about various mechanical properties of a fibre composite. In the following chapter we shall see how certain types of imperfections in a composite can be studied with the help of wave propagation. Since the propagation of an elastic wave does not normally damage a composite, it provides a convenient technique for the non-destructive testing of a fibre composite. An elastic wave for this purpose can, of course, be easily produced by standard ultrasonics generators (transducers).

In this section we shall briefly describe some experimental techniques for the study of wave propagation in fibre composites which can be used for non-destructive testing purposes. Various other non-destructive testing techniques for fibre composites have also been developed, such as acoustic emission and radiography using X-rays.

The technique of acoustic emission is based on the fact that the solids emit sound waves before failure when subjected to a high stress. An analysis of the emitted sound can yield information about certain types of damage in the composite. On the other hand, X-ray radiography is a direct observation technique and is particularly useful for studying the fibre alignment and also for locating cracks and studying their propagation. For details of these and other testing techniques see, for example, Reynolds and Hancox (1970).

One important advantage of the ultrasonic technique is that it is capable of yielding rapid and accurate results for the bulk properties, such as the elastic moduli of the composite. Moreover, as we have seen previously, a short ultrasonic wave can also be used to study several 'microscopic' features of the composite, such as the fibre–fibre force constants. There are some experimental problems in the use of the ultrasonic technique for fibre composites due to the highly anisotropic nature of the composite and attenuation of the wave pulses. However, these problems have been largely overcome in the recent developments in this field (see, for example, Reynolds, 1969, 1971 and Curtis 1969).

We shall briefly describe here three main methods of ultrasonic measurements in fibre composites.

1. *Transit time measurement*: This method has been used by several authors, such as, for example, Elvery and Nwokoye (1969) and Zimmerman and Cost (1970) (see also Curtis, 1969). This is a direct method of measurement of the ultrasonic velocity in the sense that the distance travelled by the wave and the time taken are directly measured. In this method the transmitter and the receiver are directly coupled to the sample. The measurements are made with the help of a double-beam oscillograph and a properly calibrated delay line for the initial electrical impulse. The problem of signal attenuation can be solved by using a sensitive amplifying mechanism and also by employing a large spectrum of ultrasonic frequencies. The difficulty, however, arises in the interpretation of experimental data, on account of the perturbations in the pulse caused by various internal reflections in the composite.

2. *Acoustic resonance method*: This method was suggested by Bell (1968) and is particularly suitable for low-frequency (in the region of kilohertz) ultrasonic waves. The equipment consists of a long rod which is mechanically coupled to the specimen. Pulses of shear or compressional waves are transmitted through the rod along its axis and the reflected signal is studied with the help of an oscilloscope. This method has recently been used by Brown et al. (1972) for studying the effects of surface treatment on the fibres and their adhesion to the matrix in the case of carbon-fibre-reinforced plastics.

3. *Immersion bath technique* (*Ultrasonic Goniometer*). This technique is based on the principle of total reflection at the interface of two materials with different wave propagation characteristics.

Let us consider two materials A and B, where A is the reference material having known wave propagation characteristics and B is the material whose properties have to be determined. The velocities of a wave (compressional or shear) in A and B will be denoted by C_A and C_B, where $C_B > C_A$. A wave coming from A at the angle of incidence θ_A will be refracted into B at the angle of refraction θ_B, where

$$\frac{\sin \theta_A}{C_A} = \frac{\sin \theta_B}{C_B} \tag{5.12.1}$$

and the angles are measured from the normal at the interface.

Since $C_B > C_A$, θ_B will be greater than θ_A. At the critical value of θ_A, θ_B will be $\pi/2$. Then the wave will be totally reflected in A according to the laws of reflection. By detecting the reflected wave, the critical value of θ_A can be easily measured, which from Eq. (5.12.1) gives C_B, provided of course C_A is known. Since the values of C_B are, in general, different for waves of different polarization (longitudinal or transverse), the corresponding values of θ_A for these waves will also be different.

In actual practice the reference material A is chosen to be a suitable liquid and B is a sample of suitable size from the material which is to be examined. This technique seems to be potentially the most powerful of all the ultrasonic techniques and has the great advantage of yielding very rapid and accurate results. A description of ultrasonic goniometry and its applications has been given by Bradfield (1968) and Curtis (1969). A very useful motorised version of an ultrasonic goniometer has been produced by Curtis (quoted by Reynolds and Hancox, 1970).

Several variations and combinations of the above-mentioned techniques have also been used by some authors to measure the elastic constants of composites. A combination of the transit time and immersion bath techniques has been used by Markham (1969, 1970). In this work the sample was immersed in a liquid bath between two fixed transducers. The change in the transit time was then measured for different thicknesses and orientations of the sample. All the elastic constants of the composite can be determined by analyzing the velocity measurements as a function of the sample orientation. A limitation, though not very serious, arises from the fact that the observations cannot be taken for those sample orientations in which the angle of incidence exceeds the critical values, because in this case the pulse will be totally reflected and will not reach the receiver.

More recent applications of the ultrasonic techniques have been reported by Smith (1972), who has measured the elastic constants of carbon fibre composites, and Gravel and Cost (1972), who have measured the elastic constants of the Al–Al_3–Ni unidirectionally solidified eutectic.

change in the transit time was then measured for different thicknesses and orientations of the sample. All the elastic constants of the composite can be determined by analysing the velocity measurements as a function of the sample orientation. A limitation, though not very serious, arises from the fact that the observations cannot be taken for those sample orientations in which the angle of incidence exceeds the critical values, because in this case the pulse will be totally reflected and will not reach the receiver.

More recent applications of the ultrasonic techniques have been reported by Smith (1972), who has measured the elastic constants of carbon fibre composites, and Gravel and Cost (1972), who have measured the elastic constants of the Al–Al_3Ni unidirectionally solidified eutectic.

CHAPTER **6**

Effect of Defective Fibres on Wave Propagation in Composties

6.1 Introduction

6.2 Effect of Fibre Misalignment on the Propagation of Long Waves in a Composite

6.3 Effect of Defects on Vibration Frequencies

6.4 Localized Vibration Modes in Fibre Composites

6.5 Relaxation Energy of a Debonded Fibre in a Composite

6.1 INTRODUCTION

In Chapter 5 we discussed wave propagation in a 'perfect' fibre composite. A perfect fibre composite was defined in Section 5.3. In this chapter we shall calculate the effect of the following imperfections in the model of a perfect fibre composite as given in Section 5.3: (1) misalignment of fibres, (2) mass defect, and (3) bonding defect.

As we shall see in Section 6.2, very useful information regarding the misalignment of the fibres can be obtained by using long waves travelling along the axis of the composite. The mass and bonding defects can be studied by using the short waves travelling in the plane normal to the composite axis, when, for reasons discussed in Section 5.8, the wave propagation will be most sensitive to such defects.

In order to calculate the effect of the mass and bonding defects on short-wave propagation in fibre composites, we have developed a Green function method which is an adaptation of the Green function method for lattice dynamics (for a review of the Green function method for lattice dynamics, see Kwok, 1967, and Maradudin, 1965). The Green function, which, in general, is a function of the wave frequency, gives the response of the solid to a force at that frequency. A brief review of the Green function method is given in Section 6.3 and its application to fibre composites is described in Section 6.4.

The zero frequency limit of the Green function gives the response of the solid to a static force and can be used, therefore, to study the static properties of a solid with defects (Tewary, 1969, 1973; Bullough and Tewary, 1972). This method is applied in Section 6.4 to calculate the work of fracture associated with the debonding of a fibre in a fibre composite.

6.2 EFFECT OF FIBRE MISALIGNMENT ON THE PROPAGATION OF LONG WAVES IN A COMPOSITE

In this section we shall consider the effect of misalignment of some fibres on the propagation of long waves in an otherwise perfect composite. A perfect composite has been defined in Chapter 5. We shall assume the same frame of reference as that used for the analysis given in Chapter 5, viz. the z axis along the axis of the composite. We shall assume the wave to be confined in the yz plane, so that k_1, the x component of the wave vector, is zero.

If C_p denotes the phase velocity, then E, the effective modulus of elasticity, can be defined as follows:

$$E = \rho C_p^2 \tag{6.2.1}$$

Since C_p depends upon the direction of the wave propagation relative to the fibre axis, E will be a function of the angle of the wave propagation relative to the fibre axis. If the direction of the wave propagation is measured with respect to a fixed axis in the composite, a measurement of E will yield information about the fibre alignment with respect to that axis.

Considering, for example, the p_1-wave (Eq. 5.9.6), we obtain in the limit of long waves, for a wave incident at an angle θ_k from the z axis,

$$E(\theta_k) = \tfrac{1}{2}(c_{11} + c_{44})\,k_2^2 + \tfrac{1}{2}(c_{44} + c_{33})\,k_3^2 - \left[\left\{(c_{11} - c_{44})\frac{k_2^2}{2} + (c_{44} - c_{33})\frac{k_3^2}{2}\right\}^2 + (c_{13} + c_{44})^2 k_2^2\, k_3^2\right]^{1/2} \tag{6.2.2}$$

where $k_2 = \sin\theta_k$, $k_3 = \cos\theta_k$ and, by symmetry, $E(\theta_k) = E(-\theta_k)$.

To represent the alignment of fibres, we introduce a normalized distribution function $F(\phi)$ so that the relative number of fibres aligned between the angles ϕ and $\phi + d\phi$ from the z axis is given by $F(\phi)\,d\phi$. The angle ϕ can have values in the range $-\pi/2$ to $\pi/2$.

For a composite in which all the fibres are perfectly aligned along the z axis $F(\phi)$ will have the form $\delta(\phi)$, whereas for a completely 'misaligned' composite $F(\phi)$ will be independent of ϕ. We shall show here how one can obtain information about $F(\phi)$ by a comparison of experimental and theoretical values of $E(\theta_k)$ for a composite.

The contribution from a fibre at an angle ϕ from the z axis to the modulus will be $E(\theta_k - \phi)$, so that the total effective modulus $E_e(\theta_k)$ for a wave incident at an angle θ_k from the z axis in the yz plane is given by

$$E_e(\theta_k) = \int_{-\pi/2}^{\pi/2} E_t(\theta_k - \phi)\,F(\phi)\,d\phi \tag{6.2.3}$$

where the suffices e and t denote the experimental and theoretical values of E, respectively, since the left-hand side is the one which can be measured experimentally.

Let us express the various functions in Eq. (6.2.3) in the form of a Fourier sine series in θ, as follows:

$$E_e(\theta) = \sum_n E_{en} \sin 2n\theta \tag{6.2.4}$$

$$E_t(\theta) = \sum_n E_{tn} \sin 2n\theta \tag{6.2.5}$$

and

$$F(\theta) = \sum_n F_n \sin 2n\theta \tag{6.2.6}$$

where the sum is over all integral values of n.

In practice, only a few Fourier coefficients will be required, since the Fourier series converges rapidly for functions which are not wildly varying. The coefficients E_{en} and E_{tn} can be obtained from $E_e(\theta)$ (observed values) and $E_{t\theta}$ (Eq. 6.2.2), respectively, as follows:

$$E_n = \frac{2}{\pi} \int_{-\pi/2}^{\pi/2} E(\theta) \sin 2n\,\theta \tag{6.2.7}$$

where $E(\theta)$ stands for either $E_e(\theta)$ or $E_t(\theta)$. The integration in Eq. (6.2.7) has to be performed numerically at least for E_{en}.

The coefficients F_n can be obtained in terms of E_{en} and E_{tn} with the help of Eq. (6.2.3). Since on the right of Eq. (6.2.3) we have a convolution of the functions $E_t(\theta)$ and $F(\theta)$, we get the following simple relation between the Fourier coefficients:

$$F_n = \frac{E_{en}}{E_{tn}} \tag{6.2.8}$$

Once F_n are known, the distribution function $F(\theta)$ can be constructed from Eq. (6.2.6). Thus, we see that a measurement of $E(\theta)$ can be used to derive the exact distribution function of the fibres in the composite.

As the Fourier coefficients are very sensitive to the details of the function, it is very important for this technique that $E_e(\theta)$ be accurately measured for many values of θ; otherwise this approach can give results which are wildly in error. When the measurements of $E_e(\theta)$ are not sufficiently accurate, the following simple approach may be more useful for a rough determination of $F(\phi)$.

We write $F(\theta)$ as a sum over delta functions:

$$F(\theta) = \sum_n t_n\, \delta(\theta - \theta_n) \tag{6.2.9}$$

and substitute this expression into Eq. (6.2.3) to yield

$$E_e(\theta) = \sum_n t_n E_t(\theta - \theta_n) \tag{6.2.10}$$

Physically t_n gives the number of fibres at an angle θ_n from the z axis. For a perfectly aligned sample $t_n = 1$ for $\theta_n = 0$ and zero for all other values of θ_n. For a completely misaligned sample θ_n is more or less continuously distributed in the range $-\pi/2$ to $\pi/2$ and t_n is the same for all n.

For a real case Eq. (6.2.10) can be used for a rough determination of t_n and θ_n by using a least square fitting procedure. In actual practice, one can start with two or three terms on the right in Eq. (6.2.10) and obtain the values of t_n and θ_n that give the best fit with the observed values for $E(\theta)$. The number of terms can then be gradually increased to get a satisfactory agreement between the observed and calculated values of $E(\theta)$. The final values of t_n and θ_n will give an idea of the alignment of the fibres.

We shall illustrate this technique by applying it to calculate the fibre orientation function for a real case. For this purpose we shall use the experimental results obtained by Reynolds (quoted by Tewary and Bullough, 1972) for the angular dependence of the phase velocity of long waves in a carbon fibre–epoxy composite with 50% fibre loading. These experimental results have been shown in Figures 6.1 and 6.2.

Let us consider a long wave travelling in the *yz* plane. In this plane the wave propagation cannot distinguish between tetragonal and hexagonal symmetries. The elastic constants for a hexagonal carbon fibre-epoxy resin composite with 50% fibre loading, as calculated by Heaton (quoted by Tewary and Bullough, 1972), are given below. This set of elastic constants will be referred to as Set 1.

Set 1 elastic constants (*units*: 10 GN/m^2)

$$c_{11} = 3.157$$

$$c_{12} = 0.959$$

$$c_{13} = 2.058$$

$$c_{33} = 43.720$$

$$c_{44} = 2.273$$

We have calculated $E(\theta_k)$ with the help of Eq. (6.2.2), using Set 1 elastic constants, taking $k = 0$ and $\rho = 2\ kg/m^3$. The results are shown in Figure 6.1.

We notice from Figure 6.1 that the agreement between the calculated and experimental results for the p_1 wave is rather poor. One possible explanation could be that the calculated values of the elastic constants do not agree with the actual elastic constants of the sample used for the experiments. We can, of course, determine the experimental values of the elastic constants from the measured values of the phase velocities. The elastic constants thus determined are given below and will be referred to as Set 2. The elastic constants c_{11}, c_{66}, c_{44} and c_{33} have been obtained from the measured phase velocities of the three waves for $\theta = 0$ and 90°. The elastic constant c_{13} has been determined from the measured phase velocity of the p_2 wave for $\theta = 45°$. It may be mentioned here that it is

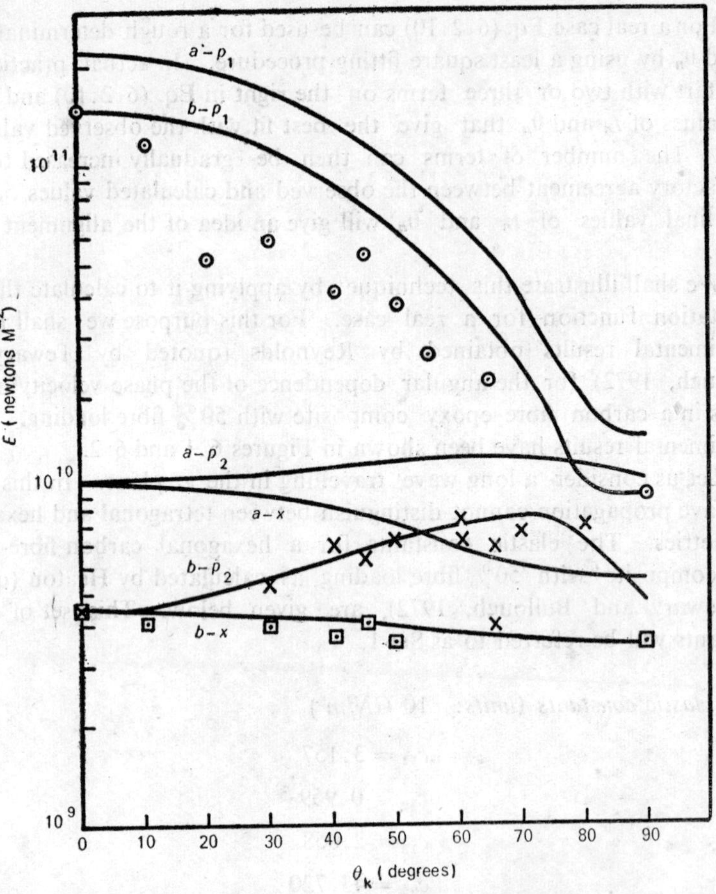

Figure 6.1 Angular dependence of the effective modulus of elasticity of a hexagonal or tetragonal composite for long elastic waves, using the elastic constants of (a) set 1 and (b) set 2. The points denote the experimental results obtained by Reynolds as quoted by Tewary and Bullough (1972): $O-p_1$ wave, $X-p_2$ wave and □—x-transverse wave.

not possible to choose a physically acceptable value so as to obtain agreement between the measured and calculated values for the velocity of the p_1 wave for any value of 0.

Set 2 elastic constants (*units*: 10 GN/m^2)

$$c_{11} = 2.00$$
$$c_{12} = 0.60$$
$$c_{13} = 1.30$$
$$c_{33} = 29.00$$
$$c_{44} = 0.90$$

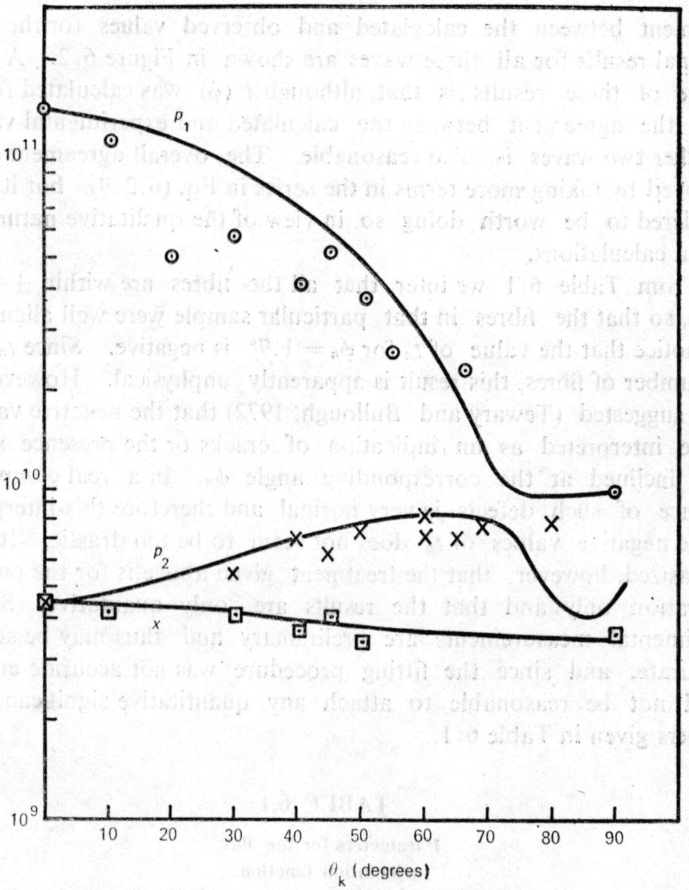

Figure 6.2 Angular dependence of the effective modulus of elasticity of a hexagonal or tetragonal composite for long elastic waves using the elastic constants in set 2 and the fibre orientation function (Eq. 6.2.9) with parameters as given in Table 6.1. The points denote the experimental results as in Figure 6.1.

The calculated values of the phase velocities for the three waves using the Set 2 elastic constants are also shown in Figure 6.1. The agreement between the calculated and the observed values is found to be satisfactory for the p_2 and the x transverse waves but is unsatisfactory for the p_1 wave. It does not seem to be possible to improve on this agreement by choice of different elastic constants.

Let us attribute this discrepancy for the p_1 wave to the misalignment of the fibres in the composite. To measure the fibre misalignments, we introduce the fibre orientation function, as discussed earlier. Taking three terms in the series in Eq. (6.2.9) and choosing the values for the six parameters t_n and ϕ_n as given in Table 6.1, we obtain a reasonable agreement between the calculated and observed values for the p_1 wave. The final results for all three waves are shown in Figure 6.2. A pleasing feature of these results is that, although $F(\phi)$ was calculated for the p_1 wave, the agreement between the calculated and experimental values for the other two waves is also reasonable. The overall agreement could be improved by taking more terms in the series in Eq. (6.2.9), but it was not considered to be worth doing so, in view of the qualitative nature of the present calculations.

From Table 6.1 we infer that all the fibres are within $\pm 5°$ of the z axis, so that the fibres in that particular sample were well aligned. We also notice that the value of t_n for $\phi_n = 1.7°$ is negative. Since t_n denotes the number of fibres, this result is apparently unphysical. However, it has been suggested (Tewary and Bullough, 1972) that the negative value of t_n can be interpreted as an indication of cracks or the presence of hollow fibres inclined at the corresponding angle ϕ_n. In a real composite the presence of such defects is very normal and therefore this interpretation for the negative values of t_n does not seem to be too drastic. It must be emphasized, however, that the treatment given above is for the purpose of illustration only and that the results are only qualitative. Since the experimental measurements are preliminary and thus may be somewhat inaccurate, and since the fitting procedure was not accurate enough, it would not be reasonable to attach any quantitative significance to the numbers given in Table 6.1.

TABLE 6.1

Parameters for the fibre orientation function

ϕ_n (degrees)	t_n
−4.6	1.26
1.7	−0.36
3.4	0.08

6.3 EFFECT OF DEFECTS ON VIBRATION FREQUENCIES

In the previous section we considered the effect of the misalignment of fibres on the propagation of long waves and how this result can be used for non-destructive testing of the 'average' fibre alignment in a composite. Now we shall examine the effect of a defective fibre on the propagation of short waves in a composite.

Two types of fibre defects will be considered here: (1) mass defect and (2) bonding defect. As mentioned in Section 6.1, the first type of defect refers to a fibre whose mass is different from the others. Similarly, the second type of defect refers to a fibre whose bonding strength is different from that of the others.

We shall assume that the concentration of the defective fibres is low, so that the total effect on the composite is linearly proportional to the number of the defective fibres. This enables us to consider a single defective fibre in an otherwise perfect composite. Thus, in our model all the fibres are well aligned along the symmetry axis and are arranged in a regular pattern which defines the symmetry of the solid. The mass and the bonding strength of all fibres is the same apart from the single defective fibre.

In this section we shall describe the general theory of the effect of the above mentioned type of defects on the vibration frequencies. Its application to a fibre composite will be given in the next section. The present treatment follows closely the theory of lattice dynamics of crystals with defects (see Maradudin, 1965, for a review).

The theory of lattice dynamics for perfect solids has already been reviewed in Section 2.2. In the Born-von Kármàn model, as described in that section, if each atom is given a displacement denoted by $\mathbf{u}\,(\mathbf{l})$, then the force on the atoms $\mathbf{F}\,(\mathbf{l})$ is given by Eq. (2.2.9). Thus, we obtain the following equation of motion:

$$m\,(\mathbf{l})\,\frac{d^2\,u_\alpha\,(\mathbf{l})}{dt^2} = -\sum_\beta \phi_{\alpha\beta}\,(\mathbf{l},\mathbf{l}')\,u_\beta\,(\mathbf{l}') \tag{6.3.1}$$

where $m\,(\mathbf{l})$ denotes the mass of the atom at site $\mathbf{l}$. If we assume a stationary wave solution of Eq. (6.3.1), so that the time development of $\mathbf{u}\,(\mathbf{l})$ can be taken to be of the form $\exp\,(i\,\omega\,t)$ we obtain the following equation

$$T_{\alpha\beta}\,(l,\,l';\,\omega^2)\,u_\beta\,(l') = 0 \tag{6.3.2}$$

where

$$T_{\alpha\beta}\,(l,\,l';\,\omega^2) = -\,M_{\alpha\beta}\,(l,\,l')\,\omega^2 + \phi_{\alpha\beta}\,(l,\,l') \tag{6.3.3}$$

$$M_{\alpha\beta}\,(l,\,l') = m\,(l)\delta_{\alpha\beta}\,\delta_{ll'} \tag{6.3.4}$$

and δ_{ij} is the well-known Kronecker delta of any two indices i and j, which is zero unless $i=j$, in which case it is unity. The presence of ω^2 in the parenthesis simply indicates that $T_{\alpha\beta}\,(\mathbf{l},\,\mathbf{l}')$ is a function of ω^2.

The object of writing the equations of motion in the form of Eq. (6.3.2) is that it allows us to use a convenient matrix notation. Let N denote the total number of the degrees of freedom in the solid, i.e. the number of distinct values which the combination of α and $\mathbf{l}$ can take. For example, in a three-dimensional solid containing n atoms, $\mathbf{l}$ can take n

values and α can take three values — x, y and z. In this case $N = 3n$. Similarly, in a two-dimensional solid containing n atoms $N = 2n$.

We can now define a N-dimensional column matrix (vector) $\mathbf{u}$ and $N \times N$ square matrices $\mathbf{T}$ (ω^2), $\mathbf{M}$ and ϕ. Typical matrix elements of these matrices are denoted by u_α $(\mathbf{l})$, $T_{\alpha\beta}$ $(\mathbf{l}, \mathbf{l}'; \omega^2)$, $M_{\alpha\beta}$ $(\mathbf{l}, \mathbf{l}')$ and $\phi_{\alpha\beta}$ $(\mathbf{l}, \mathbf{l}')$, respectively, which have been defined in Eqs. (6.3.3) and (6.3.4). It is obvious from Eq. (6.3.4) that the matrix $\mathbf{M}$ is a diagonal matrix containing the atomic masses along its diagonal. In the matrix notation, therefore, Eq. (6.3.2) is written as

$$\mathbf{T}\,(\omega^2)\,\mathbf{u} = 0 \tag{6.3.5}$$

The matrices introduced above have been defined in what is called in group theoretic language the 'space' of the lattice or the representation offered by the lattice sites. A very brief explanation of this terminology will be given here; for details reference should be made to a textbook on group theory, e.g. Mariot (1962).

Let us define a N-dimensional column matrix $\mathbf{r}$ with the matrix elements r_α $(\mathbf{l})$, where r_α $(\mathbf{l})$ denotes the α component of the radius vector of the site $\mathbf{l}$. Each quantity r_α $(\mathbf{l})$ can be regarded as a certain vector. Thus, we have a set of N vectors r_α $(\mathbf{l})$ which define the positions of all the lattice sites. These N vectors, which are called the basis vectors, are said to 'span' the space of the lattice sites and can be used to define this particular space.

The transformation properties of the matrices defined in a space are closely related to the transformation of the basis vectors under an operation of the group which defines the symmetry of the system. The basis vectors of a complete set transform among themselves under the operations of the symmetry group. The transformation law for ϕ has been given in Section 2.2. The other matrices follow a similar law.

A space defined by a smaller set of basis vectors which are part of a bigger set is called a subspace of the original space. A subspace is invariant against a transformation if all its basis vectors defining the subspace transform among themselves under that transformation. If $\mathbf{S}$ denotes the 3×3 matrix of a certain operation, then, for an invariant subspace, the vector defined by

$$r_\alpha\,(\mathbf{L}) = \sum_\beta S_{\alpha\beta}\, r_\beta\,(\mathbf{l})$$

will also be a basis vector of the same subspace.

The formulae given above are applicable to a perfect as well as an imperfect solid. In the case of a perfect solid, all the atomic masses are equal, i.e. m $(\mathbf{l})$ is independent of $\mathbf{l}$. The diagonal matrix $\mathbf{M}$ therefore becomes a scalar. Further, for a perfect solid, ϕ $(\mathbf{l}, \mathbf{l}')$ depends on $\mathbf{l}$ and $\mathbf{l}'$

only through their difference, as discussed in Section 2.2. It is then possible to define the Fourier transform $\phi(\mathbf{k})$ of ϕ, as given in Eq. (2.2.10). The 3×3 dynamical matrix $\mathbf{D}(\mathbf{k})$ is defined as in Eq. (2.2.11). Then it can be verified by taking the Fourier transform of Eq. (6.3.5) and imposing the requirement of the vanishing determinant, that the eigenvalues of $\mathbf{D}(\mathbf{k})$ give the allowed values of the frequency ω^2. The calculations have been given in the previous chapter.

In the case of a solid containing a defect, $\mathbf{M}(\mathbf{l})$ is still diagonal but no longer scalar. Moreover, in general, ϕ will also change, and will depend on $\mathbf{l}$ and $\mathbf{l}'$ separately and not merely on their difference. It is, therefore, not possible to define a simple Fourier transform of ϕ as was done in Eq. (2.2.10), and the solution of Eq. (6.3.5) cannot be obtained simply by diagonalizing a 3×3 dynamical matrix.

We shall describe a Green function method for the calculation of the frequencies from Eq. (6.3.5) in the case of a solid containing a single defect. The mass and the force constant matrices in this case will be denoted by $\mathbf{M}^*$ and ϕ^*, respectively, so that

$$\mathbf{T}^*(\omega^2) = -\mathbf{M}^*\omega^2 + \phi^* \tag{6.3.6}$$

where ω^2 have to be obtained from the following equation:

$$\mathbf{T}^*(\omega^2)\,\mathbf{u} = 0 \tag{6.3.7}$$

The matrices $\mathbf{T}$, $\mathbf{M}$ and ϕ without the star superscript will refer to the perfect solid. Let us write

$$\mathbf{T}^* = \mathbf{T} - \Delta\mathbf{T} \tag{6.3.8}$$

$$\mathbf{M}^* = \mathbf{M} - \Delta\mathbf{M} \tag{6.3.8 a}$$

and

$$\phi^* = \phi - \Delta\phi \tag{6.3.8 b}$$

where $\Delta\mathbf{T}$, $\Delta\mathbf{M}$ and $\Delta\phi$ denote the change in the corresponding matrices caused by the presence of a defect. For notational convenience the dependence of $\mathbf{T}$, $\Delta\mathbf{T}$ and $\mathbf{T}^*$ on ω^2 has not been explicitly indicated.

From Eqs. (6.3.3), (6.3.6) and (6.3.8)–(6.3.8 b) we obtain the relation

$$\Delta\mathbf{T} = -\Delta\mathbf{M}\,\omega^2 + \Delta\phi \tag{6.3.8 c}$$

With the help of Eq. (6.3.8), Eq. (6.3.7) can be written as

$$\mathbf{T}\mathbf{u} = \Delta\mathbf{T}\mathbf{u} \tag{6.3.9}$$

or

$$\mathbf{u} = \mathbf{G}\,\nabla\mathbf{T}\mathbf{u} \tag{6.3.10}$$

where

$$\mathbf{G} = \mathbf{T}^{-1} \tag{6.3.11}$$

The matrix **G** is also a $N \times N$ square matrix and its typical matrix element will be denoted by $G_{\alpha\beta}(\mathbf{l}, \mathbf{l}'; \omega^2)$. It can easily be verified that **G** is the Green function matrix defined by Eqs. (2.2.13) and (2.2.14). Before attempting to solve Eq. (6.3.9), we shall describe some of the properties of Green function which are relevant to the present discussion.

From Eq. (2.2.13), by using the representation of a matrix in terms of its eigenvalues and eigenvector, we can write

$$G_{\alpha\beta}(\mathbf{k}, \omega^2) = \frac{1}{m} \sum_j \frac{e_\alpha(\mathbf{k}, j)\, e_\beta(\mathbf{k}, j)}{-\omega^2 + \omega^2(\mathbf{k}, j)} \qquad (6.3.12)$$

where $\omega^2(\mathbf{k}, j)$ ($j=1, 2, 3$) are the three eigenvalues of $\mathbf{D}(\mathbf{k})$ corresponding to the eigenvectors $\mathbf{e}(\mathbf{k}, j)$.

According to the discussion following Eq. (2.2.10), only discrete values of **k** which are confined in the Brillouin zone are allowed in the lattice theory. Hence, a maximum frequency, which will be denoted by ω_m, must exist. Thus, the allowed frequencies are distributed in a band of frequencies from O to ω_m. The band width is ω_m (or ω_m^2 for squared frequencies). In the Debye model ω_m can be related to the Debye temperature (see Chapter 5).

We see from Eq. (6.3.12) that the Green function has singularities for $\omega^2 < \omega_m^2$ at the points $\omega^2 = \omega^2(\mathbf{k}, j)$. It is a characteristic property of the Green function that it will be singular for a frequency lying within the band of allowed frequencies in a perfect lattice. The singularity is formally treated by adding a small imaginary part in ω^2 and considering the real and imaginary parts of the Green function separately (Maradudin, 1965).

We shall be particularly interested in the case when $\omega^2 > \omega_m^2$, i.e. in frequencies which are outside the band of allowed frequencies of the perfect solid. In this case the denominator in Eq. (6.3.12) can be expanded as a power series in $\omega^2(\mathbf{k}, j)/\omega^2$, viz.

$$G_{\alpha\beta}(\mathbf{k}, \omega^2) = -\frac{1}{m\omega^2}\Big[\delta_{\alpha\beta} + \frac{1}{\omega^2}\sum_j e_\alpha(\mathbf{k}, j)\, e_\beta(\mathbf{k}, j)\, \omega^2(\mathbf{k}, j) + \frac{1}{\omega^4}\sum_j e_\alpha(\mathbf{k}, j)\, e_\beta(\mathbf{k}, j)\, \omega^4(\mathbf{k}, j) + \ldots\Big] \qquad (6.3.13)$$

where we have used the following orthonormal property of the eigenvectors:

$$\sum_j e_\alpha(\mathbf{k}, j)\, e_\beta(\mathbf{k}, j) = \delta_{\alpha\beta}$$

$\delta_{\alpha\beta}$ being the Kronecker delta, as defined earlier.

Using the following representation of the nth power of $\mathbf{D}(\mathbf{k})$ in terms of its eigenvectors and eigenvalues,

$$\left\{\mathbf{D}^n(\mathbf{k})\right\}_{\alpha\beta} = \sum_j e_\alpha(\mathbf{k}, j)\, e_\beta(\mathbf{k}, j)\, \omega^{2n}(\mathbf{k}, j) \tag{6.3.14}$$

which can be derived from elementary maxtrix algebra, we can write Eq. (6.3.13) in the following form:

$$G_{\alpha\beta}(\mathbf{k}, \omega^2) = -\frac{1}{m\omega^2}\left[\delta_{\alpha\beta} + \frac{1}{\omega^2}\left\{\mathbf{D}(\mathbf{k})\right\}_{\alpha\beta} + \frac{1}{\omega^4}\left\{\mathbf{D}^2(\mathbf{k})\right\}_{\alpha\beta} + \ldots\right]$$

or, in matrix notation,

$$\mathbf{G}(\mathbf{k}, \omega^2) = -\frac{1}{m\omega^2}\left[I + \frac{1}{m\omega^2}\phi(\mathbf{k}) + \frac{1}{m^2\omega^4}\phi^2(\mathbf{k}) + \ldots\right] \tag{6.3.15}$$

where we have used the relation between $\mathbf{D}(\mathbf{k})$ and $\phi(\mathbf{k})$ as given by Eq. (2.2.11), and $\mathbf{I}$ is the unit matrix.

Finally, by taking the inverse Fourier transform of $\mathbf{G}(\mathbf{k}, \omega^2)$, as defined in Eq. (2.2.14), we get the following series for the Green function:

$$G_{\alpha\beta}(\mathbf{l}, \mathbf{l}'; \omega^2) = -\frac{1}{m\omega^2}\left[\delta_{\alpha\beta}\,\delta_{l,l'} + \frac{1}{m\omega^2}\phi_{\alpha\beta}(\mathbf{l}, \mathbf{l}') + \frac{1}{m^2\omega^4}\sum_{\mathbf{l}'',\gamma}\phi_{\alpha\gamma}(\mathbf{l}, \mathbf{l}'')\,\phi_{\gamma\beta}(\mathbf{l}', \mathbf{l}) + \ldots\right] \tag{6.3.16}$$

for $\omega^2 \gg \omega_m^2$.

The series in Eq. (6.3.16) provides a convenient method for a calculation of the near-neighbour Green function matrix elements for frequencies which are much larger than the maximum allowed frequency of the perfect crystal. It also provides a simple physical interpretation; the first term in Eq. (6.3.16), which represents the zero-order approximation, is simply the well-known free particle Green function $1/m\omega^2$. It corresponds to the situation when the frequency ω^2 is so high that each particle can be regarded as an independent oscillator, i.e. a free particle. The higher terms in Eq. (6.3.16) give the effect of coupling of each particle with its near neighbours on the Green function. The coupling between two particles $\mathbf{l}$ and $\mathbf{l}'$ in the Born-von Kármán model is measured by the force constants $\phi_{\alpha\beta}(\mathbf{l}, \mathbf{l}')$.

Let us now return to Eq. (6.3.10). As mentioned earlier, in our model, which is valid for only a low concentration of defects, we shall assume a single fibre (or atom in the context of the present discussion) to be defective. Suppose its mass is $m-\Delta m$ and the change in its force constants is restricted to only a few of its neighbours. Assuming the defect to be at the origin ($l=0$), we have

$$\Delta M_{\alpha\beta}(\mathbf{l}, \mathbf{l}') = \Delta M\, \delta_{\alpha\beta}\, \delta_{l, l'}\, \delta_{l0} \tag{6.3.17}$$

and

$$\Delta\phi_{\alpha\beta}(\mathbf{l}, \mathbf{l}')$$

is non-vanishing only if $\mathbf{l}$ and $\mathbf{l}'$ refer to either the defect ($l=0$) or an atom with which the defect has a direct interaction.

The radius vectors of such lattice sites for which $\Delta\phi$ and $\Delta\mathbf{M}$ and therefore $\Delta\mathbf{T}$ are non-vanishing can be used to define a subspace in the sense explained earlier in this section. This subspace will be referred to as the defect space. If the solid retains any rotational symmetry about the defect site, the basis vectors of the defect space will transform among themselves under the operations of the corresponding group. The defect space is therefore an invariant subspace.

Finally, in this context we shall also define the coordination number of the defect. If the defect interacts directly with $\eta-1$ atoms, then η is called the coordination number of the defect. Since the defect and its $\eta-1$ neighbours will contribute to the matrix $\Delta\mathbf{T}$, the defect space will be defined by the radius vectors of η sites. Thus, the defect space will be of the dimension $n\eta$, where $n=3$ or 2 depending on, respectively, whether it is a three or a two-dimensional solid.

We shall now partition the $N \times N$ matrix $\Delta\mathbf{T}$ in block form, as follows:

$$\Delta T = \left(\begin{array}{c|c} \Delta T_p & 0 \\ \hline 0 & 0 \end{array}\right) \tag{6.3.18}$$

where ΔT_p is the $n\eta \times n\eta$ square matrix which contains all the non-vanishing elements of $\Delta\mathbf{T}$ (or $\Delta\mathbf{M}$ and $\Delta\phi$). $\Delta\mathbf{T}_p$ can be called the defect matrix. By hypothesis, the other block elements of $\Delta\mathbf{T}$ in Eq. (6.3.18) are zero, as denoted in that equation.

The matrices $\mathbf{G}$ and $\mathbf{u}$ in Eq. (6.3.10) can also be partitioned in the same representation as follows:

$$G = \left(\begin{array}{c|c} \mathbf{g} & \mathbf{G}' \\ \hline \tilde{\mathbf{G}}' & \mathbf{G}'' \end{array}\right) \tag{6.3.19}$$

and

$$u = \left(\begin{array}{c} \mathbf{u}_p \\ \hline \mathbf{u}' \end{array}\right) \tag{6.3.19a}$$

where $\mathbf{g}$ and $\mathbf{u}_p$ are, respectively, the Green function and the displacement (column) matrices in the defect space. The dimensions of $\mathbf{g}$, $\mathbf{G}'$, $\tilde{\mathbf{G}}'$ and $\mathbf{G}''$ are $n\eta \times n\eta$, $n\eta \times (N-n\eta)$, $(N-n\eta) \times n\eta$ and $(N-n\eta) \times (N-n\eta)$, respectively. The dimensions of the column matrices $\mathbf{u}_p$ and $\mathbf{u}'$ are $n\eta$ and $(N - n\eta)$, respectively.

Substituting the forms of $\Delta\mathbf{T}$, $\mathbf{G}$ and $\mathbf{u}$ from Eqs. (6.3.18)–(6.3.19a) into Eq. (6.3.10), we obtain the following two matrix equations:

$$\mathbf{u}_p = \mathbf{g}\,\Delta\mathbf{T}_p\,\mathbf{u}_p \tag{6.3.20}$$

and

$$\mathbf{u}' = \tilde{\mathbf{G}}'\,\Delta\mathbf{T}_p\,\mathbf{u}_p \tag{6.3.21}$$

The condition that a non-trivial solution of Eq. (6.3.20) exists requires that the determinant of the matrix $\mathbf{P} = \mathbf{I} - \mathbf{g}\,\Delta\mathbf{T}_p$ must vanish. Thus, we obtain the secular equation

$$|\,\mathbf{I} - \mathbf{g}\,\Delta\mathbf{T}_p\,| = 0 \tag{6.3.22}$$

where $|\,\mathbf{P}\,|$ denotes the determinant of the matrix $\mathbf{P}$ and $\mathbf{I}$, as before, is the unit matrix. As $\mathbf{g}$ and $\Delta\mathbf{T}_p$ are functions of ω^2, Eq. (6.3.22), being a determinantal equation of a finite order, can be solved for ω^2, which gives the frequencies perturbed by the presence of the defect.

In most cases it is a reasonable or at least the usual assumption that the defect interacts with only a few atoms, so that η, the coordination number of the defect is small. The order of the determinant in Eq. (6.3.22) in such cases is small and therefore Eq. (6.3.22) is amenable to standard techniques of solution. An additional simplification can be achieved by exploiting the symmetry of the configuration with the help of group theory. Such techniques have been discussed in several review articles and books, such as, for example, those by Maradudin (1965) and Maradudin et al. (1971).

In the general case $\mathbf{g}$ is a very complicated function of ω^2 and therefore Eq. (6.3.22) can only be solved numerically. However, if this equation has a solution for a value of ω^2 which is outside the band of allowed frequencies of the perfect solid (i.e. for $\omega^2 > \omega_m^2$), then the simple series representation of the Green function as given in Eq. (6.3.16) can be used. In this case, if only one or two terms in the series for Green function are retained, it is possible to obtain an analytical expression for ω^2.

The vibration modes whose frequencies lie outside the band of perfect lattice frequencies are called 'localized' modes. The other type of perturbed modes whose frequencies lie within the band of perfect lattice frequencies are called 'band' modes. The localized modes occur only if the values of $\Delta\mathbf{M}$ and $\Delta\phi$ satisfy certain conditions, whereas the band modes are generally present in a perturbed lattice.

The main difference between the two types of modes is that in a localized mode the atomic vibrational amplitude is very large at or near the defect site and decays very rapidly (exponentially or even faster) with distance from the defect site. These modes are thus localized near the defect. On the other hand, the atomic vibrational amplitudes have an oscillatory wave-like dependence on distance from the defect site in the case of band modes.

The localized modes, if they exist, are of particular interest because they are much easier to detect experimentally than the band modes. An interesting correspondence can be noted here between the elastic wave propagation in a solid, which can be regarded as a dynamic loading of the solid, and the static loading, in which a time-independent stress is applied to the solid. The localized modes in the former case correspond to the stress concentration near the defect in the latter case. The localized modes, if present, should therefore be important for a study of any damage in the solid produced by an intense elastic wave such as a shock wave.

Fortunately, the mathematical analysis of the localized modes is much simpler than the band modes. This is because, as has been said earlier, the frequencies of the localized modes are larger than ω_m^2 and in this case the Green function can be conveniently calculated from Eq. (6.3.16). In the application of the present theory to fibre composites, as given in the following section, we shall, therefore, consider only the localized modes.

6.4 LOCALIZED VIBRATION MODES IN FIBRE COMPOSITES

In this section we shall apply the general theory given in the previous section to study the localized vibration modes near a defective fibre in a fibre composite. Only a tetragonal composite will be considered here—the calculations for a hexagonal composite can be carried out in an analogous manner.

The model for the perfect tetragonal composite has been described in Chapter 5. The model for the defect has been explained in the previous section. Thus, we shall consider a single fibre defect in an otherwise perfect composite. The defect may be a mass defect (type I) or the bonding defect (type II), or both.

As in Chapter 5, the z axis of the coordinate system is taken to be along the fibres and the x and the y axes are taken in a plane normal to the fibres. The origin on the xy plane is taken on the defective fibre. It is obvious that in this case the composite is still invariant to all the symmetry operations given in Appendix 1, provided that these operations are applied about the defective fibre.

According to the discussion given in Chapter 5, wave propagation will be most sensitive to the fibre mass and bonding when the wave is travelling in the xy plane, i.e. when $k_3 = 0$. We shall therefore consider only wave propagation in the xy plane. In this case the composite behaves as a two-dimensional solid. The two-dimensional lattice structure of a tetragonal composite in the xy plane was shown in Figure 5.1a. The coordinates of the nearest and the next-nearest neighbour fibres relative to the origin in this case are given in Table 6.2.

TABLE 6.2

Coordinates of the near neighbour fibres in the xy plane in a tetragonal composite

Fibre Label	Coordinates in units of the lattice constant a	
	x	y
1	1	0
2	0	1
3	−1	0
4	0	−1
5	1	1
6	−1	1
7	−1	−1
8	1	1

According to Eq. (5.8.1), the dynamical matrix in this case factorizes into a 2×2 matrix $\mathbf{D}^p$ and a 1×1 matrix, i.e, a pure number D^z. The former matrix characterizes the waves with their polarization vectors in the xy plane. This corresponds to the plane strain analysis in the continuum theory. The matrix $\mathbf{D}^z$ characterizes the wave polarized in the z direction and corresponds to the anti-plane strain analysis. We shall consider these two cases separately.

Anti-plane strain modes

In this case only the zz element of the force constant matrix between two fibres is relevant. This element for fibres $\mathbf{l}$ and $\mathbf{l}'$, in our notation, is written as $\phi_{zz}(\mathbf{l}, \mathbf{l}')$. In this sub-section, for the sake of brevity, we shall omit the subscripts z.

The force constant matrices for the perfect composite are given below;

$$\phi(0, 1) = \phi(0, 2) = \phi(0, 3) = \phi(0, 4) = -F_1 \qquad (6.4.1)$$

and

$$\phi(0, 5) = \phi(0, 6) = \phi(0, 7) = \phi(0, 8) = -F_2 \qquad (6.4.1a)$$

where the numbers in parentheses refer to the fibre labels given in Table 6.2. With the help of Eq. (2.2.6) we obtain

$$\phi(0, 0) = 4F_1 + 4F_2 \equiv F_0 \qquad (6.4.2)$$

The force constants F_1 and F_2 as defined here differ from F and G in Eq. (5.8.3) only in units, since F and G were expressed in the units of the fibre mass m. The symbol G in this chapter has been used for the Green function. It should not cause any confusion with the force constant G, which was defined in the previous chapter. We shall not need a reference to this force constant again in this chapter.

By taking the Fourier transform of ϕ as defined in Eq. (2.2.10) and using Eqs. (2.2.11) and (6.4.1)–(6.4.2), it can easily be verified that the dynamical matrix for the perfect composite in this case is exactly the same as that given by Eq. (5.8.3).

Let ΔF_1 and ΔF_2 denote the change in F_1 and F_2, respectively, when a defective fibre of mass $m - \Delta m$ is placed at the origin. Again with the help of Eq. (2.2.6), we obtain

$$\Delta\phi(0, 0) = \Delta F_0 = 4(\Delta F_1 + \Delta F_2) \tag{6.4.3}$$

$$\begin{aligned}\Delta\phi(1, 1) &= \Delta\phi(2, 2) = \Delta\phi(3, 3) = \Delta\phi(4, 4)\\ &= \Delta F_1\end{aligned} \tag{6.4.3a}$$

and

$$\begin{aligned}\Delta\phi(5, 5) &= \Delta\phi(6, 6) = \Delta\phi(7, 7) = \Delta\phi(8, 8)\\ &= \Delta F_2\end{aligned} \tag{6.4.3b}$$

The above relations have been given in order to clarify the nature of the change in the force constants. In the actual calculations we shall assume that $\Delta F_2 = 0$. Thus, the defect interaction is restricted to the nearest neighbour of the defect only, whereas the fibre–fibre interaction in the perfect composite was assumed to extend up to the second neighbours.

In the present case the defective fibre and its four nearest neighbours constitute the defect space, which was defined in Eq. (6.3.18). The coordination number of the defect is 5, which is also the dimensionality of the defect space, since each fibre has only one degree of freedom, i.e. along the z axis. The order of the determinantal equation (6.3.22) is 5×5, which has to be solved for a calculation of the perturbed frequencies.

Since we are interested here only in the localized modes, we can use the series representation of the Green function as given in Eq. (6.3.16). Assuming that ω^2 is much larger than ω_m^2, we shall retain only the first two terms of this series. Then the elements of the Green function and their symbols can be written as follows:

$$G(0, 0) = -\frac{1}{\mu}\left(1 + \frac{F_0}{\mu}\right) \equiv g_0 \tag{6.4.4}$$

$$\begin{aligned}G(0, 1) &= G(0, 2) = G(0, 3) = G(0, 4)\\ &= \frac{F_1}{\mu^2} \equiv g_1\end{aligned} \tag{6.4.4a}$$

and

$$G(0,5) = G(0,6) = G(0,7) = G(0,8)$$

$$= \frac{F_2}{\mu^2} \equiv g_2 \qquad (6.4.4b)$$

where

$$\mu = m\omega^2 \qquad (6.4.5)$$

It should be emphasized that G and ϕ are properties of the perfect lattice. On account of the translation symmetry, therefore, as discussed in Section 2.2, they are independent of the choice of the origin. They depend upon the coordinates of the two atoms only through their difference. Thus, $G(\mathbf{l}, \mathbf{l})$ and $\phi(\mathbf{l}, \mathbf{l})$ for any value of $\mathbf{l}$ are equal to g_0 and E_0, respectively. Similarly, $G(0,1) = G(2,5) = G(4,8)$, etc. Precisely analogous relations are valid for $\phi(\mathbf{l}, \mathbf{l}')$. This, of course, does not apply to $\Delta\phi(\mathbf{l}, \mathbf{l}')$, which is a property of the defective composite in which the translation symmetry has been destroyed.

We shall now write the explicit forms of $\Delta\mathbf{T}_p$ and g matrix as follows:

$\Delta\mathbf{T}_p$ matrix: (6.4.6)

		1	2	3	4
0	$-\Delta m\,\omega^2 + \Delta F_0$	$-\Delta F_1$	$-\Delta F_1$	$-\Delta F_1$	$-\Delta F_1$
1	$-\Delta F_1$	ΔF_1	0	0	0
2	$-\Delta F_1$	0	ΔF_1	0	0
3	$-\Delta F_1$	0	0	ΔF_1	0
4	$-\Delta F_1$	0	0	0	ΔF_1

g matrix: (6.4.7)

		1	2	3	4
0	g_0	g_1	g_1	g_1	g_1
1	g_1	g_0	g_2	0	g_2
2	g_1	g_2	g_0	g_2	0
3	g_1	0	g_2	g_0	g_2
4	g_1	g_2	0	g_2	g_0

In the above matrices the numbers 0–4 above the horizontal and on the left of the vertical lines refer to the fibre levels as defined in Table 6.2.

It is now quite straightforward to construct the determinant of the matrix $\mathbf{I} - \mathbf{g}\,\Delta\mathbf{T}_p$ and solve the determinantal equation for ω^2.

It is not, however, essential to solve the 5×5 determinantal equation. As has been suggested earlier, a considerable simplification can be achieved with the help of group theory. A description of group theory and the group theoretical technique for the simplification of the determinantal equation is outside the scope of this book, and therefore only the results will be given here. Readers who are not familiar with this field may refer to, for example, Mariot (1962) and Callaway (1964) for group theory and Maradudin (1965) for a group theoretical technique for the solution of defect dynamics problem.

The present two-dimensional configuration is invariant against the operations xy, $\bar{x}\bar{y}$, $\bar{y}x$, $y\bar{x}$, $\bar{x}y$, $x\bar{y}$, $\bar{y}\bar{x}$ and yx, where the overhead bars denote the minus sign. These eight operators constitute what is known as the group Δ (for the group theoretical nomenclature which has been used here, see Callaway, 1964). This group has five irreducible representations, which are denoted by Δ_1, Δ_2, Δ_3, Δ_4 and Δ_5. The first four are one-dimensional (non-degenerate) representations, whereas the last—Δ_5—is a two-dimensional (doubly degenerate) representation.

As mentioned in the previous section and as can be verified in the present case, the defect space as defined by the origin and four nearest-neighbour fibre sites is an invariant subspace of the group Δ. A representation of the group Δ can therefore be constructed in the defect space. This will be a reducible representation.

With the help of the standard character analysis it can be shown that this representation contains the irreducible representations Δ_1, Δ_2 and Δ_5, 2, 1 and 1 times, respectively. With the help of the vectors transforming according to these irreducible representations, the 5×5 matrices $\mathbf{g}$ and $\Delta\mathbf{T}_p$ can be reduced to the block diagonal form, which will contain one 2×2 block matrix corresponding to Δ_1, one 1×1 block matrix corresponding to Δ_2 and two identical 1×1 block matrices corresponding to Δ_5. Thus, with the help of group theory we have been able to reduce the problem of solving a 5×5 determinantal equation to a 2×2 determinantal equation and two simple algebraic equations.

A further simplification can be achieved by noting that the representation Δ_1 has a non-vanishing component at the origin. For the localized modes, therefore, only the Δ_1 representation is relevant. In this case we do not have to solve the remaining two algebraic equations, which correspond to the representations Δ_2 and Δ_5.

We shall now give the reduced 2×2 matrices for $\Delta\mathbf{T}_p$ and $\mathbf{g}$ in the Δ_1 representation.

Reduced $\Delta \mathbf{T}_p$ matrix (denoted by $\Delta \mathbf{T}_r$):

$$\Delta \mathbf{T}_r = \begin{pmatrix} -\Delta m\omega^2 + \Delta F_0 & -2\,\Delta F_1 \\ -2\,\Delta F_1 & \Delta F_1 \end{pmatrix} \tag{6.4.8}$$

Reduced $\mathbf{g}$ matrix (denoted by $\mathbf{g}_r$):

$$\mathbf{g}_r = \begin{pmatrix} g_0 & 2g_1 \\ 2g_1 & g_0 + 2g_2 \end{pmatrix} \tag{6.4.9}$$

The reduced 2×2 determinantal equation which has to be solved for ω^2 instead of the original 5×5 determinantal equation is given below:

$$\begin{vmatrix} 1 + g_0\,\Delta m\omega^2 - 4\,\Delta F_1\,(g_0 - g_1) & 2\Delta F_1\,(g_0 - g_1) \\ 2g_1\,\Delta m\omega^2 + 2\Delta F_1\,(g_0 + 2g_2 - 4g_1) & 1 - \Delta F_1\,(g_0 + 2g_2 - 4g_1) \end{vmatrix} = 0 \tag{6.4.10}$$

where we have used the relation $\Delta F_0 = 4\Delta F_1$, obtained from Eq. (6.4.3) by putting $\Delta F_2 = 0$.

Although Eq. (6.4.10) can be solved analytically for ω^2 using Eqs. (6.4.4)–(6.4.4b) for the Green function elements, it is more instructive to consider the approximate solutions of Eq. (6.4.10), which are much simpler. These approximate solutions have been discussed below.

Mass defect only. First we shall consider the case when only the mass of the defective fibre is different and its bonding strength is the same as for the other fibres. In this case $\Delta F_1 = 0$ and only the first diagonal term in the determinant in Eq. (6.4.10) has to be considered.

Substituting for g_0 from Eq. (6.4.4) in Eq. (6.4.10) with $\Delta F_1 = 0$, we obtain

$$1 - \frac{\Delta m}{m}\left(1 + \frac{F_0}{m\omega^2}\right) = 0$$

or

$$\omega^2 = \frac{\Delta m}{m} - \frac{F_0}{m - \Delta m} \tag{6.4.11}$$

Equation (6.4.11) gives the frequency of the localized mode near a defective fibre on the assumption that its mass but not the bonding strength is different from that of the other fibres.

From Eq. (6.4.11) we can immediately derive the condition for the occurrence of the localized modes, which is that Δm must be positive, since otherwise ω will be imaginary. A positive Δm means that the defective fibre must be lighter than the other fibres. Thus, we derive the conclusion that if the bonding strength of all the fibres is the same, localized modes are possible in a composite only if the defective fibre has a lower mass. This condition may be satisfied in a real composite containing some hollow but well-bonded fibres.

The other inference which we can derive from Eq. (6.4.11) regarding the nature of the localized modes is that the value of ω^2 increases with Δm. Thus, we infer that the lighter the defective fibre, the stronger the localized mode.

Bonding defect only. Next we shall consider the case when $\Delta m = 0$ but ΔF_1 is non-vanishing. We shall make a further assumption that ΔF_1 is so small that its second power can be neglected in comparison with unity. With this assumption of linearity in ΔF_1, Eq. (6.4.10) reduces to the following two equations:

$$1 + \frac{4\Delta F_1}{m\omega^2}\left(1 + \frac{F_0 + F_1}{m\omega^2}\right) = 0 \tag{6.4.12}$$

and

$$1 + \frac{\Delta F_1}{m\omega^2}\left(1 + \frac{F_0 - 2F_2 + 4F_1}{m\omega^2}\right) = 0 \tag{6.4.13}$$

These are quadratic equations in ω^2 and their solutions can easily be written down. The solutions of Eqs. (6.4.12) aud (6.4.13), in that order, are given below:

$$\omega^2 = -\frac{2\Delta F_1}{m}\left[1 \pm \left(1 - \frac{F_0 + F_1}{\Delta F_1}\right)^{1/2}\right] \tag{6.4.14}$$

$$\omega^2 = -\frac{\Delta F_1}{2m}\left[1 \pm \left\{1 - \frac{4(F_0 - 2F_2 + 4F_1)}{\Delta F_1}\right\}^{1/2}\right] \tag{6.4.15}$$

We notice from the above equations that the necessary condition for ω to be real is that ΔF_1 must be negative. (The lower sign in each case will lead to an imaginary value of ω; only the upper signs are to be included.) A negative value of ΔF_1 implies an increased force constant of the defective fibres with its neighbours and therefore an increased bonding strength. Thus, we derive the conclusion that if all the fibres in a composite have equal mass, localized modes will occur near a fibre which is more strongly bonded to the matrix than the other fibres.

Both mass and bonding defects present together. Finally, in the present anti-plane strain analysis we shall consider the case when both Δm and ΔF_1 are non-vanishing. For simplicity, we shall consider only the first diagonal term of Eq. (6.4.10). This is not really a good approximation but should be adequate for the present purpose of a qualitative discussion. More realistic results can be obtained by properly solving Eq. (6.4.10), which is straightforward though quite involved.

We shall make a further approximation that ω^2 is so large that we can neglect the terms of order ω^{-4}. With the help of these rather drastic approximations we obtain the following simple result for the frequency of the localized mode:

$$\omega^2 = \frac{\Delta m}{m} \cdot \frac{F_0}{m - \Delta m}\left[1 - \frac{4 m \Delta F_1}{\Delta m F_0}\right] \tag{6.4.16}$$

For $\Delta F_1 = 0$ the result in Eq. (6.4.16) reduces to that in Eq. (6.4.11), as expected. In the limit $\Delta m = 0$ the correspondence between Eq. (6.4.16) and Eq. (6.4.14) is not so direct because of the different nature of approximations used in the derivation of these two equations.

In order to establish the correspondence between ω^2 as given by Eq. (6.4.16) with $\Delta m = 0$ and that derived in the previous sub-section for $\Delta m = 0$, we have to neglect the ω^{-4} term in Eq. (6.4.12). The resulting value of ω^2 then agrees with that given by Eq. (6.4.16), which is equal to $-4 \Delta F_1/m$ (for $\Delta m = 0$).

From Eq. (6.4.16) we notice that a sufficient condition for ω^2 to be real is that Δm be positive and ΔF_1 be negative. The general conclusion, therefore, is that the localized modes will occur in a composite containing a defective fibre which is lighter but more strongly bonded to the matrix than the others.

This, however, is not the necessary condition. A more restrictive condition for the occurrence of the localized modes can be obtained from Eq. (6.4.16) as follows. For ω^2 to be positive

$$\Delta m \left(1 - \frac{m}{\Delta m} \frac{\Delta F_1}{F_1}\right) > 0 \tag{6.4.17}$$

where we have put $F_0 \approx 4 F_1$, which is valid if F_2 is small (see Eq. 6.4.2).

The localized modes can therefore occur for $\Delta m > 0$, provided that

$$\frac{\Delta m}{m} > \frac{\Delta F_1}{F_1} \tag{6.4.18}$$

and for $\Delta m < 0$, if

$$\Delta F_1 < 0 \tag{6.4.19}$$

and, in addition,

$$\frac{|\Delta m|}{m} < \frac{|\Delta F_1|}{F_1} \tag{6.4.19a}$$

Plane strain modes

The analysis of wave propagation in the xy plane when the polarization vectors are also in the xy plane corresponds to the plane strain analysis in the continuum theory. The calculation of the localized mode frequencies in this case follows the same lines as in the anti-plane strain case but is more complicated, because of an additional degree of freedom.

As described at the end of Section 5.8, the force constants between two fibres in this case have to be regarded as 2 × 2 matrices on account of their coupling in the x and y directions. The force constants between the fibres **l** and **l′** will therefore be denoted by $\phi_{\alpha\beta}$ (**l**, **l′**), where α and β can take two values each ($\alpha, \beta = x, y$). These matrices for the perfect composite on the basis of the same model as was used in the previous section can be written as follows:

$$\phi(0, 1) = \phi(0, 3) = -\begin{pmatrix} A_1 & 0 \\ 0 & A_2 \end{pmatrix} \tag{6.4.20}$$

$$\phi(0, 2) = \phi(0, 4) = -\begin{pmatrix} A_2 & 0 \\ 0 & A_1 \end{pmatrix} \tag{6.4.20a}$$

$$\phi(0, 5) = \phi(0, 7) = -\begin{pmatrix} B_1 & B_2 \\ B_2 & B_1 \end{pmatrix} \tag{6.4.20b}$$

and

$$\phi(0, 6) = \phi(0, 8) = -\begin{pmatrix} B_1 & -B_2 \\ -B_2 & B_1 \end{pmatrix} \tag{5.4.20c}$$

where the numbers in parentheses refer to the fibre labels as defined in Table 6.2. Using Eq. (2.2.6), we obtain

$$\phi(0, 0) = \begin{pmatrix} A & 0 \\ 0 & A \end{pmatrix} \tag{6.4.21}$$

where

$$A = 2\,(A_1 + A_2 + 2\,B_1) \tag{6.4.22}$$

The force constants defined in the above equations are the same as those defined at the end of Section 5.8 except for the units, since in that section the force constants were expressed in units of mass.

As in the previous section, the defective fibre is located at the origin of coordinates. The mass of the defective fibre is again assumed to be $m - \Delta m$, so that the matrix elements of $\Delta \mathbf{M}$, the change in the diagonal matrix of masses, are the same as in the anti-plane strain case and are given by Eq. (6.3.17).

If we assume, as before, that $\Delta\phi$ is non-vanishing only for the four nearest neighbours of the defective fibre, the coordination number of the defect will again be 5. However, in this case each fibre has two degrees of freedom and therefore the defect space is of dimension 10. Hence the order of the determinantal equation which has to be solved for the perturbed vibration mode frequencies will be 10. However, with the help of group theoretical techniques, as indicated in the previous case, the 10 × 10 can be reduced to some lower-order determinants and only a 3 × 3 determinantal equation has to be solved for the localized modes.

As may be expected intuitively, the qualitative behaviour of the localized mode frequencies and the nature of their dependence on Δm and $\Delta \phi$ are similar to those observed in the anti-plane strain case. In view of the fact that no experimental results are available as yet on the propagation of short waves in fibre composites, there is not a strong enough incentive to perform a numerically accurate calculation of the localized mode frequencies. At present, therefore, a qualitative discussion of the effect of defects on the vibration frequencies of a fibre composite should be of sufficient interest.

For the purpose of illustration we shall consider only the mass defect in the present case of the plane strain modes and assume $\Delta \phi = 0$. This assumption results in a very substantial simplification of the problem. With this assumption the coordination number of the defect is just 1 and the defect space, which only consists of the origin of the coordinates, is 2. The matrices $\Delta \mathbf{T}_p$ and $\mathbf{g}$ are 2×2 diagonal, which have been given below:

$$\Delta \mathbf{T}_p = \begin{pmatrix} -\Delta m \omega^2 & 0 \\ 0 & -\Delta m \omega^2 \end{pmatrix} \tag{6.4.23}$$

and

$$\mathbf{g} = \begin{pmatrix} g_0 & 0 \\ 0 & g_0 \end{pmatrix} \tag{6.4.24}$$

where

$$g_0 = G(0, 0).$$

By using the series representation of the Green function for the localized modes as given in Eq. (6.3.16) and in Eq. (6.4.21) for $\phi(0, 0)$, we obtain

$$g_0 = -\frac{1}{m\omega^2}\left(1 + \frac{A}{m\omega^2}\right) \tag{6.4.25}$$

The determinant in Eq. (6.3.22) in the present case is diagonal with equal diagonal elements. The local mode frequency is therefore doubly degenerate and is given by

$$\omega^2 = \frac{\Delta m}{m} \cdot \frac{A}{m - \Delta m} \tag{6.4.26}$$

Equation (6.4.26), as expected, is exactly similar to Eq. (6.4.11) (apart from the difference in the force constants which enter these equations). The qualitative inferences which were derived for the localized modes near a mass defect in the anti-plane strain case are therefore also valid in the present case.

The constant F_0 in Eq. (6.4.11) is a measure of the bonding strength of the fibres in the perfect composite in the z direction, whereas the constant A in Eq. (6.4.26) measures the bonding strength on the xy plane. It may be recalled from Section 5.8 that A actually defines the restoring force on a fibre if it is displaced in the x or y direction by a unit amount. There is no universally agreed definition of the bonding strength but A can be accepted as a measure of the bonding strength of the fibres in a perfect composite.

Finally, we wish to point out that the preceding analysis indicates that a measurement of the localized mode frequencies can yield useful information about the mass and the bonding defects of the fibres in a composite. For example, the presence of hollow fibres and/or strongly bound fibres can be detected by this method. In principle, an analysis of the band modes can also yield similar information but the localized modes are much easier to detect experimentally than the band modes.

A study of the localized modes can be carried out only by using short waves. Long waves, as shown in Section 6.2, can be used to study the average state of the fibre alignment in a composite. Thus, the results of this chapter, together with those derived in the preceding chapter, should be quite useful in developing non-destructive testing techniques for fibre composites based on the use of elastic waves.

It should be emphasized that the theory of the localized modes as given here is based upon the harmonic approximation. The limitations of the harmonic theory have already been discussed in Chapter 5. The important point to bear in mind before applying the results of this section is that the attenuation of the wave caused by anharmonic effects has been totally neglected. This effect may be especially significant in the case of localized modes, because of the large amplitude associated with these modes at or near the defect. Nevertheless the results derived in this section should provide a useful order of magnitude estimate of the localized mode frequencies.

6.5 RELAXATION ENERGY OF A DEBONDED FIBRE IN A COMPOSITE

So far in this chapter we have discussed the effect of a defective fibre on the dynamics of a fibre composite. In this section we shall consider the static properties of a composite containing a defective fibre. In particular, we shall calculate the contribution of the fibre–fibre interaction on the energy required to debond a fibre from the matrix. This calculation is therefore relevant to the study of fracture of a composite on the basis of the Griffith theories discussed in Chapter 4. This contribution, called the relaxation energy, was referred to in Section 4.4.

The reason why this present section has been included in this chapter and not in Chapter 4 is that the technique used here for the calculation of the relaxation energy has been derived from that used in the dynamical problem. This has been possible because of the fact that the static behaviour of a solid can be regarded as a certain limiting case of its dynamical behaviour. This correspondence between the statics and the dynamics of the ordinary solids was exploited by Tewary (1969) to develop a formalism for lattice statics. The treatment given here is essentially an adaptation of Tewary's (1969) method of lattice statics (see also Bullough and Tewary, 1972; Tewary, 1973).

Let us first define the problem and describe the model which we shall use. Let us consider a perfect fibre composite in the same sense as used in the preceding two sections. If a static (time-independent) displacement $u_\alpha(\mathbf{l})$ ($\alpha = x$ or y, assuming the same frame of reference as that defined earlier in this chapter) in the xy plane is given to each fibre, then the change in the potential energy of the composite can be written as a Taylor series in $u_\alpha(\mathbf{l})$, as given in Eq. (2.2.5). The concept of potential energy of a composite in terms of the force constants has already been introduced in Chapter 5.

For a perfect composite $\phi_\alpha(\mathbf{l}) = 0$ for all $\mathbf{l}$ and α according to restriction (1) in Section 2.2, following Eq. (2.2.5). This restriction arises from the translation symmetry of the perfect composite combined with the condition of the invariance of the potential energy of the composite against rigid body translations. The translation symmetry requires $\phi_\alpha(\mathbf{l})$ to be the same for all $\mathbf{l}$, whereas the latter condition require $\sum_{\mathbf{l}} \phi_\alpha(\mathbf{l}) = 0$. The only way these two requirements can be satisfied together is to make $\phi_\alpha(\mathbf{l}) = 0$ for all α and $\mathbf{l}$.

Physically the vanishing of $\phi_\alpha(\mathbf{l})$ can be understood as follows: suppose that the interaction between fibres separated by a vector distance $\mathbf{r}$ can be denoted by an effective potential $V(r)$. We shall assume that $V(r)$ is a central potential, i.e. it depends only on the magnitude of r and not its direction. This assumption has been made for the sake of simplicity and is not essential for the present argument. It will however, be found useful at a latter stage.

It can be verified by using the definition of the first-order Taylor coefficient that

$$f_\alpha(l) = -\left(\frac{\partial V(r)}{\partial r_\alpha}\right)_{\mathbf{r}=\mathbf{r}(\mathbf{l})} \tag{6.5.1}$$

The coefficient $\phi_\alpha(\mathbf{l})$ can thus be regarded as the α component of the force on the fibre at site $\mathbf{l}$. Because of the symmetrical arrangement of the fibres around each site, the net force on each fibre is zero; therefore $\phi_\alpha(\mathbf{l})$ for each $\mathbf{l}$ is zero.

The preceding discussion is valid only for a perfect composite. Now consider the effect of a single defective fibre, which can be conveniently located at the origin of the coordinates. We shall assume that the defect is weak, so that only the first derivative of the potential at a near-neighbour site is different and its second derivative is not affected. The second derivative of $V(r)$ defines the force constants $\phi_{\alpha\beta}(0, \mathbf{l})$ as follows:

$$\phi_{\alpha\beta}(0, \mathbf{l}) = -\left(\frac{\partial^2 V(r)}{\partial r_\alpha \, \partial r_\beta}\right)_{\mathbf{r}=\mathbf{r}(\mathbf{l})} \tag{6.5.2}$$

Since by hypothesis the force constants between the defective fibre and a perfect fibre are the same as the force constants between two perfect fibres, the force constants between any two fibres can be obtained from Eq. (6.5.2). It may be recalled that in the case of a perfect composite $\phi_{\alpha\beta}(\mathbf{l}, \mathbf{l}')$ will depend on $\mathbf{l}$ and $\mathbf{l}'$ only through their difference, on account of translation symmetry.

Since the first derivative of the potential due to the defective fibre at any fibre site, say $\mathbf{l}$, differs from the perfect fibres around $\mathbf{l}$, the net force at the site will not vanish and $\phi_\alpha(\mathbf{l})$ will survive. Thus, $\phi_\alpha(\mathbf{l})$ can be identified as the force at the site $\mathbf{l}$ due to the defective fibre at the origin.

Under the influence of force $\phi_\alpha(\mathbf{l})$, the fibre at site $\mathbf{l}$ will be displaced. A displacement of this fibre will exert additional forces on its neighbours, which will also be displaced. Each fibre will thus affect its neighbours and as a result all the fibres in the composite will suffer a displacement. This process can be referred to as fibre relaxation in the composite, by analogy with a similar phenomenon in ordinary solids which is called atomic relaxation. The change in energy associated with the relaxation process can then be called the relaxation energy.

In the preceding discussion we have assumed that the presence of the defective fibre does not induce any changes in the force constants. This assumption is reasonable only for a weak defect. However, it is easy to visualize that this assumption was not essential to the argument given above. The main point to note is that the incorporation of a single defective fibre in an otherwise perfect composite changes the energy of the composite.

We shall now proceed to calculate the relaxation energy, which will be denoted by E_r. Let $u_\alpha(\mathbf{l})$ denote the displacement field caused by the defective fibre located at the origin. Let $f_\alpha(\mathbf{l})$ denote the α component of the force at the site $\mathbf{l}$ and $\overset{*}{\phi}_{\alpha\beta}(\mathbf{l}, \mathbf{l}')$ denote the force constant between the fibres at $\mathbf{l}$ and $\mathbf{l}'$. The star superscript in the force constants is meant to indicate that the effect of the defect on the force constants has been included. The force constants for the perfect lattice will be denoted by $\phi_{\alpha\beta}(\mathbf{l}, \mathbf{l}')$. The approximation used for the 'weak' defect in the qualitative argument given earlier implies

$$\overset{*}{\phi}_{\alpha\beta}(\mathbf{l}, \mathbf{l}') = \phi_{\alpha\beta}(\mathbf{l}, \mathbf{l}') \tag{6.5.3}$$

Using the Taylor series expansion of the energy as given in Eq. (2.2.5), the energy Φ associated with the displacement field $u_\alpha(\mathbf{l})$ can be written as

$$\Phi = -\sum_{\alpha, l} f_\alpha(l)\, u_\alpha(l) + \tfrac{1}{2} \sum_{\substack{\alpha, l \\ \beta, l'}} \phi^*_{\alpha\beta}(l, l')\, u_\alpha(l)\, u_\beta(l') \tag{6.5.4}$$

where, in the spirit of harmonic approximation, we have neglected the cubic and higher powers of $u_\alpha(l)$.

The equilibrium values of $u_\alpha(l)$ which define the stable configuration of the composite in the presence of a defective fibre are obtained by minimizing Φ with respect to $u_\alpha(l)$, i.e. as the solution of the set of equations

$$\frac{\partial \Phi}{\partial u_\alpha(l)} = -f_\alpha(l) + \sum_{\beta, l'} \phi^*_{\alpha\beta}(l, l')\, u_\beta(l')$$

which can be written in the form

$$f_\alpha(l) = \sum_{\beta, l'} \phi^*_{\alpha\beta}(l, l')\, u_\beta(l') \tag{6.5.5}$$

As in Section 6.3, we can define a $N \times N$ square matrix ϕ and N-dimensional column matrices $\mathbf{f}$ and $\mathbf{u}$, where N is the total number of degrees of freedom in the composite.

In the matrix notation Eq. (6.5.5) can be written as

$$\mathbf{f} = \phi^* \mathbf{u} \tag{6.5.6}$$

or

$$\mathbf{u} = \mathbf{G}^* \mathbf{f} \tag{6.5.7}$$

where the matrix $\mathbf{G}^*$ is defined to be the inverse of ϕ^*, viz.

$$\mathbf{G}^* = \{\phi^*\}^{-1} \tag{6.5.8}$$

The displacement field as given by Eq. (6.5.7) defines the configuration of the minimum energy. The value of this energy, i.e. the minimum of Φ, is the relaxation energy. Substituting for $\mathbf{u}$ from Eq. (6.5.7) in Eq. (6.5.4) we obtain for E_r

$$E_r = (\Phi)_{\min} = -\tilde{\mathbf{f}}\mathbf{u} + \tfrac{1}{2}\tilde{\mathbf{u}}\phi^*\mathbf{u}$$

$$= -\tfrac{1}{2}\tilde{\mathbf{f}}\mathbf{u} \tag{6.5.9}$$

$$= -\tfrac{1}{2}\tilde{\mathbf{f}}\mathbf{G}^*\mathbf{f} \tag{6.5.10}$$

In order to understand the significance of $\mathbf{G}^*$, let us write ϕ^* in the following form:

$$\phi^* = \phi - \Delta\phi \tag{6.5.11}$$

where $\Delta\phi$ denotes the contribution of the defect to the force constant matrix. From Eq. (6.5.8) and (6.5.11) we obtain

$$\mathbf{G}^* = (\mathbf{I} - \mathbf{G}_0 \,\Delta\phi)^{-1} \mathbf{G}_0 \qquad (6.5.12)$$

or

$$\mathbf{G}^* = \mathbf{G}_0 + \mathbf{g}_0 \,\Delta\phi \,\mathbf{G}^* \qquad (6.5.13)$$

where

$$\mathbf{G}_0 = \{\phi\}^{-1} \qquad (6.5.14)$$

and **I** is a unit matrix.

From Eqs. (6.3.3), (6.3.11) and (6.5.14) we find that

$$G_0 = \lim_{\omega \to 0} \{T\}^{-1} = \lim_{\omega \to 0} G \qquad (6.5.15)$$

which means that $\mathbf{G}_0$ is the zero frequency limit of the Green function **G** which was introduced in Section 6.3.

The Green function $\mathbf{G}_0$ defined in this section is called the static Green function to distinguish it from **G**, which may be called the phonon Green function. An important difference between $\mathbf{G}_0$ and **G** is that, unlike **G**, $\mathbf{G}_0$ is independent of the fibre mass.

It is obvious that $\mathbf{G}_0$, like **G**, is a property of the perfect composite and does not depend upon the defect. The quantity $\mathbf{G}^*$ as defined in Eq. (6.5.8) is quite similar to $\mathbf{G}_0$ but for the fact that $\mathbf{G}^*$ also contains a contribution $\Delta\phi$ from the defect. $\mathbf{G}^*$ is therefore called the static defect Green function. $\mathbf{G}^*$ also has a dynamical analogue but it does not concern us here. In fact, in this section we shall deal with only the static Green functions. Henceforth, for the sake of brevity, we shall omit the prefix 'static' and call $\mathbf{G}_0$ and $\mathbf{G}^*$ simply the Green functions.

Equation (6.5.13) giving a relation between the defect and the perfect Green functions, is sometimes called the Dyson equation on account of its similarity to the well-known Dyson equation in quantum field theory, A series representation of $\mathbf{G}^*$ in terms of $\mathbf{G}_0$ and $\Delta\phi$ can immediately be derived from the Dyson equation and is given below:

$$\mathbf{G}^* = \mathbf{G}_0 + \mathbf{G}_0 \,\Delta\phi \mathbf{G}_0 + \mathbf{G}_0 \,\Delta\phi \,\mathbf{G}_0 \,\Delta\phi \mathbf{G}_0 + \ldots \qquad (6.5.16)$$

This series may be quite convenient for a calculation of $\mathbf{G}^*$ in the case of a weak defect, i.e. when $\Delta\phi$ is small. However, in most cases of practical interest $\mathbf{G}^*$ can be calculated exactly in a fairly simple manner. These methods have been discussed in considerable detail by Tewary (1969), Bullough and Tewary (1972) and Tewary (1973) for ordinary solids and are also applicable to fibre composites. In the present context, as mentioned earlier, we are not interested in numerically very accurate results. For our present purpose of a qualitative discussion the zero-order approximation in Eq (6.5.16) is quite sufficient, so that we can take

$$\mathbf{G}^* \approx \mathbf{G}_0 \qquad (6.5.17)$$

We may remind ourselves that the series representation given in

Eq. (6.3.16) is not valid for $\mathbf{G}_0$, since in the present case ω^2 is zero. In general, it is not possible to obtain a closed analytical expression for $\mathbf{G}_0$ and it has to be calculated numerically. Numerical methods for the calculation of $\mathbf{G}_0$ have also been given by Tewary (1969).

It is, however, possible to obtain a useful asymptotic form for $\mathbf{G}_0$ by using the correspondence between the lattice theory and the continuum theory. This correspondence was indicated in Chapter 2 and was used in the dynamic case to obtain some relations between the force constants and the elastic constants in Chapter 5. With the help of this correspondence it can be shown (see, for example, Tewary, 1973) that the Green function obtained on the basis of the continuum model is the leading term in the asymptotic expansion of the discrete lattice Green function.

The continuum model Green functions are well known (see, for example, Eshelby, 1956). For a three-dimensional solid, the Green function has a $1/r$ dependence on the distance r (for large r). In the case of a two-dimensional solid the Green function has an asymptotic behaviour $\log r$ for plane strain modes and $\tan^{-1} y/x$ for anti-plane strain modes.

The discussion so far in this section has been rather formal. Now let us apply this method to a particular case.

We shall calculate the relaxation energy associated with a debonded fibre in an otherwise perfect composite. We shall assume the model which was used for the dynamic case in Section 6.4, i.e. a tetragonal composite with a single defective fibre. Only the plane strain modes have to be considered.

The force constants between the fibres in the perfect composite in the plane strain modes have been given in Eqs. (6.4.20) and (6.4.21). We shall make a further approximation that B_1 and B_2 are zero, which means that only the nearest-neighbour interactions are allowed.

The matrix $\phi(\mathbf{k})$ in this case is a 2×2 matrix and is a diagonal matrix, viz.

$$\phi(\mathbf{k}) = \begin{pmatrix} \omega_1^2(\mathbf{k}) & 0 \\ 0 & \omega_2^2(\mathbf{k}) \end{pmatrix} \tag{6.5.18}$$

where

$$\omega_1^2(\mathbf{k}) = 2A_1(1 - \cos k_1) + 2A_2(1 - \cos k_2) \tag{6.5.19}$$

and

$$\omega_2^2(\mathbf{k}) = 2A_1(1 - \cos k_2) + 2A_2(1 - \cos k_1) \tag{6.5.19a}$$

The allowed $\mathbf{k}$ vectors are distributed in the Brillouin zone, which has been defined in Section 5.8.

The Fourier transform of the Green function matrix will be denoted by $\mathbf{G}_0(\mathbf{k})$ It is also a 2×2 diagonal matrix and can be obtained very simply in the present model by taking the inverse of $\phi(\mathbf{k})$ (see Eq. 2.2.13). We obtain

$$G_0(\mathbf{k}) = \begin{pmatrix} \dfrac{1}{\omega_1^2(\mathbf{k})} & 0 \\ 0 & \dfrac{1}{\omega_2^2(\mathbf{k})} \end{pmatrix} \tag{6.5.20}$$

The defect in the present model is a debonded fibre which is characterized by the forces f_α (**l**) and the condition

$$\phi^*_{\alpha\beta}(0, \mathbf{l}') = 0 \text{ (for all } \mathbf{l}') \tag{6.5.21}$$

since a debonded fibre cannot interact with any other fibre. From Eq. (6.5.21) and (6.5.11) we obtain

$$\Delta\phi_{\alpha\beta}(0, \mathbf{l}) = \phi_{\alpha\beta}(0, \mathbf{l}) \tag{6.5.22}$$

We see that $\Delta\phi$ in the present case is large and therefore the approximation in Eq. (6.5.17) is not valid. We can, therefore, hope to get only an order-of-magnitude estimate of E_r. It may be mentioned here that the approximation in Eq. (6.5.17) is not wildly wrong even in the present case, as has been verified in the case of a vacancy in a Born–von Kármán lattice (See Tewary, 1969).

The most general form of the net forces caused by the debonded fibre at the origin on its nearest neighbour fibres, subject to the symmetry restrictions, are given below:

$$\left.\begin{aligned} f(0) &= \begin{pmatrix} 0 \\ 0 \end{pmatrix} \\ f(1) = -f(3) &= \begin{pmatrix} \xi \\ 0 \end{pmatrix} \\ f(2) = -f(4) &= \begin{pmatrix} 0 \\ \xi \end{pmatrix} \end{aligned}\right\} \tag{6.5.23}$$

where the numbers following f refer to the fibre labels as defined in Table 6.2. The value of ξ in Eqs. (6.5.23) is as yet undetermined.

The form of the displacement matrix elements $\mathbf{u}_\alpha$ (**l**) will be similar to that of f_α (**l**), since both **u** and **f** transform like a vector. We can therefore write

$$\left.\begin{aligned} u(0) &= \begin{pmatrix} 0 \\ 0 \end{pmatrix} \\ u(1) = -u(3) &= \begin{pmatrix} \zeta \\ 0 \end{pmatrix} \\ \text{and} \quad u(2) = -u(4) &= \begin{pmatrix} 0 \\ \zeta \end{pmatrix} \end{aligned}\right\} \tag{6.5.24}$$

where ζ can be determined, if required, from Eq. (6.5.7).

The Fourier transforms of **u** and **f** will be denoted by **u (k)** and **f (k)**, respectively, and are given below:

$$\mathbf{u}\,(\mathbf{k}) = 2\,i \begin{pmatrix} u_1 & (\mathbf{k}) \\ u_2 & (\mathbf{k}) \end{pmatrix} \tag{6.5.25}$$

and

$$f\,(\mathbf{k}) = 2\,i\,\xi \begin{pmatrix} \sin & k_1 \\ \sin & k_2 \end{pmatrix} \tag{6.5.26}$$

where i is the square root of -1 and u_1 **(k)** and u_2 **(k)** have to be determined.

Taking the Fourier transform of both sides of Eq. (6.5.7) (assuming $\mathbf{G}^* = \mathbf{G}$), we obtain

$$\mathbf{u}\,(\mathbf{k}) = \mathbf{G}\,(\mathbf{k})\;\mathbf{F}\,(\mathbf{k}) \tag{6.5.27}$$

which gives, with the help of Eqs. (6.5.20). (6.5.25) and (6.5.26),

$$u_1\,(\mathbf{k}) = \frac{\xi \sin k_1}{\omega_1^2\,(\mathbf{k})} \tag{6.5.28}$$

and

$$u_2\,(\mathbf{k}) = \frac{\xi \sin k_2}{\omega_2^2\,(\mathbf{k})} \tag{6.5.28a}$$

Finally, the relaxation energy E_r as defined by Eq. (6.5.9) is given by:

$$E_r = -\frac{2\xi^2}{\rho_k \Omega_k} \sum_{\mathbf{k}} \left[\frac{\sin^2 k_1}{\omega_1^2\,(\mathbf{k})} + \frac{\sin^2 k_2}{\omega_2^2\,(\mathbf{k})} \right] \tag{6.5.29}$$

where ρ_k denotes the density of the allowed **k** vectors in the Brillouin zone and Ω_k is the volume of the Brillouin zone.

The sum over **k** in Eq. (6.5.29) has to be evaluated numerically. However, an approximate value of this sum can be obtained with the help of Born's continuum approximation. In this approximation we expand the sine and the cosine functions of **k** and retain only up to the second-order term, viz.

$$\sin^2 k_{1,\,2} = k_{1,\,2}^2$$

and

$$\cos k_{1,\,2} = 1 - \frac{k_{1,\,2}^2}{2}$$

Substituting these values in Eq. (6.5.29) and using (6.5.19) for $\omega_1^2\,(\mathbf{k})$ and $\omega_2^2\,(\mathbf{k})$, we obtain:

$$E_r = -\frac{8\xi^2}{\Omega_k} \int_0^{\pi}\!\!\int_0^{\pi} \left[\frac{k_1^2}{A_1 k_1^2 + A_2 k_2^2} + \frac{k_2^2}{A_1 k_2^2 + A_2 k_1^2} \right] dk_1\, dk_2 \tag{6.5.30}$$

where we have replaced the summation by the two-dimensional integration over k_1 and k_2. The extra factor 4 in Eq. (6.5.30) has appeared because we have restricted the integration to only the positive values of k_1 and k_2.

Although the integral in Eq. (6.5.30) can be evaluated exactly, we shall make another approximation which reduces the integral to a particularly simple form. This approximation consists of putting

$$A_1 = A_2 = A \tag{6.5.31}$$

in Eq. (6.5.30), which yields the final result for the relaxation energy

$$E_r = -\frac{2\xi^2}{A} \tag{6.5.32}$$

since

$$\Omega_k = 4\pi^2$$

It should be emphasized that the approximation $A_1 = A_2$ in Eq. (6.5.31) is particularly unphysical, since it makes the composite mechanically unstable. However, the actual value of E_r does not seem to be highly sensitive to this approximation. Our purpose at present is only to obtain an order-of-magnitude estimate of E_r. We should therefore forgive ourselves for not being rigorous, particularly since this approximation has the very pleasing feature of yielding a very simple expression for E_r, as given by Eq. (6.5.32). The disturbing question of our model not being mechanically stable can be avoided by assuming that A is an average value of A_1 and A_2 rather than forcing A_1 and A_2 to be equal. Then the approximation made in reaching Eq. (6.5.32) consists of replacing A_1 and A_2 by their average value A, which can be defined as follows:

$$A = \frac{A_1 + A_2}{2} \tag{6.5.33}$$

Now we shall obtain a crude estimate of E_r assuming that A_1 and A_2 are known. First an estimate of ξ will be obtained by using the effective potential concept, which was introduced earlier in this section.

If $V(r)$ denotes the central potential due to a fibre in the perfect composite, then the effective net potential due to a debonded fibre is given by $-V(r)$. This can be verified by simple physical arguments. However, an interesting way of arriving at this result is to simulate a debonded fibre (which is defined as a fibre which does not interact with any other fibre) by putting a source of an equal but negative potential, i.e. $-V(r)$, on top of a fibre in the perfect composite. If the potential due to the debonded fibre which has been assumed to be located at the origin of coordinates is taken to be $-V(r)$, then the force due to this fibre at its nearest-neighbour sites can be obtained from Eq. (6.5.1) and is given below:

$$\left.\begin{aligned} f(1) = -f(3) &= \begin{pmatrix} V_2 \\ 0 \end{pmatrix} \\ f(2) = -f(4) &= \begin{pmatrix} 0 \\ V_2 \end{pmatrix} \end{aligned}\right\} \tag{6.5.34}$$

and

where

$$V_2 = \left(\frac{dV(r)}{dr}\right)_{r=1} \tag{6.5.35}$$

and the derivative has been evaluated at the nearest-neighbour position, for which $r = 1$ (see Table 6.2).

Comparing Eqs. (6.5.34) and (6.5.23), we obtain

$$\xi = V_2 \tag{6.5.36}$$

The force constants between the fibres in the perfect composite can also be obtained in terms of the derivatives of $V(r)$ with the help of Eq. (6.5.2). Thus, we get the following matrix for the force constants between two nearest neighbour fibres:

$$\phi(0, 1) = \phi(0, 3)$$

$$= -\begin{pmatrix} V_1 & 0 \\ 0 & V_2/a \end{pmatrix} \tag{6.5.37}$$

where

$$V_1 = \left(\frac{d^2V(r)}{dr^2}\right)_{r=1} \tag{6.5.38}$$

and a is the lattice constant as defined in Table 6.2.

Comparing Eqs. (6.5.37) and (6.4.20), we obtain

$$A_1 = V_1/a \tag{6.5.39}$$

and

$$A_2 = V_2/a \tag{6.5.40}$$

Finally, from Eqs. (6.5.36) and (6.5.40) we obtain the required relation between ξ and the force constants, i.e.

$$\xi = A_2 a \tag{6.5.41}$$

It should be emphasized that such a relation between the force due to a debonded fibre and the force constants between fibres in a perfect composite could be derived only because we had assumed the interaction potential to be a central potential. This assumption is probably reasonable for a transversely isotropic composite but not in a general case.

Now we require an estimate of A_1 and A_2. From Eqs. (5.6.13) we find that $A_1 = ac_{11}$ and $A_2 = ac_{66}$ by setting the second-neighbour force constant in that equation to zero. The factor 'a' has been introduced to take care of the units. For c_{11} and c_{66} we use the values given in Set 2 of the elastic constants in Section 6.2. These values, of course, were obtained for a hexagonal composite but we can use them in the present case, since our purpose is to obtain only an order-of-magnitude estimate.

Taking $a = 10^{-6}$ m, we obtain

$$A_1 \approx 2 \times 10^4 \text{ N/m}$$

$$A_2 \approx 10^4 \text{ N/m}$$

$$A \approx 1.5 \times 10^4 \text{ N/m}$$

and

$$\xi \approx 10^{-2} \text{ N}$$

Then from Eq. (6.5.32) we obtain the following estimate for the relaxation energy associated with the debonding of a fibre:

$$E_r \approx -0.1 \times 10^{-7} \text{ J/fibre}$$

To express this energy as the more familiar surface energy, we have to divide it by the area per fibre, i.e. the area of the two-dimensional square unit cell for the lattice structure shown in Figure 5.1a, which is equal to a^2. Taking a to be approximately equal to a micron as before and using the fact that in the linear approximation the energy of two defective fibres is additive, we obtain the following order-of-magnitude estimate for the relaxation energy per unit area:

$$E_r = 10n_d \text{ kJ/m}^2 \tag{6.5.42}$$

where n_d is the number density of the debonded fibre, i.e. the number of debonded fibres for each perfect fibre. This value is of the same order of magnitude as the experimentally observed value (see Phillips and Tetelman, 1972). Thus, we see that the fibre–fibre interaction makes a very important contribution to the work of fracture and therefore must be included in a realistic calculation.

Unfortunately, at present we do not have accurately measured values of the fibre–fibre force constants. They can, in principle, be measured from the dispersion of short waves as described in Chapter 5. If reliable values of the fibre–fibre force constants are available, then the Green function method as described in this chapter can give a quite accurate value of the fibre debonding energy.

Appendices

APPENDIX 1
OPERATIONS OF THE TETRAHEDRAL POINT GROUP

On the basis of the cartesian axes, as shown in Figures 5.1 and 5.2 the matrix representations of the tetrahedral point group for the composite are given below:

1. $$\begin{pmatrix} 1 & 0 & 0 \\ 0 & 1 & 0 \\ 0 & 0 & 1 \end{pmatrix}$$

2. $$\begin{pmatrix} 1 & 0 & 0 \\ 0 & 1 & 0 \\ 0 & 0 & -1 \end{pmatrix}$$

3. $$\begin{pmatrix} 1 & 0 & 0 \\ 0 & -1 & 0 \\ 0 & 0 & 1 \end{pmatrix}$$

4. $$\begin{pmatrix} 1 & 0 & 0 \\ 0 & -1 & 0 \\ 0 & 0 & -1 \end{pmatrix}$$

5. $$\begin{pmatrix} -1 & 0 & 0 \\ 0 & 1 & 0 \\ 0 & 0 & 1 \end{pmatrix}$$

6 $$\begin{pmatrix} -1 & 0 & 0 \\ 0 & 1 & 0 \\ 0 & 0 & -1 \end{pmatrix}$$

7. $$\begin{pmatrix} -1 & 0 & 0 \\ 0 & -1 & 0 \\ 0 & 0 & 1 \end{pmatrix}$$

8. $$\begin{pmatrix} -1 & 0 & 0 \\ 0 & -1 & 0 \\ 0 & 0 & -1 \end{pmatrix}$$

9. $$\begin{pmatrix} 0 & 1 & 0 \\ 1 & 0 & 0 \\ 0 & 0 & 1 \end{pmatrix}$$

10. $$\begin{pmatrix} 0 & 1 & 0 \\ 1 & 0 & 0 \\ 0 & 0 & -1 \end{pmatrix}$$

11. $$\begin{pmatrix} 0 & 1 & 0 \\ -1 & 0 & 0 \\ 0 & 0 & 1 \end{pmatrix}$$

12. $$\begin{pmatrix} 0 & 1 & 0 \\ -1 & 0 & 0 \\ 0 & 0 & -1 \end{pmatrix}$$

$$13.\quad \begin{pmatrix} 0 & -1 & 0 \\ 1 & 0 & 0 \\ 0 & 0 & 1 \end{pmatrix} \qquad 15.\quad \begin{pmatrix} 0 & -1 & 0 \\ -1 & 0 & 0 \\ 0 & 0 & 1 \end{pmatrix}$$

$$14.\quad \begin{pmatrix} 0 & -1 & 0 \\ 1 & 0 & 0 \\ 0 & 0 & -1 \end{pmatrix} \qquad 16.\quad \begin{pmatrix} 0 & -1 & 0 \\ -1 & 0 & 0 \\ 0 & 0 & -1 \end{pmatrix}$$

APPENDIX 2

OPERATIONS OF THE HEXAGONAL POINT GROUP

On the basis of the cartesian axes, as shown in Figures 5.1 and 5.2 the matrix representations of the hexagonal point group for the composite are given below:

$$1.\ \begin{pmatrix} 1 & 0 & 0 \\ 0 & 1 & 0 \\ 0 & 0 & 1 \end{pmatrix} \qquad 7.\ \tfrac{1}{2}\begin{pmatrix} -1 & \sqrt{3} & 0 \\ \sqrt{3} & 1 & 0 \\ 0 & 0 & 2 \end{pmatrix}$$

$$2.\ \begin{pmatrix} 1 & 0 & 0 \\ 0 & 1 & 0 \\ 0 & 0 & -1 \end{pmatrix} \qquad 8.\ \tfrac{1}{2}\begin{pmatrix} -1 & \sqrt{3} & 0 \\ \sqrt{3} & 1 & 0 \\ 0 & 0 & -2 \end{pmatrix}$$

$$3.\ \begin{pmatrix} 1 & 0 & 0 \\ 0 & -1 & 0 \\ 0 & 0 & 1 \end{pmatrix} \qquad 9.\ \tfrac{1}{2}\begin{pmatrix} 1 & -\sqrt{3} & 0 \\ \sqrt{3} & 1 & 0 \\ 0 & 0 & 2 \end{pmatrix}$$

$$4.\ \begin{pmatrix} 1 & 0 & 0 \\ 0 & -1 & 0 \\ 0 & 0 & -1 \end{pmatrix} \qquad 10.\ \tfrac{1}{2}\begin{pmatrix} 1 & -\sqrt{3} & 0 \\ \sqrt{3} & 1 & 0 \\ 0 & 0 & -2 \end{pmatrix}$$

$$5.\ \tfrac{1}{2}\begin{pmatrix} 1 & \sqrt{3} & 0 \\ \sqrt{3} & -1 & 0 \\ 0 & 0 & 2 \end{pmatrix} \qquad 11.\ \tfrac{1}{2}\begin{pmatrix} -1 & -\sqrt{3} & 0 \\ \sqrt{3} & -1 & 0 \\ 0 & 0 & 2 \end{pmatrix}$$

$$6.\ \tfrac{1}{2}\begin{pmatrix} 1 & \sqrt{3} & 0 \\ \sqrt{3} & -1 & 0 \\ 0 & 0 & -2 \end{pmatrix} \qquad 12.\ \tfrac{1}{2}\begin{pmatrix} -1 & -\sqrt{3} & 0 \\ \sqrt{3} & -1 & 0 \\ 0 & 0 & -2 \end{pmatrix}$$

13. $\begin{pmatrix} -1 & 0 & 0 \\ 0 & -1 & 0 \\ 0 & 0 & 1 \end{pmatrix}$

19. $\frac{1}{2}\begin{pmatrix} 1 & -\sqrt{3} & 0 \\ -\sqrt{3} & -1 & 0 \\ 0 & 0 & 2 \end{pmatrix}$

14. $\begin{pmatrix} -1 & 0 & 0 \\ 0 & -1 & 0 \\ 0 & 0 & -1 \end{pmatrix}$

20. $\frac{1}{2}\begin{pmatrix} 1 & -\sqrt{3} & 0 \\ -\sqrt{3} & -1 & 0 \\ 0 & 0 & -2 \end{pmatrix}$

15. $\begin{pmatrix} -1 & 0 & 0 \\ 0 & 1 & 0 \\ 0 & 0 & 1 \end{pmatrix}$

21. $\frac{1}{2}\begin{pmatrix} -1 & \sqrt{3} & 0 \\ -\sqrt{3} & -1 & 0 \\ 0 & 0 & 2 \end{pmatrix}$

16. $\begin{pmatrix} -1 & 0 & 0 \\ 0 & 1 & 0 \\ 0 & 0 & -1 \end{pmatrix}$

22. $\frac{1}{2}\begin{pmatrix} -1 & \sqrt{3} & 0 \\ -\sqrt{3} & -1 & 0 \\ 0 & 0 & -2 \end{pmatrix}$

17. $\frac{1}{2}\begin{pmatrix} -1 & -\sqrt{3} & 0 \\ -\sqrt{3} & 1 & 0 \\ 0 & 0 & 2 \end{pmatrix}$

23. $\frac{1}{2}\begin{pmatrix} 1 & \sqrt{3} & 0 \\ -\sqrt{3} & 1 & 0 \\ 0 & 0 & 2 \end{pmatrix}$

18. $\frac{1}{2}\begin{pmatrix} -1 & -\sqrt{3} & 0 \\ -\sqrt{3} & 1 & 0 \\ 0 & 0 & -2 \end{pmatrix}$

24. $\frac{1}{2}\begin{pmatrix} 1 & \sqrt{3} & 0 \\ -\sqrt{3} & 1 & 0 \\ 0 & 0 & -2 \end{pmatrix}$

References

Abarcar, R.B. and Cunniff, P.F. (1972). J. Comp. Mat. 6, 504.

Abolinsh, D.S. (1965). Mekhanika Polimerov 1, 28.

Abrahams, M. and Dimmock, J. (1971). Plastics and Polymers 39, 187.

Achenbach, J.D. and Herrmann, G. (1967), North-western University, Structural Mech. Lab. Report No. 67-3, AD–657461.

Adams, D.F. (1970). J. Comp. Mat. 4, 310.

Adams, D.F. (1974). J. Comp. Mat. 8, 320.

Adams, D.F. (1976). Mat. Science & Engg. 23, 55.

Adams, D.F. and Doner, D.R. (1967a). J. Comp. Mat. 1, 4.

Adams, D.F. and Doner. D.R. (1967b). J. Comp. Mat. 1, 152.

Adams, D.F., Doner. D.R. and Thomas, R.L. (1967). AFML Report TR–67–96; AD–654065.

Adams, D.F. and Tsai, S.W. (1969). J. Comp. Mat. 3, 368.

Adams, D.M. (1969). Plastics and Polymers 37, 385.

Adams, R.D., Flitcroft, J.E., Hancox, N.L. and Reynolds, W.N. (1973). J. Comp. Mat. 7, 68.

Adams, R.D., Fox, M.A.O., Flood, R.J.L., Friend, R.J. and Hewitt, R.L. (1969). J. Comp. Mat, **3**, 594.
Agarwal, B.D. and Dally, J.W. (1975). J. Mat. Science **10**, 193.
Ahmad, I. and Barranco, J.M. (1970). Metall. Trans. **1**, 989.
Allison, I.M. and Holloway, L.C. (1967). Br. J. Appl. Phys. **18**, 979.
Allred, R.E., Hoover, W.R. and Horak, J.A. (1974). J. Comp. Mat. **8**, 15.
Allred, R.E. and Schuster, D.M. (1973). J. Mat. Science, **8**, 245.
Amirbayat, J. (1971). Fibre Science and Tech. **3**, 309.
Amirbayat, J. and Hearle, J.W.S. (1969). Fibre Science and Tech. **2**, 143.
Amirbayat, J. and Hearle, J.W.S. (1970a). Fibre Science and Tech, **2**, 233.
Amirbayat, J. and Hearle, J.W.S. (1970b). Fibre Science and Tech. **3**, 147.
Anderson, E. (1973). J. Mat. Science **8**, 676.
Argon, A.S. (1972). Treat. Mat. Science and Tech. (Academic Press, New York; Ed.—H. Herman), **1**, 79.
Arridge, R.G.C. (1965). Br. J. Appl. Phys. **16**. 1181.
Asamoah. N.K. and Wood, W.G. (1970). J. Strain Analysis, **5**, 88.
Ashton, J.E. (1969a). J. Comp. Mat. **3**, 355.
Ashton, J.E. (1969b). J. Comp. Mat. **3**, 470.
Ashton, J.E. and Love, T.S, (1969). J. Comp. Mat. 3, 230.
Ashton, J.E. and Waddoups, M.E. (1969). J. Comp. Mat. 3, 148.
Atkins, A.G. (1975a). J. Mat. Science **10**, 819.
Atkins, A.G. (1975b). J. Aircraft **12**, 850.
Aveston, J. and Kelly, A. (1973). J. Mat. Science **8**, 352.
Auzukains, Ya. V., Bulavs, F. Ya. and Gunyaev, G.M. (1974). Soviet Polymer Mechanics **9**, 22.
Bader, M.G. (1971). Design Engineering; Feb. issue, p. 59.
Bader, M.G. and Ellis, R.M. (1974). Composites **5**, 253.
Baker, A.A. (1968). J. Mat. Science **3**, 412.
Baker, A.A. (1975). Materials Science & Engineering **17**, 177.
Baker, A.A., Braddick, D.M. and Jackson, P.W. (1972). J. Mat. Science **7**, 747.
Baker, A.A. and Cratchley, D. (1964). Appl. Materials Res. **3**, 215.
Baker, A.A., Mason, J.E. and Cratchley, D. (1966). J. Mat. Science **1**, 229.
Baldwin, D.H. and Sierakowski, R.L. (1975). Composites **6**, 30.
Barker, L.M. (1971). J. Comp. Mat. **5**, 140.
Barnet, F.R. and Norr, M.K. (1976). Composites **7**, 92.
Bartenev, G.M. and Motorina, L.I. (1971). Glass and Ceramics **28**, 185.
Bates, J.J. (1973). Carbon Fibres in Engineering. (McGraw-Hill, U.K.; Editor–M. Langley), p. 194.
Batson, R.G. and Hyde, J.H. (1922). Mechanical Testing (Chapman and Hall, London).
Beaumont, P.W.R. and Harris, B. (1971). Proc. Int. Conf. on Carbon Fibres, Paper 49 (Plastics Inst. London).
Beaumont, P.W.R. and Harris, B. (1972). J. Mat. Science **7**, 1265.
Beaumont, R.A. (1954). Mechanical Testing of Metallic Materials (Pitman, London) III Edition.
Beaumont, P.W.R. and Phillips, D.C. (1972a). J. Comp. Mat. **6**, 32.
Beaumont, P.W.R. and Phillips, D.C. (1972b). J. Mat. Science **7**, 682.

Beckwith, T.G. and Buck, N.L. (1961). Mechanical Measurements. (Addison-Wesley; Calif.).
Bedwell, M. (1973). Carbon Fibres in Engineering (McGraw-Hill, U.K.; Editor–M. Langley), p. 160.
Behrendt, D.R. (1976). NASA Technical Memorandum, NASA TM–X–73402.
Behrens, E. (1967a). J. Acoust. Soc. Am. **42**, 367.
Behrens, E. (1967b). J. Acoust. Soc. Am. **42**, 378.
Behrens, E. and Kremheller, A. (1969). Non-destructive Testing **2**, 55.
Bell, J.F.W. (1968). Ultrasonics **6**, 11.
Benveniste, Y. and Weitsman, Y. (1974). Acta Mechanics **19**, 179.
Berg, K.R. (1967). AIAA/ASME VIII Structures, Structural Dynamics and Materials Conference; Palm Springs, California; p. 706.
Berg, C.A. (1971). Fibre Science & Tech. **3**, 261.
Berg, C.A., Melton, R., Kalnin, I. and Dunn, T. (1971). J. Mat. Science **6**, 683.
Berg, C.A. and Salama, M. (1972). J. Materials **2**, 216.
Berg. C.A. and Salama, M. (1973). Fibre Science & Tech. **6**, 125.
Bert, C.W. (1974). Composites **5**, 20.
Bert, C.W. and Francis, P.H. (1974). A.I.A.A. Journal **12**, 1173.
Bhattacharyya, S. and Parikh, N.M. (1970). Metall. Transact. **1**, 1437.
Bhattacharyya, S. and Parikh, N.M. (1971). Fibre Science & Tech. **3**, 209.
Bishop, P.H.H. (1966). Ministry of Aviation, Farnborough (U.K.). Report TR 66245.
Bloom, J.M. and Wilson, H.B. Jnr. (1967). J. Comp. Mat. **1**, 268.
Bocker-Pedersen, O. (1974). J. Mat. Science **9**, 948.
Boller, K.H. (1964). Modern Plast. **41**, 145.
Bolotin, V.V. (1965a). Mekhanika Polimerov **1**, 27.
Bolotin, V.V. (1965b). Izv. Akad. Nauk. CSSR, Mekh. **1**, 74.
Bolotin, V.V. (1966). Mekhanika Polimerov **2**, 11.
Bomford, M.J. (1968). Ph.D. Thesis, Cambridge University.
Born, M. and Huang, K. (1954). Dynamical Theory of Crystal Lattices. (Oxford University Press, London).
Bose, S.K. and Mal, A.K. (1974). J. Mech. Phys. Solids **22**, 217.
Boucher, S. (1975). Revue-M-Mecanique **21**, 243.
Braddick, D.M., Jackson, P.W. and Walker, P.J. (1971). J. Mat. Science **6**, 419.
Bradfield, G. (1968). Non-destructive Testing **1**, 165.
Brandmaier, H.E. (1969). J. Comp. Mat. **3**, 728.
Brody, H. and Ward, I.M. (1971). Polym. Engineering & Science **11**, 139.
Broutman, L.J. (Editor) (1974). Composite Materials: Vol 5: Fracture and Fatigue (Academic Press, New York).
Broutman, L.J. and Krock, R.H. (1967). Modern Composite Materials (Addison-Wesley, Massachussetts).
Brown, C.G., Hancox, N.L. and Reynolds, W.N. (1972). AERE Harwell Industrial Report NDT–47.
Brown, J.H. (1973). Magazine of Cement Research **25**, 31.
Brown, J.Q. (Jnr.) (1961). J. Soc. Plast. Engineers **17**, 989.
Budianski, B. (1965). J. Mech. Phys. Solids **13**, 223.
Bulavs, F. Ya. and Birze, A.N. (1975). Soviet Polymer Mechanics **10**, 198.
Bullock, R.E. (1974). J. Comp. Mat. **8**, 200.
Bullough, R. and Tewary, V.K. (1972). Interatomic Potentials and Simulation of Lattice Defects (Plenum, New York; Eds.–P.C. Gehlen, J.R. Beeler Jnr. and R.I. Jaffee), p. 155.
Burke, J.J., Reed, N.L. and Weiss, V. (1966). Strengthening Mechanisms; Metals and Ceramics (Syracuse Univ. Press, New York).

Butcher, B.R. (1976a). Composites **7**, 12.
Butcher, B.R. (1976b). Composites **7**, 81.
Callaway, J. (1964). Energy Band Theory (Academic Press, New York).
Carrara, A.S. and McGarry, F.J. (1968). J. Comp. Mat. **2**, 222.
Chamis, C.C. (1967). Ph.D. Thesis (Case Western Reserve University).
Chamis, C.C. (Ed.) (1975). Composite Materials Vol 7: Structural Design and Analysis (Academic Press, New York).
Chamis, C.C., Lark, R.F. and Sullivan, T.L. (1975). NASA Technical Note, NASA TN D–7879.
Chamis, C.C. and Sendeckyj, G.P. (1968). J. Comp. Mat. **2**, 332.
Chamis. C.C. and Sinclair, J.H. (1976). NASA Technical Note, NASA TN–D–8215.
Chamis, C.C. and Sullivan, T.L. (1976). NASA Technical Memorandum NASA–TM X–71825.
Chan, H.C. and Patterson, W.A. (1972). J. Mat. Science **7**, 856.
Chang, C.I., Conway, H.D. and Weaver, T.C. (1972). Fibre Science and Tech. **5**, 143.
Chaplin, C.R. (1974). J. Mat. Science **9**, 329.
Chappell, M.J. and Millman, R.S. (1974). J. Mat. Science **9**, 1933.
Chappell, M.J., Morley, J.G. and Martin, A. (1975). J. Phys. D (Applied Phys.) **8**, 1071.
Chawla, K.K, and Metzger, M. (1972). J. Mat. Science **7**, 34.
Chen, C.H. (1970). J. Appl. Mech. **37**, 198.
Chen, C.H. and Cheng, S. (1967). J. Comp. Mat. **1**, 30.
Chen, E.P. and Sih, G.C. (1971). J. Comp. Mat. **5**, 12.
Chen, P.E. (1971). Polym. Engineering and Science **11**, 51.
Chen, P.E. and Lewis, T.B. (1970). Polym. Engineering and Science **10**, 43.
Cheskis, H.P. and Heckel, R.W. (1970). Mettal. Transact. **1**, 1931.
Chew, P.E. and Lin, J.M. (1969). Materials Res. and Standards **9**, 29.
Chiang, F.P. and Slepetz, J. (1973). J. Comp. Mat. **7**, 134.
Chiao, T.T. and Moore, R.L. (1971). J. Comp. Mat. **5**, 124.
Chiao, C.C., Sherry, R.J. and Chiao, T.T. (1976). Composites **7**, 107.
Chou, P.C., Carleore, J. and Hsu, C.M. (1972). J. Comp. Mat. **6**, 80.
Chou, P.C., McNamee, B.M. and Chou, D.K. (1973). J. Comp. Mat. **7**, 22.
Chow, T.S. and Hermans, J.J. (1969). J. Com. Mat. **3**, 382.
Christensen, R.M. and Waals, F.M. (1972). J. Comp. Mat. **6**, 518.
Christensen, R.M. (1976). Int. J. Solids & Structures **12**, 537.
Clausen, W.E. and Leissa, A.W. (1967). AFML Report TR–67–15; AD-656431.
Cline, H.E. (1966). Gen. Electric Report No. 66–G.C. 0332.
Colclough, W.J. and Russel, J.G. (1972). Aeronautical J. **76**, 53.
Coleman, B.D. (1958). J. Mech. Phys. Solids **7**, 60.
Collins, M.F. and Haywood, B.C. (1969). Carbon **7**, 663.
Collings, T.A. (1974). Composites **5**, 108.
Conway, H.D. and Chang, C.I. (1971). Fibre Science & Tech. **3**, 249.
Cook, J. (1968). J. Phys. D. (Appl. Phys.), **1**, 799.
Cooper, G.A. (1966). J. Mech. Phys. Solids **14**, 103.
Cooper, G.A. (1971). Rev. Phys. in Tech. **2**, 49.
Cooper, G.A., Gladman, D.G. and Sillwood, J.M. (1974). J. Mat. Science **9**, 835.
Cooper, G.A. and Kelly, A. (1967). J. Mech. Phys. Solids **15**, 279.
Cooper, G.A. and Kelly, A. (1969). Interfaces in Composites, ASTM Special Publication No. STP–452.
Cooper, G.A. and Sillwood, J.M. (1972). J. Mat. Science **7**, 325.
Cooper, R.E. (1970). J. Mech. Phys. Solids **18**, 179.
Copley, S.M. (1974). J. Adhesion **6**, 139.

Cottrell, A.H. (1964). Proc. Roy. Soc. A**282**, 2.
Courtney, T.H. and Wulff, J. (1966). J. Mat. Science **1**, 383.
Cox, H.L. (1952). Br. J. Appl. Phys. **2**, 72.
Craig, W.H. and Courtney, T.H. (1975). J. Mat. Science **10**, 1119.
Crane, R.L. and Tressler, R.E. (1971). J. Comp. Mat. **5**, 537.
Cratchley, D. (1963). Powder Metall. **11**, 59.
Cratchley, D. (1965). Metall. Rev. **10**, 79.
Crisfield, M.A. (1971). Inst. Civil Engineers Proc. **48**, 413.
Crivelli-Visconti, I. and Cooper, G.A. (1969). Nature **221**, 754.
Cruse, T.A. (1973). J. Comp. Mat. **7**, 218.
Cruse, T.A. and Stout, M.G. (1973). J. Comp. Mat. **7**, 272.
Curtis, G.J. (1969). Ultrasonics for Industry (Iliffe Books, London), p. 4.
Curtis, G.J., Milne, J.M. and Reynolds W.N. (1968). Nature **220**, 1024.
Dally, J.W. and Broutman, L.J. (1967). J. Comp. Mat. **1**, 424.
Daniel, I.M. (1970). J. Comp. Mat. **4**, 178.
Daniel, I.M. and Rowlands, R.E. (1971). J. Comp. Mat. **5**, 250.
Daniels, B.K., Harakas, N.K. and Jackson, R.C. (1971). Fibre Science & Tech. **3**, 187.
Daniels, H.E. (1945). Proc. Roy. Soc. A**138**, 405.
Darlington, M.W. and McGinley, P.L. (1975). J. Mat. Science **10**, 906.
Darlington, M.W., McGinley, P.L. and Smith, G.R. (1976). J. Mat. Science **11**, 877.
Davidge, R.W. and Phillips, D.C. (1972). J. Mat. Science **7**, 1308.
Davies, W.E.A. (1971). J. Phys. D. (Appl. Phys.) **4**, 1325.
Davis, J.H. (1971). Plastics and Polymers **39**, 137.
Davis, R.O. Jnr. and Wu, J.H. (1972). J. Comp. Mat. **6**, 126.
Davydov, S.D., Zaretskii-Feoktistov, G.G. and Sudakov, V.V. (1975). Soviet Polymer Mechanics **9**, 867.
Dean, A.V. (1967). J. Inst. Metals **95**, 77.
De Vekey, R.C. (1974). J. Mat. Science **9**, 1898.
Dexter, R.R. and Singer J. (Editors) (1974). Proc. 9th Congress of Int. Council of Aeronautical Sciences, Haifa 1974. (Weizmann Sciences Press of Israel).
Dibenedetto, A.T., Gauchel, J.V., Thomas, R.L. and Barlow, J.W. (1972). J. Materials **7**, 211.
Dibenedetto, A.T. and Wambach, A.D. (1972). Int. J. of Polymeric Materials **1**, 159.
Dietz, A.G.H. (1954). Fibre Glass Reinforced Plastics (Reinhold, New York; Ed.— R.H. Sonneborn).
Diggwa, A.D.S. and Norman, R.H. (1972). Plastics & Polymers **40**, 263.
Dimmock, J. and Abrahams, M. (1969). Composites **1**, 87.
Donald, I.W. and McMillan, P.W. (1976). J. Mat. Science **11**, 949.
Dow, N.F. (1963). Gen. Elect. Rep. No. R63–SD61.
Dow, N.F. (1966). Proc. SPI XXI Annual Conference, Sec. 5–B.
Dow, N.F., Rosen, B.W. and Hashin, Z. (1966). NASA Report No. CR–492.
Drumheller, D.S. and Lundergan, C.D. (1975). Int. J. Solids and Structures **11**, 75.
Drumheller, D.S. and Norwood, F.R. (1975). Int. J. Solids and Structures **11**, 53.
Dudek, T.J. (1970). J. Comp. Mat. **4**, 232.
Dudukalenko, V.V., Ivanishcheva, O.I. and Legenya, B.I. (1974). J. Appl. Mech. & Tech. Phys. (Translation of Zhurnal Prikladnoi Mekhaniki i Tekhnicheskoi Fiziki) **14**, 272.
Durelli, A.J., Parks, V.J., Feng, H.C. and Chiang, F. (1970). Proc. Int. Conf. on Mechanics of Composite Materials, Philadelphia (Pergamon Press, Oxford. Ed.– F.W. Wendt).

Dvorak, G.J., Rao, M.S.M. and Tarn, J.Q. (1973). J. Comp. Mat. **7**, 194.
Dynes, P.J. and Kaelble, D.H. (1974). J. Adhesion **6**, 195.
Ekvall, J.C. (1961). Proceedings of ASME Aviation Conference, Los Angeles, California.
Ekvall, J.C. (1966). AIAA/ASME VII Structures, Structural Dynamics and Materials Conference, Palm Springs, California; p. 250.
Ellis, C.D. and Harris, B. (1973). J. Comp. Mat. **7**, 76.
Ellison, E.G. and Harris, G.B. (1966). Appl. Materials Res. **5**, 33.
Elvery, R.H. and Nwokoye, D.N. (1969). Symposium on NDT of Concerete and Timber (Institution of Civil Engineers), p. 69.
Epstein, B. (1948). J. Appl. Phys. **19**, 140.
Eshelby, J.D. (1956). Solid State Physics (Academic Press, New York; Eds.—F. Seitz and D. Turnbull), **3**, 79.
Eshelby, J.D. (1971). Science Prog. **59**, 161.
Ezekiel, H.M. (1971). Fibre Science & Tech. **3**, 243.
Fedorov, F.I. (1968). Theory of Elastic Waves in Crystals (Plenum Press, New York).
Felbeck, D.K. (1968). Introduction to Strengthening Mechanisms (Prentice-Hall, New Jersey).
Felix, M.P. (1974). J. Com. Mat. **8**, 275.
Fenner, A.J. (1965). Mechanical Testing of Materials (George Newnes Ltd., London).
de Ferran, E.M. and Harris, B. (1970). J. Comp. Mat. **4**, 62.
Fichter, W.B. (1969). NASA Technical Note No. D–5433.
Fitz-Randolph, J., Phillips, D.C, Beaumont, P.W.R. and Tetelman, A.S. (1972). J. Mat. Science **7**, 289,
Fokin, A.G. and Shermergor, T.D. (1967). Mekhanika Tverdogo Tela **1**, 129.
Ford, H. and Alexander, J.M. (1963). Advanced Mechanics of Materials (Longmans, London).
Forsyth, P.J.E., George, R.N. and Ryder, D.A. (1964). Appl. Materials Res. **3**, 223.
Foye, R.L. (1972). J. Comp. Mat. **6**, 293.
Foye, R.L. (1973a). J. Comp. Mat. **7**, 178.
Foye, R.L. (1973b). J. Comp. Mat. **7**, 310.
Frank, F.C. (1970). Proc. Roy. Soc. **A319**, 127.
Franklin, H.G. (1970). Fibre Science & Tech. **2**, 241.
Freeman, M.A.R., Day, W.H. and Swanson, S.A.U. (1971). Medical and Biological Eng. **9**, 619.
Friedrich, E., Pompe, W. and Kopjov, I.M. (1974). J. Mat. Science **9**, 1911.
Fuji, T., Mizukawa, K. and Markawa, Z. (1972). Proc. XV Japan Cong. on Materials Research, p. 154.
Garg, S.K., Svalbonas, V. and Gurtman, G.A. (1973). Analysis of Structural Composite Materials (Marcel Dekker Inc., New York).
Garmong, G. (1972). Metall. Trans. **3**, 1919.
Garmong, G. and Shephard, L.A. (1971). Metall. Trans. **2**, 175.
Garmong, G. and Thompson, R.B. (1973). Metall. Trans. **4**, 863.
Gebauer, J., Hasselman, D.P.H. and Thomas, D.A. (1972). J. Am. Ceramic Soc. **55**, 175.
George, F.D., Ford, J.A. and Salkind, M.J. (1968). Metal Matrix Composites ASTM Special Tech. Publ. No. 438, p. 59.
Gerberich, W.W. and Zackay, V.F. (1972). Metall. Trans. **3**, 747.
Gill, R.M. (1972). Carbon Fibres in Composite Materials (published for the Plastics Institute by Iliffe Books, London).
Gillis, P.P. (1970). Fibre Science and Technology **2**, 193.

Goggin, P.R. (1973). J. Mat. Science **8**, 233.
Gordon, J.E. (1952). J. Roy. Aero Soc. September issue, p. 704.
Gravel, J.V. and Cost, J.R. (1972). Metall. Trans. **3**, 1973.
Greszczuk, L.B. (1965). AIAA/ASME VI Structures, Structural Dynamics and Materials Conference,, Palm Springs, California, p. 285.
Greszczuk, L.B. (1969). Interfaces in Composites. ASTM Special Tech. Publ. 452, p. 42.
Greszczuk, L.B. (1975). A.I.A.A. Journal **13**, 1311.
Grigolyuk, E.I. and Fil' shtinskii, L.A. (1965). Uprugost i Plastichnost (Moscow), p. 7.
Gücer, D.E. and Gurland, J. (1962). J. Mech. Phys. Solids **10**, 356.
Guess, T.R. and Hoover, W.R. (1973). J. Comp. Mat. **7**, 2.
Gurev, A.V., Gokhberg, Ya. A. and Fedorov, V.I. (1975). Industrial Laboratory (Translation of Zavodskaya Laboratoriya) **41**, 742.
Hackett, R.M. (1971). Polym. Engineering and Science **11**, 220.
Haener, J. and Ashbaugh, N. (1967). J. Comp. Mat. **1**, 54.
Hahn, H.T. (1975). J. Com. Mat. **9** 316.
Hahn, H.T. and Tsai, S.W. (1973). J. Comp. Mat. **7**, 102.
Hahn, H.T. and Tsai, S.W. (1974). J. Comp. Mat. **8**, 160.
Halpin, J.C. (1969). J. Comp. Mat. **3**, 732.
Halpin, J.C., Jerine, D. and Whitney, J.M. (1971). J. Comp. Mat. **5**, 36.
Halpin, J.C. and Pagano, N.J. (1969). J. Comp. Mat. **3**, 720.
Halpin, J.C. and Tsai, S.W. (1969). AFML Report TR–67–423.
Ham, R.L. and Place, T.A. (1966). J. Mech. Phys. Solids **14**, 271.
Hamilton, R.G. and Berg, C.A. (1973). Fibre Science & Tech. **6**, 55.
Hamstad, M.A. and Chiao, T.T. (1973). J. Comp. Mat. **7**, 320.
Hanasaki, S. and Hasegawa, Y. (1974). J. Comp. Mat. **8**, 306.
Hancock, P. and Cuthbartson, R.C. (1970). J. Mat. Science **5**, 762.
Hancox, N.L. (1975). J. Mat. Science **10**, 234.
Hardy, N.E. (1969). Engineering Materials and Design **12**, 1325.
Harris, B. (1973). Carbon Fibres in Engineering (McGraw-Hill, U.K.; Ed.—M. Langley), p. 1.
Harris, B. and Bunsell, A.R. (1975). Composites **6**, 197.
Harrison, N.L. (1971). Fibre Science & Tech. **4**, 101.
Harrison, N.L. (1972). Fibre Science & Tech. **5**, 197.
Harrison, N.L. (1973). Fibre Science & Tech. **6**, 25.
Hashin, Z. (1962). J. Appl. Mech. **29**, 143.
Hashin, Z. (1965). J. Mech. Phys. Solids **13**, 119.
Hashin, Z. (1972). Theory of fibre reinforced materials. NASA Contract Report No. CR–1974.
Hashin, Z. and Rosen, B.W. (1964). J. Appl. Mech. **31**, 223.
Hashin, Z. and Shtrikman, S. (1961). J. Franklin Inst. **271**, 336.
Hashin, Z. and Shtrikman, S. (1962a). J. Mech. Phys. Solids **10**, 335.
Hashin, Z. and Shtrikman, S. (1962b) J. Mech. Phys. Solids **10**, 343.
Hashin, Z. and Shtrikman, S. (1963). J. Mech. Phys. Solids **11**, 143.
Hazell, E.A. (1970). Composites **1**, 362.
Hazell, E.A. (1971). Composites **2**, 110.
Heaton, M.D. (1968). J. Phys. D. (Appl. Phys.) **1**, 1039.
Heaton, M.D. (1970). J. Phys. D. (Appl. Phys.) **3**, 672.
Hecker, S.S., Hamilton, C.H. and Ebot, L.J. (1970). J. of Materials 5, 868.
Hedgepeth, J.M. (1961). NASA Tech. Report No. TN–D–882.
Hedgepeth, J.M. and Van Dyke, P. (1967). J. Comb. Mat. **1**, 294.
Helfet, J.L. and Harris, B. (1972). J. Mat. Science **7**, 494,

Herakovich, C.T. (Ed.) (1975). Proc. A.S.M.E. Winter Annual Meeting, Houston, Texas, Dec. 1975.

Hermans, J.J. (1967). Proc. Roy. Acad. Amsterdam **B70**, 1.

Herring, H.W., Lytton, J.L. and Steele, J.H. Jnr. (1973). Metall. Trans. **4**, 807.

Herrmann, G. and Achenbach, J.D. (1967a). AIAA/ASME VIII Structures, Structural Dynamics and Materials Conference, Palm Springs, California, p. 112.

Herrmann, G. and Achenbach, J.D. (1967b). Northwestern University, Structral Mech. Lab. Report No. 67–7.

Herrmann, L.R. and Pister, K. S. (1963). Proc. ASME Annual Meeting, Philadelphia; Report ASME P.N. 63WA–239.

Hewitt, R.L. and de Malherbe, M.C. (1970). J. Comp. Mat. **4**, 280.

Hietman, P.W., Shephard, L.A. and Courtney, T.H. (1973). J. Mech. Phys. of Solids **21**, 75.

Hill, R. (1963). J. Mech. Phys. Solids **11**, 357.

Hill, R. (1964a). J. Mech. Phys. Solids **12**, 199,

Hill, R. (1964b). J. Mech. Phys. Solids **12**, 213.

Hill, R. (1965a). J. Mech. Phys. Solids **13**, 189.

Hill, R. (1965b). J. Mech. Phys. Solids **13**, 213.

Hing, P. and Groves, G.W. (1972). J. Mat. Science **7**, 427.

Hlavacek, M. (1975). Int. J. Solids and Structures **11**, 199.

Holiday, L. (1966). Composite Materials (Elsevier, New York).

Holmes, M. and Al-Khayatt, Q. J. (1975). Composites **6**, 157.

Holmes, B.S. and Tsou, F.K. (1972). J. Appl. Phys. **43**, 957.

Hoover, W.R. and Guess, T.R. (1973). J. Comp. Mat, **7**, 334.

Huang, W. (1971). J. Comp. Mat. **5**, 320.

Hughes, D.D. and Way, J.L. (1973). Composites **4**, 167.

Huntington, H.B. (1958). Solid State Physics (Academic Press, New York; Ed. F. Seitz and D. Turnbull) **7**, 213.

Hutchinson, J.W. (1970). Proc. Roy. Soc. **A319**, 247.

Iremonger, M.J. and Wood, W.G. (1967). J. Strain Analysis **2**, 239.

Iremonger, M.J. and Wood, W.G. (1969), J. Strain Analysis **4**, 121.

Iremonger, M.J. and Wood, W.G. (1970). J. Strain Analysis **5**, 212.

Isakson, G. and Levy, A. (1971). J. Comp. Mat. **5**, 273.

Ismar, H. (1971). Zeits. für Metallkunde **62**, 359.

Jackson, P.W., Baker, A.A. and Braddick, D.M. (1971). J. Mat. Science **6**, 427.

Jackson, P.W., Braddick, D.M. and Walker, P.J. (1972). Fibre Science & Tech. **5**, 219.

Jackson, P.W. and Cratchley, D. (1966). J. Mech. Phys. Solids **14**, 49.

James, A.M. and Vanghn, R.L. (1976). Composites **7**, 73.

Janssen, D.M., Datta, S.K. and Jahsman, W.E. (1972). J. Mech. Phys. Solids **20**, 1.

Jech, R.W., McDanels, D.L. and Weeton, J.W. (1959). Proc. VI Sagamore Conf. ASTIA No. AD:161.443, p. 116.

Jech, R.W. and Weber, E.P. (1959). Reactive Metals. Ed.—W.R. Clough (Interscience, New York), p. 109.

Johnson, W., Phillips, L.N. and Watt, W. (1968). The Production of Carbon Fibres; British Patent No. 1,110,791.

Jones, B.F. (1971). J. Mat. Science **6**, 1225.

Jones, B.F. and Wilkins, B.J.S. (1972). Fibre Science & Tech. **5**, 315.

Jones R.M. (1974). A.I.A.A. Journal **12**, 112.

Jones, W.R. and Johnson, J.W. (1971). Carbon **9**, 645.

Kafka, V. (1974). Acta Technica CSAV **19**, 87.

Karpinos, D.M. and Bespyatyi, V.A. (1971). Sov. Powd. Met. & Metal Ceramics **10**, **310**.

Karpinos, D.M. and Tuchunskii, L.I. (1968a). Soviet Powder Metallurgy **9**, 735.
Karpinos, D.M. and Tuchunskii, L.I. (1968b). Soviet Powder Metallurgy **11**, 901.
Kawamura, K. and Jenkins, G.M. (1970). J. Mat. Science **5**, 262.
Kedward, K.T. and Hindle, G.R. (1970). J. Strain Analysis **5**, 309.
Kellerer, H., Matera, R. and Piatti, G. (1972), Aluminium **48**, 657.
Kelly, A. (1964). Proc. Roy. Soc. **A282**, 63.
Kelly, A. (1966). Strong Solids (Clarendon Press, Oxford).
Kelly, A. (1970). Proc. Roy. Soc. **A319**, 95.
Kelly, A. and Davies, G.J. (1965). Metall. Rev. **10**, 37.
Kelly, A. and Lilholt, H. (1969). Phil. Mag. **20**, 164.
Kelly, A. and Tyson, W.R. (1965a). High Strength Materials (John Wiley, New York; Ed.—V.F. Zackay), p. 578.
Kelly, A. and Tyson, W.R. (1965b). J. Mech. Phys. Solids **13**, 329.
Kelly, A. and Tyson, W.R. (1966). J. Mech. Phys. Solids **14**, 177.
Kendall, D.P. (1972). J. Materials **7**, 430.
Kendall, K. (1975). J. Mat. Science **10**, 1011.
Kendall, K. (1976). J. Mat. Science **11**, 1267.
Kerner, E.H. (1956). Proc. Phys. Soc. **B69**, 808.
Khachaturyan, A.G. and Shatalov, G.A. (1971). Phys. of Metals and Metallography **31**, 1.
Khoroshun, L.P. (1966a). Prikladnaia Mekhanika **2**, 14.
Khoroshun, L.P. (1966b). Mekhanika Polimerov **2**, 365.
Khoroshun, L.P. (1966c). Prikladnaia Mekhanika **2**, 99.
Khoroshun, L.P, (1968). Mekhanika Polimerov **4**, 78.
Khoroshun, L.P. and Shevchenko, S.N. (1975). Prikladnaya Mekhanika **11**, 104.
Kicher, T.P. and Stevenson, J.F. (1967). Integrated approach to research and design methods for carbon composite materials. U.S. Air Force Cantract No. AF 33 (615)-3110 Prog. Report No. 8.
Kilchinskii, A,A, (1965). Prikladnaia Mekhanika **1**, 65.
Kilchinskii, A.A. (1966). Thermal stresses in elements of construction (Naukova Dumka, Kiev), **6**, 123.
King, J.L. (1972), J. Strain Analysis **7**, 146.
Kittel, C. (1976), Introduction to Solid State Physics (John Wiley, New York, 5th ed.).
Knibbs, R.H. and Morris, J.B. (1974). Composites **5**, 209.
Knight, M. and Hahn, H.T. (1975). J. Comp. Mat. **9**, 77.
Ko, H.Y. and Sture, S. (1974). J. Comp. Mat. **8**, 178.
Koeneman, J.B. (1970). Ph.D. Thesis, Case Western Reserve University.
Kohn, W. (1974). J. Appl. Mech. Trans. ASME Series E **41**, 97.
Kohn, W. (1975). J. Appl. Mech. (Trans. ASME, Series E) **42**, 159.
Kolevatov, Yu. A., Olkhovich-Novosadyuk, N.A. and Tynnyi, A.N. (1976). Soviet Materials Science **10**, 536.
Konish, H.J. Jr., Swedlow, J.L. and Cruse, T.A. (1972). J. Comp. Mat. **6**, 114.
Koss, D.A. and Copley, S.M. (1971). Metall. Trans. **2**, 1557.
Kotchick, D.M., Hink, R.C. and Tressler, R.E. (1975). J. Comp. Mat. **9**, 327.
Kothari, L.S. and Singwi, K.S. (1957). Phys. Rev. **106**, 230.
Kothari, L.S. and Tewary, V.K. (1963a). J. Chem. Phys. **38**, 417.
Kothari, L.S. and Tewary, V.K. (1963b). Phys. Letters **6**, 248.
Krautkramer, J. and Krautkramer, H. (1969). Ultrasonic Testing of Materials (Springer-Verlag, Berlin).
Kreider, K.G. and Leverant, G.R. (1966). X SAMPE Symposium on Advanced Fibrous Reinforced Composites. (Hollywood, U.S.A.).
Kreider, K.G. (Ed.) (1974). Composite Materials Vol 4: Metallic Matrix Composites (Academic Press, New York).

Kreider, K.G. and Marciano, M. (1969). Trans. Metall. Soc. AIME **245**, 1279.
Krummheuer, W.R. and Alexander, H.A. (1971). Zeits. für Metallkunde **62**, 129.
Kulkarni, S.V., Rice, J.S. and Rosen, B.W. (1975). Composites **6**, 217.
Kulkarni, S.V., Rosen, B.W. and Zweben, C. (1973). J. Comp. Mat. **7**, 387.
Kwok, P.C.K. (1967). Solid State Physics (Academic Press, New York; Eds.—F. Seitz and D. Turnbull) **20**, 213.
Lager, J.R. and June, R.R. (1969). J. Comp. Mat. **3**, 48.
Landau, L.D. and Lifshitz, E.M. (1970). Theory of Elasticity (Pergamon Press, New York).
Lange, F.F. and Radford, K.C. (1971). J. Mat. Science **6**, 1197.
Langley, M. (1970). The Chartered Mechanical Engineer **17**, 56.
Langley, M. (Editor) (1973). Carbon Fibres in Engineering (McGraw-Hill, U.K.).
Larder, R.A. and Beadle, C.W. (1976). J. Comp. Mat. **10**, 21.
Lavendel, E.E. and Kalinka, Yu. A. (1974). Mekhanika Polimerov **7**, 916.
Lavengood, R.E. (1972). Polym. Engineering & Science **12**, 48.
Lavengood, R.E. and Gulbransen, L.B. (1969). Polym. Engineering & Science **9**, 365.
Lavengood, R.E. and Ishai, O. (1971). Polym. Engineering & Science **11**, 226.
Lawrence, P. (1972). J. Mat. Science **7**, 1.
Laws, V. (1971). J. Phys. D. (Appl. Phys.) **4**, 1737.
Leach, P. and Ashbee, K.H.G. (1974). Composites **5**, 67.
Lee, R.E. and Harris, S.J. (1974). J. Mat. Science **9**, 359.
Leigh, R.S., Szigeti, B. and Tewary, V.K. (1971). Proc. Roy. Soc. **A320**, 505.
Lekhnitskii, S.G. (1963). Theory of Elasticity of an Anisotropic Body (Holden-Day) p. 184.
Liberman, M.L. and Noles, G.T. (1972). J. Mat. Science **7**, 654.
Lifshitz, J.M. and Rotern, A. (1972). J. Mat. Science **7**, 861.
Lin, J.M., Chen, P.E. and Dibenedetto, A.T. (1971). Polym. Engineering & Science **11**, 344.
Lin, T.H. (1967). J. Comp. Mat. **1**, 144.
Lin, T.H., Salinas, D. and Ito, Y.M. (1972). J. Comp. Mat. **6**, 48.
Livesey, R. (1971). The Engineer **233**, 42.
Lloyd, D.J. and Tangri, K. (1974). J. Mat. Science **9**, 482.
Lomakin, V.A. and Koltunov, M.A. (1965). Polymer Mechanics (English Translation of Mekhanika Polimerov) **1**, 79.
Lou, Y.C. and Schapery, R.A. (1971). J. Comp. Mat. **5**, 208.
Love, A.E.H. (1944). A Treatise on Theory of Elasticity. (Dover Publications, New York).
Lundergan, C.D. and Drumheller, D.S. (1971). J. Appl. Phys. **42**, 669.
MacMillan, N.H. (1972). J. Mat. Science **7**, 239.
Majumdar, S. and McLaughlin, P.V. (1975). Int. J. Solids & Structures **11**, 777.
Mallick, P.K. and Broutman, L.J. (1974). J. Mat. Science **9**, 1420.
Mann, R.K. and Campbell, W.K. (1975). J. Textile Institute **66**, 15.
Mansfield, E.H. (1976a). Aeronautical Research Council, London. Reports and Memoranda, R & M No. 3782.
Mansfield, E.H. (1976b). Aeronautical Research Council, (London) Current Papers C.P. No. 1347.
Mansfield, E.H. and Purslow, D. (1976). Aeronautical Research Council, London; Current papers No. 1339.
Maradudin, A.A. (1965). Reports on Progress in Physics **28**, 331.
Maradudin, A.A., Montroll, E.W., Weiss, G.H. and Ipatova. I.P. (1971). Theory of Lattice Dynamics in the Harmonic Approximation. Solid St. Phys. Suppl. 3, II edition (Academic Press, New York and London; Eds.—H. Ehrenreich, F. Seitz and D. Turnbull).

Mariot, L. (1962). Group Theory and Solid State Physics. (Prentice Hall Int., London).
Markham, M.F. (1969). Ultrasonics for Industry (Illiffe Books, London) p. 1.
Markham, M.F. (1970). Composites 1, 145.
Markham, M.F. and Dawson, D. (1975). Composites 6, 173.
Marston, T.U., Atkins, A.G. and Felbeck, D.K. (1974). J. Mat. Science 9, 447.
Marom, G. and White, E.F.T. (1972). J. Mat. Science 7, 1299.
McClung, R.W. (1974). Annual Review of Materials Science (Annual Reviews Inc., California, U.S.A.; Ed. R.A. Huggins) 4, 1.
McDanels, D.L. and Signorelli, R.A. (1976). NASA Technical Memorandum, NASA TMX-71875.
McDanels, D.L., Signorelli, R.A. and Weeton, J.W. (1967). Fibre Strengthened Metallic Composites (ASTM. Philadelphia, U.S.A.) p. 124,
McKinney, J.M. (1972). J. Comp. Mat. 6, 164.
McLean, D. and Read, B.E. (1975). J. Mat. Science 10, 481.
Mau, S.T , Tong, P. and Pian, T.H.H. (1972). J. Comp. Mat. 6, 304.
Medredev, M.Z. (1975). Soviet Polymer Mechanics 9, 919.
Metcalfe, A.G. (Ed.) (1974). Composite Materials Vol 1: Metal Matrix Composites and Interfaces (Academic Press, New York).
Middleton, B.D. (1968-1970). Materials Research & Stand. 10, 29.
Mileiko, S.T. (1969). J. Mat. Science 4, 974.
Mileiko, S.T. (1970). J. Mat. Science 5, 254.
Mileiko, S.T. and Khvostunkov, A.A. (1973). J. Appl. Mech. & Tech. Phys. (Translation of Zhurnal Prikladnoi Mekhaniki i Tekhnicheskoi) 12, 628.
Millman, R.S. and Morley J.G. (1976). Materials Science & Engineering 23, 1.
Moehlenpath, A.E., Ishai, O. and Dibenedetto, A.T. (1971). Polym. Engineering & Science 11, 129.
Molyneux, M. (1973). Carbon Fibres in Engineering. (McGraw-Hill, U.K. Editor —M. Langley) p. 46.
Moon, F.C. (1972). J. Comp. Mat. 6, 62.
Moore, R.L. and Lepper, J.K. (1974). J. Testing and Evaluation 2, 173.
Morley, J.G. (1970). Proc. Roy. Soc. A 319, 117.
Morley, J.G. (1971). Composites 2, 80.
Morris, A.W.H. and Smith, R.S. (1971). Fibre Science and Tech. 3, 219.
Morris, A.W.H. and Steigerwald, E.A. (1967). Trans. Metall. Soc. AIME 239, 730.
Morton, J. and Groves, G.W. (1974). J. Mat. Science 9, 1436.
Morton, J. and Groves, G.W. (1975). J. Mat. Science 10, 170.
Motavkin, A.V., Kalinka, Yu. A. and Teleshov, V.A. (1975). Soviet Polymer Mechanics 10, 37.
Mukherjee, S. and Lee, E.H. (1975). Computers & Structures 5, 279.
Mullin, J.V. and Knoell, A.C. (1970). Mat. Res. & Stand. 10, 16.
Mullin, J.V. and Mazzio, V.F. (1972a). J. Comp. Mat. 6, 268.
Mullin, J.V. and Mazzio, V.F. (1972b). J. Mech. Phys. of Solids 20, 391.
Musgrave, M.J.P. (1954a). Proc. Roy. Soc. A226, 339.
Musgrave, M.J.P. (1954b). Proc. Roy. Soc. A226, 356.
Musgrave, M.J.P. (1970). J. Mech. Phys. Solids 18, 207.
Nelson, R.B. and Navi, P. (1975). J. Acoust. Soc. America 57, 773.
Nemat-Nasser, S. and Fu, F.C.L. (1974). J. Appl. Mechanics, Trans. of the ASME, Series E 41, 288.
Nemat-Nasser, S.. Fu, F.C.L. and Mingawa, S. (1975). Int. J. Solids and Structures 11, 617.
Nevadunsky, J.J., Lucas, J.J. and Salkind. M.J. (1975). J. Comp. Mat. 9, 394.
Nicolai, L. and Narkis, M. (1971). Polym. Engineering & Science 11, 3.

Nielson, L.E. (1970). J. Appl. Phys. **41**, 4626.
Nikolaev, V.P. (1974). Mekhanika Polimerov **7**, 984.
Nixdorf, J., Rochow, H. and Seidelmann, U. (1971). Aluminium **47**, 550.
Nosarev, A.V. (1967). Mekhanika Polimerov **3**, 858.
Noton, B.R. (Ed.) (1974). Comosite Materials Vol 3: Engineering Applications of Composites (Academic Press, New York).
Nye, I.F. (1957). Physical Properties of Crystals. (Oxford University Press, London).
Ogden, R.W. (1974). J. Mech. Phys. Solids **22**, 541.
Ogorkiewicz, R.M. (1971). Composites **2**, 29.
Ogorkiewicz, R.M. and Sayigh, A.A.M. (1971). J. Strain Analysis **6**, 226.
Oplinger, D.W., Parker, B.S. and Chiang, F.P. (1974). Experimental Mechanics **14**, 347.
Outwater, J.O. and Carnes, W.O. (1967). NASA Report No. AD–659363 (Contract No. DAA–21–67–C–0041).
Outwater, J.O. and Murphy, M.C. (1969). XXVI Annual Conf.—Reinforced Plastics and Composites Division of Society of Plastics Industry, Payer 11–C.
Outwater, J.Q. Jnr. (1956). Modern Plastics **33**, 156.
Owen, D.R.J. (1972). Fibre Science and Tech. **5**, 37.
Owen, D.R.J., Holbeche, J. and Zienkiewicz, O.C. (1969). Fibre Science and Tech. **1**, 185.
Owen, D.R.J. and Lyness, J.F. (1972). Fibre Science and Tech. **5**, 129.
Owen, M.J. and Bishop, P.T. (1973). J. Comp. Mat. **7**, 146.
Owen, M.J. and Morris, S. (1970a). Mod. Plastics **47**, 158.
Owen, M.J. and Morris, S. (1970b). XXV Annual Conf.—Reinforced Plastics and Composites Division of Society of Plastics Industry, Paper 8–E.
Owen, M.J. and Morris. S. (1972). Plastics & Polymers **40**, 209.
Owen, M.J. and Rose, R.G. (1972). Plastics & Polymers **40**, 325.
Pagano, N.J. (1969). J. Comp. Mat. **3**, 398.
Pagano, N.J. (1970). J. Comp. Mat. **4**, 20.
Pagano, N.J. (1974a). J. Comp. Mat. **8**, 214.
Pagano, N.J. (1974b). J. Comp. Mat. **8**, 65.
Pagano, N.J. (1975). J. Comp. Mat. **9**, 67,
Pao, Y.C. and Maheshwari, M.N. (1974). Computer Methods in Applied Mechanics & Engineering **3**, 305.
Parratt, N.J. (1972). Fibre-Reinforced Materials Technology (Van Nostrand Reinhold, London).
Partsevskii, V.V. (1973). Mekhanika Polimerov **6**, 279.
Paul, B. (1960). Trans. Metall. Soc. AIME **218**, 36.
Perry, A.J., de Lamotte, E. and Phillips, K. (1970). J. Mat. Science **5**, 945.
Perry, A.J., Phillips, K. and de Lamotte, E. (1971). Fibre Science & Tech. **3**, 317.
Perry, J.L. and Adams, D.F. (1975). Composites **6**, 166.
Peterson, J.M. and Hermans, J.J. (1969). J. Comp. Mat. **3**, 338.
Petit, P.H. (1969). ASTM Report No. STP–460.
Petrasek, D.W., Signorelli, R.A. and Weeton, J.W. (1967). NASA Tech. Note D–3886.
Phillips, D.C. (1972). J. Mat. Science **7**, 1175.
Phillips, D.C. (1974). J. Mat. Science **9**, 1847.
Phillips, D.C. and Tetelman, A.S. (1972). Composites **3**, 216.
Phillips, L.N. (1976). Composites **7**, 7.
Phillips, K., Perry, A.J., Holloy, G.E., de Lamotte, E., Hintermann, H.E. and Gass, H. (1971). J. Mat. Science **6**, 270.
Pickett, G. (1965). AFML Report TR–220; AD–473790.

Pickett, G. and Johnson, M.W. (1996). AFML Report TR-65220 Pt. 2: AD-646216.
Piehler, H.R. (1965). Trans. Metall. Soc. AIME **233**, 12.
Pierce, F.J. (1926). J. Text. Inst. **17**, 355.
Piggott, M.R. (1970). J. Mat. Science **5**, 669.
Piggott, M.R. (1974). J. Mat. Science **9**, 494.
Pink, E. and Campbell, J.D. (1974). J. Mat. Science **9**, 658.
Pinnel, M.R. and Lawley, A. (1970). Metall. Trans. **1**, 1337.
Pipes, R.B. and Cole, B.W. (1973). J. Comp. Mat. **7**, 246.
Pipes, R.B. and Daniel, I.M. (1971). J. Comp. Mat. **5**, 255.
Pipes, R.B. and Pagano, N.J. (1970). J. Comp. Mat. **4**, 538.
Pipes, R.B. and Pagano, N.J. (1974). J. Appl. Mech., Trans. of the ASME, Series E **41**, 668.
Pipkin, A.C. (1975). Quarterly J. Mech. & Appl. Maths. **28**, 271.
Plueddemann, E.P. (Editor) (1975). Composite Materials Vol 6: Interfaces in Polymer Matrix Composites (Academic Press, New York).
Pompe, W. and Schultrich, B. (1974). Annalen der Physik **31**, 101.
Prager, W. (1969). Trans. ASME **36**, 542.
Prewo, K.M. and Kreider, K.G. (1972) Metall. Trans. **3**, 2201.
Prosen, S.P. (1970). Fibre Science & Tech. **3**, 81.
Protopopov, K.G. and Piskunov. N.V. (1975). Industrial Laboratory (Translation of Zavodskaya Laboratoriya) **40**, 1526.
Quackenbush, N.E. and Thomas, R.L. (1967). Philco-Ford; AD 820492.
Rabinovich, A.L. (1964). Soviet Phys. Doklady **8**, 1233.
Rabinovich, A.L. and Verkhovskii, I.A. (1964). Inzhenerny Zhurnal **4**, 90.
Rauch, H.W. Snr., Sutton, W.H. and McCreight, L.R. (1966). Ceramic Fibres and Fibrous Composite Materials (Academic Press, New York).
Reed, R.P. and Munson, D.E. (1972). J. Comp. Mat. **6**, 232.
Reifsnider, K.L. and Kelly, C.E. (1972). J. Mat. Science **7**, 1.
Reynolds, W.N. (1969). Plastics and Polymers **37**, 155.
Reynolds, W.N. (1971). Proc. International Conf. on Carbon Fibres, their Composites and Applications (published by the Plastics Institute, London) paper no. 52.
Reynolds, W.N. and Hancox, N.L. (1970). Proc. Seventh International Reinforced Plastics Conference (British Plastics Federation) paper no. 12.
Rhodes, M.D. and Mikulas, M.M. (1975). NASA Technical Memorandum, NASA TM X-72771.
Riley, V.R. (1968). J. Comp. Mat. **2**, 436.
Roderick, G.L. and Whitcomb, J.D. (1975). J. Comp. Mat. **9**, 391.
Romualdi, J.P. (1974). Materials Science & Engineering **15**, 31.
Rose, J.L. and Mortimer, R.W. (1974). J. Comp. Mat. **8**, 191.
Rose, J.L., Wang, A.S.D. and Deska, E.W. (1974). J. Comp. Mat. **8**, 419.
Rosen, B.W. (1964). AIAA J. **2**, 1965.
Rosen, B.W. (1965). Fibre Composite Materials (American Soc. for Metals, Ohio, U.S.A.)
Rosen. B.W. (1970a), Proc. Roy. Soc. **A319**, 79.
Rosen, B.W. (1970b). Proc. Int. Conf. on Mechanics of Composite Materials (Pergamon, Oxford. Ed.—F.W. Wendt).
Rosen, B.W. (1972). J. Comp. Mat. **6**, 552.
Rosen, B.W. (1973). Composites **4**, 16.
Rosen, B.W., Dow, N.F. and Hashin, Z. (1964). NASA Report CR-31.
Rosen, B.W. and Shu, L.S. (1971). J. Comp. Mat. **5**, 279.
Ross, C.A. and Sierakowski, R.L. (1975). Shock and Vibration Digest **7**, 96.
Rotem, A. and Baruch, J. (1974), J. Mat. Science **9**, 1789.
Sachse, W. (1974). J. Comp. Mat. **8**, 378.

Sahu, S. and Broutman, L.J. (1972). Polym. Engineering & Science **12**, 91.

Sakurada, I., Ita, T. and Nakamae, K. (1964). Bull. Inst. Chem. Res. Kyoto University **42**, 77.

Sarian, S. (1973). J. Mat. Science **8**, 251.

Sayers. K.H. and Harris, B. (1973). J. Comp. Mat. **7**, 129.

Schapery, R.A. (1967). J. Comp. Mat. **1**, 228.

Schneider, G.J. (1972). Fibre Science & Tech. **5**, 29.

Schrager, M. and Carey, J. (1970). Polym. Engineering & Science **10**, 369.

Schuerch, H. (1966). AIAA J. **4**, 102.

Schultz. A.B. and Tsai, S.W. (1968). J. Comp. Mat. **2**, 368.

Schumann, W., Wuthrich, W. and Teichmann, G. (1972). J. Comp. Mat. **6**, 536.

Schuster, D.M. and Scala, E. (1963). J. Metals **15**, 697.

Schuster, D.M and Scala, E. (1964). Trans. Metall. Soc. AIME **230**, 1635.

Sendeckyj, G.P. (1970). J. Comp. Mat. **4**, 500.

Sendeckyj, G.P. (1971). J. Comp. Mat. **5**, 82.

Sendeckyj, G.P. (Ed.) (1974). Composite Materials Vol 2: Mechanics (Academic Press, New York).

Sendeckyj, G.P. and Ing–Wu, Yu (1971). J. Comp. Mat. **5**, 533.

Shaffer, B.W. (1964). AIAA J. **2**, 348.

Shorshorov, M.K. (1976). Composites **7**, 17.

Sierakowski, R.L., Hemp, G. and Hockstad, A. (1971). J. Comp. Mat. **5**, 417.

Sih, G.C. and Chen, E.P. (1973). J. Comp. Mat. **7**, 230.

Sih, G.C, Paris, P.C. and Irwin, G.R. (1965). Int. J. Fracture Mech. **1**, 189.

de Silva, A.R.T. (1968). J. Mech. Phys. Solids **16**, 169.

Skelton, J. (1970). Materials Res. & Stand. **10**, 20.

Skudra, A.M. (1974a). Soviet Polymer Mechanics **8**, 482.

Skudra, A.M. (1974b). Soviet Polymer Mechanics **8**, 57.

Skudra, A.M. and Bulavs, F. Ya (1973). Soviet Polymer Mechanics **7**, 221.

Small, L. (1960) Hardness—Theory and Practice Vol I (Service Diamand Tool Co., London).

Smith J.C. (1974). J. Research of US National Bureau of Standards **78A**, 355.

Smith, R.E. (1972). J. Appl. Phys. **43**, 2555.

Smith, T.R. and Owen, M.J. (1969a). Mod. Plastics **46**, 124.

Smith, T.R. and Owen, M.J. (1969b). Mod. Plastics **46**, 128.

Soldatov, M.M. (1975). Soviet Polymer Mechanics **9**, 791.

Spencer, A.J.M. (1965). Int. J. Mech. Sci. **7**, 197.

Spencer, A.J.M. (1972). Deformation of Fibre-reinforced Materials (Clarendon Press, Oxford).

Spencer, A.J. M., Moss, R.L. and Rogers, T.G. (1975). J. Elasticity **5**, 287.

Stone, D.E.W. (1969). J. Strain Analysis **4**, 88.

Stowell, E.Z. and Liu, T S. (1961). J. Mech. Phys. Solids **9**, 242.

Strelyaev, V.S. and Sachkovskaya, L.L. (1974). Industrial Laboratory (Translation of Zavodskaya Laboratoriya) **39**, 1379.

Stuhrke, W.F. (1968). Metal Matrix Composites; ASTM Special Technical Publication 438 (Philadelphia).

Sumsion, H.T. (1976). J. Spacecrafts & Rockets **13**, 150.

Sun, C.T., Feng, W.H. and Koh, S.L. (1974). Int. J. Engineering Science **12**, 919.

Sutherland, H.J. and Lingle, R. (1972). J. Comp. Mat. **6**, 490.

Swift, D.G. (1975). J. Phys. D (Appl. Phys.) **8**, 223.

Symm, G.T. (1970). J. Comp. Mat. **4**, 426.

Tabor, D. (1951). The Hardness of Metals (Clarendon Press, Oxford).

Tardiff, G. Jnr. (1973), Eng. Fract. Mech. **5**, 1.

Tarnopolsky, Yu., Portnov, G. and Zhigun, I. (1967), Mekhanika Polimerov **3**, 243.

Tauchert, T.R. (1971). J. Comp. Mat. **5**, 456.
Tetlow, R. (1973). Carbon Fibres in Engineering. (McGraw-Hill, U.K. Ed.—M. Langley) p. 108.
Tewary, V.K. (1969). AERE Harwell Report T.P. 388.
Tewary, V.K. (1973) AERE Harwell Report T.P. 548. (Published in Advances in Phys. **22**, 757.).
Tewary, V.K. and Bullough, R. (1971). J. Phys. D. (Appl. Phys.) **4**, L5.
Tewary, V.K. and Bullough, R. (1972). Treat. Mater. Science & Techn. (Academic Press, New York; Ed.—H. Herman) **1**, 115.
Theocaris, P. and Paipetis, S.A. (1975). J. Comp. Mat. **9**, 244.
Thomason, P.F. (1972a). J. Mech. Phys. Solids **20**, 19.
Thomason, P.F. (1972b). J. Mech. Phys. Solids **20**, 153.
Thornton, J.S. and Thomas, A.D. Jnr. (1972). Metall. Trans. **3**, 637.
Timoshenko, S. and Goodier, J.N. (1951). Theory of Elasticity (McGraw-Hill, New York).
Trachte, K.L. and Dibenedetto, A.T. (1971). Int. J. Polymeric Materials **1**, 75.
Treloar, L.R.G. (1960). Polymer **1**, 95.
Tsai, S.W. (1964). NASA Report CR–71.
Tsai, S.W. (1965). NASA Contract Report No. CR–224.
Tsai, S.W. and Pagano, N.J. (1968), Composite Materials Workshop (Technomic Publishing Co. Inc. Stanford, Connecticut; Eds.—S.W. Tsai, J.C. Halpin and N.J. Pagano) p. 233.
Tsai, S.W., Springer, G.S. and Schultz, A.B. (1963). Proc. XIV International Astronautical Congress, p. 387.
Tunik, A.L. and Tomashevskii, V.T. (1974). Mekhanika Polimerov **7**, 893.
Turhan, D. (1975). Middle East Technical University Journal of Pure and Applied Sciences **8**, 213.
Turner, S. (1973). Mechanical Testing of Plastics, Iliffe, London.
Tyson. W.R. (1964). Ph.D. Thesis, Cambridge University.
Tyson, W.R. and Davies, G.J. (1965). Br. J. Appl. Phys. **16**, 199.
Umanskii, E.S. (1969). Soviet Powder Metallurgy **1**, 80.
Varatharajulu, V.K. and Kayer, J.S. (1969). Fibre Science & Tech. **2**, 9.
Varschavsky, A. (1972). J. Mat. Science **7**, 159.
Veltri, R.D. and Galasso, F.S. (1971). J. Am. Ceramic Soc. **54**, 319.
Vennett, R.M., Wolf, S.M. and Levett, A.P. (1970). Metal. Trans. **1**, 1569.
Waddoups. M.E., Eisenmann, J.R. and Kaminski, B.E. (1971). J. Comp. Mat. **5**, 446.
Wall, L.D. Jnr. and Card, M,F. (1971). NASA Tech. Note D–6140.
Walpole, L.J. (1969). J. Mech. Phys. Solids **17**, 235.
Wang, F.F.V. (1970). Materials Science & Engineering **7**, 103.
Wan., J.T. (1969). J. Comp. Mat. **3**, 590.
Watt, W. (1970). Proc. Roy. Soc. A**319**, 17.
Watt, W., Phillips, L.N. and Johnson, W. (1966). Engineer **221**, 815.
Weaver, C.W. and Williams, J.G. (1975). J. Mat. Science **10**, 1323.
Weibull, W. (1939). Ing. Vetenskaps Handl. No. 151.
Weibull, W. (1951). J. Appl. Mech. **18**, 293,
Weidmann, G.W and Ogorkiewicz, R.M. (1974). Composites **5**, 117.
Weitsman, Y. and Aboudi, J. (1975). Israel J. Technology **13**, 39.
Whitney, J.M. (1966). Textile Res. J. **36**, 765.
Whitney, J.M. (1967). J. Comp. Mat. **1**, 188.
Whitney, J.M. (1969a) J. Comp. Mat. **3**, 534.
Whitney, J.M. (1969b). J. Comp. Mat. **3**, 715.
Whitney, J.M. (1969c). J. Comp. Mat. **3**, 359.

Whitney, J.M. (1971). J. Comp. Mat. **5**, 340.
Whitney, J.M. (1972). J. Comp. Mat. **6**, 426.
Whitney, J.M. and Riley, M.B. (1966). AIAA Journal **4**, 1537.
Whitney, J.M.. Stansbarger, D.L. and Howell, H.B. (1971). J. Comp. Mat. **5**, 24.
Whitney, W. and Kimmel, R.M. (1972). Nature (Phys. Sciences) **237**, 93.
Wilcox, B.A. and Clauer, A.H. (1969). Trans. Metall. Soc. AIME **245**, 935.
Wilson, H.B. Jnr. and Hill, J.L. (1965). Rohm and Hass Special Report No. S–50, AD–468 596.
Wolson, E.A. and Parsons, B. (1969). Fibre Science & Tech. **2**, 155.
Wright, P.K. and Ebert, L.J. (1972). Metall. Trans. **3**, 1645.
Wu, E.M. (1967). J. Appl. Mech. Trans. ASME Series E, **34**, 967.
Wu, T.T. (1965). J. Appl. Mech. **32**. 211.
Yeh, R.H.T. (1970). J. Appl. Phys. **41**, 3553.
Yeh, R.H.T. (1971). J. Appl. Phys. **42**, 1101.
Yue, A.S.. Crossman, F.W., Vidoz, A.E. and Jacobsen, M.I. (1968). Trans. Metall. Soc. AIME, **242**, 2441.
Zackay, V.F. (1965). High Strength Materials (John Wiley, New York).
Zecca, A.R. and Hay, D.R. (1970). J. Comp. Mat. **4**, 556.
Zienkiewicz, O.C. (1971). The Finite Element Method in Engineering Science (McGraw-Hill, London).
Zimmerman, J.E. and Cost, J.R. (1970). J. Acoust. Soc. Amer. **47**, 795.
Zweben, C. (1968). Proc. AIAA, VI Aerospace Science Meeting, Paper No. 68–173.
Zweben, C. (1969). J. Comp. Mat. **3**, 713.
Zweben, C. (1971). J. Mech. Phys. Solids **19**, 103.
Zweben, C. (1972). Eng. Frac. Mech. **4**, 1.
Zweben, C. and Rosen, B.W. (1969). Proc. AIAA, VII Aerospace Sciences Meeting. Paper No. 69–123.
Zweben, C. and Rosen, B.W. (1970). J. Mech. Phys. Solids **18**, 189.

Subject Index

Author Index